IRRITATING EXPERIMENTS

HALLER'S CONCEPT AND THE EUROPEAN CONTROVERSY ON IRRITABILITY AND SENSIBILITY, 1750–90

THE WELLCOME SERIES IN THE HISTORY OF MEDICINE

Forthcoming Titles:

Emblematic Monsters:
The Description and Interpretation of
Human Birth Defects in Europe
Alan W. Bates

New Medical Challenges during the Scottish Enlightenment
Guenter B. Risse

The Wellcome Series in the History of Medicine series editors are
V. Nutton, C.J. Lawrence and M. Neve.
Please send all queries regarding the series to Michael Laycock,
The Wellcome Trust Centre for the History of Medicine at UCL,
210 Euston Road, London NW1 2BE, UK.

www.ucl.ac.uk/histmed

IRRITATING EXPERIMENTS

HALLER'S CONCEPT AND THE EUROPEAN CONTROVERSY ON IRRITABILITY AND SENSIBILITY, 1750–90

Hubert Steinke

Rodopi

Amsterdam – New York, NY 2005

First published in 2005
by Editions Rodopi B. V., Amsterdam – New York, NY 2005.

Design and Typesetting by Michael Laycock,
The Wellcome Trust Centre for the History of Medicine at UCL.
Printed and bound in The Netherlands by Editions Rodopi B.V.,
Amsterdam – New York, NY 2005.

Index by Indexing Specialists (UK) Ltd.

British Library Cataloguing in Publication Data
A catalogue record for this book is available from the British Library
ISBN: 90-420-1852-6

'Irritating Experiments –
Haller's Concept and the European Controversy
on Irritability and Sensibility, 1750–90'
Amsterdam – New York, NY:
Rodopi. – ill.
(Clio Medica 76 / ISSN 0045-7183;
The Wellcome Series in the History of Medicine)

Front cover:

Frontispiece to the *Mémoires sur la Nature Sensible et Irritable des Parties du Corps Animal*

Printed in The Netherlands

Contents

List of Figures

Abbreviations

Corr. Bonnet	O. Sonntag (ed.), *The Correspondence between Albrecht von Haller and Charles Bonnet* (Bern, Stuttgart, Vienna: Huber, 1983).
Corr. Caldani	E. Hintzsche (ed.), *Albrecht von Haller - Marc Antonio Caldani: Briefwechsel 1756-1776* (Bern and Stuttgart: Huber, 1966).
Corr. Tissot	E. Hintzsche (ed.), *Albrecht von Hallers Briefe an Auguste Tissot, 1754-1777* (Bern, Stuttgart, Vienna: Huber 1977).
De Partibus	A. von Haller, 'De Partibus Corporis Humani Sensilibus et Irritabilibus. Die 22. April. & die 6. Maii 1752', *Commentarii Societatis Regiae Scientiarum Gottingensis* (Vol 2, Göttingen: Vandenhoeck, 1753), 114–58.
Elementa	A. von Haller, *Elementa Physiologiae Corporis Humani* (8 vols, Lausanne and Bern: Bousquet, Arnet, Grasset, Societas typographica, 1757–66).
Encyclopédie	D. Diderot and J.l-R. d'Alembert (eds), *Encyclopédie, ou Dictionnaire Raisonnée des Sciences, des Arts et des Métiers* (17 vols, Paris: Briasson etc., 1751–65).
Encyclopédie, Suppl.	*Supplément à l'Encyclopédie, ou Dictionnaire Raisonnée des Sciences, des Arts et des Métiers, par une Société de Gens de Lettres* (4 vols, Amsterdam: Rey, 1776–7).
Encyclopédie d'Yverdon	*Encyclopédie, ou Dictionnaire Universelle Raisonnée des Connaissances Humaines: Mis en Ordre par M. de Félice* (42 vols, 6 vols of suppls., Yverdon: s.n., 1770–9).
Epistolae	A. von Haller (ed.), *Epistolarum ab Eruditis Viris ad Alb. Hallerum Scriptarum. Pars 1, Latinae,* (6 vols, Bern: Societas typographica, 1773–5).
GGA	*Göttingische Zeitung von Gelehrten Sachen* (1739–52), *Göttingische Anzeigen von Gelehrten Sachen* (1753–).
Haller Papers	Burgerbibliothek Bern, Nachlass Albrecht von Haller.

Irritable parts	A. von Haller, *A Dissertation on the Sensible and Irritable Parts of Animals.* Transl. from the Latin [by Donald Monro]; with a Preface by M. Tissot (London: Nourse, 1755).
Mémoires	A. von Haller (ed.), *Mémoires sur la Nature Sensible et Irritable des Parties du Corps Animal* (4 vols, Lausanne: Bousquet, Arnet, 1756–60).
Opuscoli	G. Bartolomeo Fabri (ed.), *Sulla Insensitività ed Irritabilità Halleriana: Opuscoli di vari Autori* (4 vols. Bologna: Corciolani, Colli, 1757–9).
Praelectiones	A. Haller (ed.), *Hermanni Boerhaave... Praelectiones Academicae in Proprias Institutiones Rei Medicae* (7 vols, Göttingen: Vandenhoeck, 1739–44).
Zimmermann Papers	Niedersächsische Landesbibliothek Hanover, MS XLII 1933.

Acknowledgements

This book is a revised version of my PhD thesis submitted to Oxford University in 2003. It originated in the study of Haller's manuscripts and, particularly, his letter collection. For many years I have been working on the Haller Project in Bern (www.haller.unibe.ch) and had privileged access to the Haller papers. The Haller Project has established an extensive database including the biographies of Haller's correspondents, the content of his correspondence, the catalogue of his library and a primary and secondary bibliography documenting his work and life. Without this excellent resource at hand I would not have been able to deal with my topic in the manner I have done. Stefan Haechler and Martin Stuber built up the database and discussed Haller and the Republic of Letters with me. Barbara Braun, Kathrin Ott, Luc Lienhard, Claudia Profos and David Krebs helped in locating manuscripts and in improving the database. Urs Boschung, the director of the project and an expert in all Hallerian studies, has been, and still is, a supporting and encouraging tutor. His cautious way of dealing with historical problems has repeatedly prompted me to verify questionable assumptions.

Having decided to embark on my project, I first went to Renato Mazzolini in Trent – who had envisaged a study on irritability and sensibility a long time ago. In his great generosity, he shared not only his thoughts but also his whole collection of papers relating to the topic with me. Maria Teresa Monti has supported me in Milan, where I consulted Haller's library, now part of the Biblioteca Nazionale Braidense. In Oxford, Laurence Brockliss has been less a supervisor than a friendly adviser. With his broad knowledge of early modern history and medicine, he helped me to return to the key issues whenever I was in danger of going astray. Without him, this book would be different; any defects, however, are my own.

I have presented various aspects of the thesis on different occasions in Bern, Zurich, Lausanne, Geneva, Dublin, London and Oxford. Vincent Barras, Marino Buscaglia, Roberto Fantini, Robert Fox, Andreas-Holger Maehle, Wolfgang Pross, George Rousseau, Elizabeth Williams and others have provided critical and stimulating comments. So did the anonymous referees of the *Clio Medica* series. The staff of several libraries have been most helpful with their speedy delivery of documents. In particular, I would like to mention those of the Burgerbibliothek Bern, the University Libraries of Bern, Göttingen and Oxford, the British and the Wellcome Library in London, the Niedersächsische Landesbibliothek Hanover and the Bibliothèque de l'Académie de Médecine in Paris. Whenever I was in need, Pia Burkhalter from the Bern Institute of the History of Medicine procured the material desired. Lucy Hay reviewed the whole text for language errors. Michael Laycock at The Wellcome Trust Centre for the History of Medicine at UCL has been a careful and supporting editor. Last but not least, the Swiss National Science Foundation, the Roche Research Foundation and the Albrecht von Haller-Stiftung made this study possible through their generous financial aid.

I wish to thank all the persons and institutions mentioned for their help, encouragement and contribution.

This book is dedicated to my parents.

Introduction

Albrecht von Haller (1708–1777), the Swiss polymath, is best known in the history of medicine for his concept of irritability and sensibility. His orations *De Partibus Sensilibus et Irritabilibus*, delivered in 1752 and published in 1753, caused a European controversy about the function of nerves and muscles and about the properties of the living body in general. They were translated within two years into French, English, German, Italian and Swedish, and have since then been considered a classic of medical literature.[1] No general history of medicine skips Haller's contribution to physiology or 'animal economy', as it was often called in these days. Haller claimed to have proven by animal experiments that only the muscular fibre possesses the ability of contraction, which he called irritability and which was responsible for movement. From this property he strictly distinguished sensibility, responsible for sensual impression and inherent only in the nerves and the parts furnished with nerves. Thus he challenged the traditional, mechanical – mainly Boerhaavian – model on three main points. First, Haller postulated a force inherent in the muscular fibre and independent of the nerves and the soul. Second and partly as a result of this, he separated – conceptually and physically – the two properties of movement and sense perception. Third, and again in part resulting therefrom, he established a strict correlation between structure and function, not on the level of corpuscules or elementary particles, however, but on the level of compound structures, ie. the muscular and nervous fibres. Several well-balanced, general descriptions of Haller's concept have been published.[2] Maria Teresa Monti and especially François Duchesneau have furnished detailed and illuminating conceptual analyses of his theory.[3] Duchesneau, Roselyne Rey and others have located it within the general development of physiological models and have worked out its differences from the theories of leading mechanists (Boerhaave, Caldani), animists (Whytt), semi-vitalists (Fontana) and vitalists (Bordeu, Barthez, Wolff, Blumenbach, Hunter).[4] The notions of irritability and sensibility of many minor authors have been presented in brief summaries by Jörg Jantzen and some other scholars.[5]

The contributions of Duchesneau and Rey reflect the epistemological turn in the history of biological sciences, initiated by George Canguilhem.[6] Like Canguilhem, they focus on the structure of concepts rather than on the broad outline of theories and describe them as models answering a specific set of problems. Nevertheless, like the traditional descriptions of the history of physiology, they are mainly dealing with abstract ideas. The new science studies emerging in the 1980s, however, have stressed that scientific and medical research and the establishment of knowledge are not only theoretical but practical and social processes as well. Ian Hacking, Frederic Holmes and David Gooding have described scientific research as an investigative procedure, during which new concepts are created thanks to a combination of careful practical examination and theoretical conceptualization.[7] Somewhat more sociologically orientated, Bruno Latour, Steve Woolgar, Karin Knorr-Cetina, Hans-Jörg Rheinberger, Michael Hagner, Steven Shapin and Simon Schaffer have argued for the social construction of knowledge in the laboratory and in the scientific community at large.[8] My aim is to follow the traditions of both the historico-epistemological analysis of physiological theories and the newer science studies. I will, therefore, on the one hand examine the content and structure of Haller's concept – and, to a smaller extent, those of some other eighteenth-century physiologists – and on the other hand describe Haller's practice of research and the practical issues and social factors in the subsequent debate. But rather than using the historical case to argue for a specific model of how science works – which is the case for most science studies – I would like to employ the approach of modern science scholars to shed some new light on the development, modulation and reception of Haller's concept of irritability and sensibility. This study deals with the social construction of medical knowledge. Thus it acknowledges the existence of both social elements and the independent life of nerves, muscles, dogs, cats, and human bodies as parts in the controversy.[9] It is less concerned with broad sociological theories than a reconstruction of relations between objects, thoughts, persons, and institutions located in a specific historical setting.

In the first part I will discuss the development of the new concept. As the term 'development' suggests, Haller's theory should be considered as the result of an ongoing research process. The experimental investigations in the years 1750–52 lie at the heart of this development that provoked the essential turn in Haller's conception, a turn that led to the rejection of all earlier notions of movement and sensation – including his own. There was no single experiment that might be called 'crucial' but rather the whole experimental process, including practical exploration and theoretical reflection. In my description, I will draw on the laboratory notebooks of Haller and his pupil Johann Georg Zimmermann which, quite surprisingly, have never previously been studied (Chapter 2). By this, I hope to attract attention to the animal experimentation that lies at the core of Hallerian physiology but that has been neglected compared with the theoretical aspects of his work.[10] Haller, of course, did not develop his new concept in a space void of theories. Several major authors in the seventeenth century – notably Harvey, Glisson, Baglivi and Borelli – put forward notions of animal motion which were inconsistent with a purely mechanical description of animal economy (Chapter 1, under *The heritage of the seventeenth century*). However, as I will argue, the Dutch school of Boerhaave and his pupils was of greater importance for Haller. The epistemological approach – predominant in modern research on the history of eighteenth-century physiology – focuses on the general conceptual frame of major theories and tends to neglect the continuous development of models. Not only do the models of major authors change, but also 'minor' authors with less elaborate theories may have furnished important new insights. Boerhaave, who represents the mechanical model par excellence, modified his concept over time and, in his late years, whilst maintaining the general mechanical framework, even assumed the existence of an innate bodily property responsible for movement. Bernhard Siegfried Albinus, Frederik Winter and others of Boerhaave's pupils developed this idea even before their master and made it an important pillar in their physiological research (Chapter 1, *Boerhaave and the Leiden school*). Haller's early notions of animal motion were very similar and presumably heavily indebted to those of his Dutch colleagues. New evidence gained from animal experimentation, however, suddenly cleared his early, rather vague ideas and led him to a new understanding of animal motion and sensation. But his formulation of 1752 was not a definitive statement. Even more than Boerhaave's, Haller's notions changed. His later statements often indicate not only adjustments but also important modifications of his initial position (Chapter 3). And these changes were, in part, a reaction to the shift in physiological thought that his work – and that of others – had brought about.

In the second part of the book, I describe the reception of Haller's work and address the chief aspects of the controversy it provoked. Modern scholars agree that Haller's work was a main point of reference in the physiological thought of his time. François Azouvi, for instance, states that it 'dominates' Haller's century and that it 'constitutes undoubtfully the monument upon which all physiologists of the second half of the century decided, for or against Haller.'[11] But what does 'dominant' mean? Azouvi is an author mainly interested in the structure of theories. For him, Haller's theory was dominant because it offered a new and convincing concept that all serious contemporary physiologists had to deal with. This should not, however, lead us to the notion, prevalent in modern general histories of medicine, that 'Haller's concepts of irritability and sensibility achieved widespread acclaim' or 'widespread acceptance.'[12] Quite generally, we do not know to what extent the ideas of the most original theorists, as displayed for instance in Duchesneau's *La Physiologie des Lumières*, were shared by general practitioners and the wider Republic of Letters. Esprit Calvet (1728–1810), professor of medicine at Avignon, for instance, developed notions of sensation and motion peculiar to himself and seems not to have been disturbed by the latest discussions in Montpellier, Paris and elsewhere, even though he knew of Haller's work.[13] There seems to have been a wide variety of notions and ideas, which is only insufficiently described by the analysis of the theories of a dozen original thinkers. To draw a map of eighteenth-century physiological beliefs would require an extensive study of medical dissertations, lecture notes and handbooks throughout Europe. Such a study I have not been able to undertake, not even for the restricted domain of neuromuscular physiology. What I have tried to do, however, is to demonstrate representative modes of reception, appropriation and rejection of Haller's specific notions of irritability and sensibility. The diversity of these models is great but most of them agreed on one point: Haller's ideas had to be rejected. We have to keep in mind the essential points that distinguished Haller's concept of irritability and sensibility in particular and his physiology in general from that of others. These were on the one hand his stress on animal experimentation as the main method of physiological research, and on the other hand the strict separation of irritability and sensibility, based on the idea of the correlation between (compound) structure and (organic) function. Neither of these ideas gained considerable acceptance. Of course, Haller's concept fostered the general notion of innate bodily faculties and thereby was used to discredit the strictly mechanist and animist models. But this notion was not particular to Haller; it was postulated before him by Dutch (Albinus, Winter, Gaub and others) and French (La Caze, Bordeu) vitalists and by many authors who have to be considered his opponents. To describe the increasing acceptance of vital properties as a support of Haller's ideas is a misunderstanding of what his physiology was about. During the exposition of the whole controversy on irritability and sensibility – which will deal with a variety of other aspects – several explanations of why Haller was important and nevertheless rejected will emerge. It had to do with the performance of animal experiments (Chapter 4), the interpretative openness of Haller's theory and the complexity of his physiology (Chapter 5), the status of surgery, the search for new pathological models (Chapter 6, under *Pathology and the practice of medicine and surgery*) and the culture of criticism (Chapter 6, *The review journals and the culture of criticism*).

There is one element of the debate that the reader probably expects to be treated but which is not: the moral aspect of vivisection. There are two reasons for this. First, the general positions of medical, scientific, theological and philosophical authors of the eighteenth century have been well documented by Andreas-Holger Maehle.[14] And secondly, moral questions were of only marginal importance in the debate that was dominated by medical and other authors interested in animal economy. Of course, Haller's experiments were extremely painful and he called them 'cruelties which I detest myself', but he was convinced that the contribution to 'the benefit of mankind and the necessity' would justify their performance.[15] Those who rejected animal experiments did so, as we will see, because they doubted their validity. The cruelties, in their view, were only unacceptable because they did not help to gain further insight into the operations of the human body. Ethical arguments were put forward only very rarely and they were not an issue which had great influence on the acceptance or rejection of Haller's theory. Haller's ideal of natural science – and that of many of his contemporaries – was that of physico-theology, which conceived the world as a complex structure created by God. The scientist's duty was to detect its inherent laws and patterns, knowing well that the last things could not be fathomed. Research, including animal experimentation, thus had a theological element and was approved and even demanded by God in so far as it served mankind.

As this study deals with Haller's specific physiological ideas and their reception, Haller himself and his views on a variety of issues such as animal experimentation, research in general or professional criticism, will always be a major point of reference. It thus seems appropriate to add a short biographical sketch of the Swiss scholar.[16] Born in 1708 and growing up in Bern, he studied medicine in Tübingen and especially in Leiden with Herman Boerhaave and Bernard Siegfried Albinus. After graduation in 1727, he visited England briefly, completed his anatomical and surgical studies in Paris and learnt the foundations of higher mathematics from Johannes Bernoulli in Basel. From 1729 to 1736 he worked as a physician in Bern and published some first minor works in anatomy and botany. He achieved his first fame, however, with his *Essay of Swiss Poems* (*Versuch Schweizerischer Gedichte*), first published in 1732. This small booklet served as the model for descriptive and philosophical poetry for the next generation and made Haller the most highly esteemed German poet of the 1730s and 1740s. In 1736, he was called as professor of anatomy, botany and surgery to the newly established University of Göttingen, where he stayed until 1753. In this period of intense scientific activity, he developed his main areas of research and laid the foundations of later works. In 1742, he published a massive flora of Switzerland and was soon acknowledged as one of the leading botanists and as the most important opponent of Linnean nomenclature. As an anatomist, he focused on the vascular system and set the new standard in this particular branch with his *Icones Anatomicae* (8 parts, 1743–56). Haller's main interest, however, was physiology. His first major work was the edition of Boerhaave's lectures on physiology, which he enlarged by an extensive commentary and thus transformed into a critical assessment of the actual state of physiological knowledge (7 vols., 1739–44). In 1747, he published his own, short textbook of physiology, which ran through four original editions and was presumably the most widely used of all his scientific works. The orations on irritability and sensibility (1752) marked a major point in the development of his particular synthesis of physiology. In recognition of his scientific contributions, Haller was ennobled by the emperor in 1749. More importantly, his standing was confirmed by membership of the main European academies, ie. those of Uppsala (1733), London (1739), Stockholm (1747), Berlin (1749), Bologna (1751) and Paris (1754). In 1751, he was elected perpetual president of the newly founded Royal Academy of Sciences of Göttingen (Göttingen was part of the Hanoverian empire). Besides his scientific research and various other literary activities, Haller was busy as chief editor (1747–53) of the *Göttingische Gelehrte Anzeigen*, the leading German review journal, for which he penned some 9,000 reviews in the years 1747–77. Setting his hopes on a political career and aiming to secure the social and economic position of his family in Bernese patrician society, Haller returned to his home town in 1753. He was never, however, elected to the Small Council (*Kleiner Rat*), the seat of political power. After some years in a modest position in Bern (*Rathausammann*), he was elected director of the salt mines in Roche in the French part of the Bernese territories, where he could implement some of the agricultural reforms he promoted. In 1764, he returned to Bern and continued to work in various Bernese municipal bodies such as the Economic Committee and the Medical Council. Haller's return to Switzerland was not a farewell to the Republic of Letters. He continued to maintain his vast correspondence, of which 3,700 letters to and 13,300 from 1,200 persons have survived.[17] And he did not relent in his scientific activity. He proceeded with his embryological investigations, already started in Göttingen, and published his major works on the development of the chicken embryo in 1758 and 1767. His opus magnum, the *Elementa physiologiae*, appeared in eight volumes over a period of ten years (1757–66). Haller presented his views on anatomy and physiology to the wider public in the Yverdon and the supplements to the Paris *Encylopédie* (1772–77), for which he wrote some two hundred articles. A second, considerably revised and enlarged edition of his Swiss flora, was published in 1768. Remote from major centres of academia, he continued to build up his large library with more than 23,000 titles, mostly belonging to the medical, botanical and natural sciences.[18] The last decade of his life Haller devoted to the edition of critically commented bibliographies of botany, anatomy, physiology, surgery and the practice of medicine. In ten volumes, he presented and discussed some 50,000 works from all branches of medicine. Besides that, he wrote three novels on the principles of government and religious works against the French freethinkers, notably Voltaire. Haller obtained perhaps the greatest satisfaction of his life in July 1777, half a year before his death, when the emperor Joseph II – on his 'incognito' voyage through Europe – declined to visit the *philosophe* in Ferney but called upon our scholar in Bern.

Notes

1. C. Richet published a French edition in the series *Les Maîtres de la Science* (Paris: Masson, 1892), K. Sudhoff edited a German translation in the series *Klassiker der Medizin* (Leipzig: Barth 1922, reprinted Leipzig: Zentralantiquariat der Deutschen Demokratischen Republik, 1968) and O. Temkin published the English text in the *Bulletin of the Institute of the History of Medicine*, 4 (1936), 651–99. A reprint of the Latin and English text appeared in *The Classics of Neurology & Neurosurgery Library* (New York: Gryphon, 1992).
2. H. Buess, 'Zur Entwicklung der Irritabilitätslehre', *Festschrift für Jacques Brodbeck-Sandreuter zu Seinem 60. Geburtstag* (Basel: [n.p.], 1942), 299–333; G. Rudolph, 'Hallers Lehre von der Irritabilität und Sensibilität', in K.E. Rothschuh (ed.), *Von Boerhaave bis Berger. Die Entwicklung der kontinentalen Physiologie im 18. und 19. Jahrhundert mit besonderer Berücksichtigung der Neurophysiologie* (Stuttgart: Fischer, 1964), 14–34; R. Toellner, 'Anima et irritabilitas. Hallers Abwehr von Animismus und Materialismus', *Sudhoffs Archiv*, 51 (1967), 130–44; idem, 'Albrecht von Haller (1708–1777)', in D. von Engelhardt and F. Hartmann (eds), *Klassiker der Medizin* (Munich: Beck, 1991), 245–61; U. Boschung, 'Neurophysiologische Grundlagenforschung: "Irritabilität" und "Sensibilität" bei Albrecht von Haller', in H. Schott (ed.), *Meilensteine der Medizin* (Dortmund: Harenberg, 1996), 242–9. English descriptions are usually confined to a brief summary; the most detailed analysis is my paper 'Haller's Concept of Irritability and Sensibility and its Reception in France', *La lettre de la Maison française d'Oxford*, 14 (2001), 37–69 (largely a shortened version of ch. 3 of this book).
3. M.T. Monti, *Congettura ed Esperienza nella Fisiologia di Haller: La Riforma*

dell'Anatomia Animata e il Sistema della Generazione (Florence: Olschki, 1990); M.T. Monti, 'Les Dynamismes du Corps et les Forces du Vivant dans la Physiologie de Haller', in G. Cimino and F. Duchesneau (eds), *Vitalisms from Haller to the Cell Theory* (Florence: Olschki, 1997), 41–66; F. Duchesneau, *La Physiologie des Lumières: Empirisme, Modèles, Théories* (The Hague: Nijhoff, 1982), esp. Chapters 5 and 6. See also T.S. Hall, *Ideas of Life and Matter: Studies in the History of General Physiology, 600 B.C. – 1900 A.D.* (2 vols., Chicago: University of Chicago Press, 1969).

4. Duchesneau, *op. cit.* (note 3); F. Duchesneau, 'Vitalism in Late 18th-Century Physiology: The Cases of Barthez, Blumenbach, and John Hunter', in W.F. Byum and R. Porter (eds), *William Hunter and the Eighteenth Century Medical World* (Cambridge: Cambridge University Press, 1985), 259–95; R. Rey, *Naissance et Développement du Vitalisme en France de la Deuxième Moitié du 18e Siècle à la Fin du Premier Empire* (Oxford: Voltaire Foundation, 2000); R.K. French, *Robert Whytt, the Soul, and Medicine* (London: Wellcome, 1969); R.G. Mazzolini and G. Ongaro (eds), *Epistolario di Felice Fontana, I: Carteggio con Leopoldo Marc'Antonio Caldani 1758–94* (Trent: Società di Studi Trentini di Scienze Storiche, 1980); G. Ongaro, 'L'irritabilità Halleriana in Italia', in G. Zanchin and L. Premuda (eds), *Lo Sviluppo Storico della Neurologia Italiana: Lo Studio delle Fonti* (Padua: La Garangola, 1990), 113–25; A. Dini, *Vita e Organismo: Le Origini della Fisiologia Sperimentale in Italia* (Florence: Olschki, 1991), chapters 1 and 2.
5. J. Jantzen, 'Physiologische Theorien', in J. Jantzen, M. Durner and F. Moiso, *Wissenschaftshistorischer Bericht zu Schellings Naturphilosophischen Schriften 1797–1800* (Stuttgart: Frommann-Holzboog, 1994), 375–668 (see esp. 'Theorien der Irritabilität und Sensibilität', 375–498); A.G. Weber, *Commentatio de Initiis ac Progressibus Irritabilitatis cum Historia Sensibilitatis atque Irritabilitatis Partium Morbosae* (Halle: Litteris Orphanotrophei, 1783); F. Freschi, *Storia della Medicina in Aggiunta e Continuazione a quella di Curzio Sprengel*, VII/I (Florence: Speranza, 1845); for the Netherlands, see G. Van der Waa, '*De Irritabilitate*': *Een Onderzoek naar de Betekenis van het Irritabiliteitsbegrip in de Geschiedenis van de Achttiende-Eeuwse Nederlandse Fysiologie* (Rotterdam: Erasmus, 1992).
6. G.Canguilhem, *La Formation du Concept de Réflexe aux XVIIe et XVIIIe Siècles* (2nd edn, Paris: Vrin, 1977; first edn Paris: Presses universitaires de France, 1955).
7. I. Hacking, *Representing and Intervening: Introductory Topics in the Philosophy of Natural Science* (Cambridge: Cambridge University Press, 1983); F.L. Holmes, *Lavoisier and the Chemistry of Life. An Exploration of Scientific Creativity* (Madison: University of Wisconsin Press, 1985); D. Gooding, *Experiment and the Making of Meaning: Human Agency in Scientific*

Observation and Experiment (Dordrecht: Kluwer, 1990).

8. B. Latour and S. Woolgar, *Laboratory Life: The Construction of Scientific Facts* (Beverly Hills: Sage, 1979); K. Knorr-Cetina, *The Manufacture of Knowledge: An Essay on the Constructivist and Contextual Nature of Science* (Oxford: Pergamon, 1981); H.R. Rheinberger and M. Hagner (eds), *Die Experimentalisierung des Lebens: Experimentalsysteme in den Biologischen Wissenschaften, 1850/1950* (Berlin: Akademie-Verlag, 1993); S. Shapin and S. Schaffer, *Leviathan and the Air-Pump: Hobbes, Boyle, and the Experimental Life* (Princeton: Princeton University Press, 1985); B. Latour, *Science in Action: How to Follow Scientists and Engineers through Society* (Cambridge, Mass.: Harvard University Press, 1987).
9. This is the position of most social constructivists; see Ludmilla Jordanova, 'The Social Construction of Medical Knowledge', *Social History of Medicine*, 8 (1995), 361–81.
10. Haller's experimental investigations have, however, been marvellously studied in the area of embryology by M.T. Monti. See her edition of Haller's *Commentarius de Formatione Cordis in Pullo* (Basel: Schwabe, 2000).
11. F. Azouvi, 'Le Vitalisme en Allemagne, en France et en Grande-Bretagne au XVIIIe Siècle', *La Lettre de la Maison Française d'Oxford*, 14 (2001), 9–19: 10.
12. R. Porter, 'Medical Science', in R. Porter (ed.), *Cambridge Illustrated History of Medicine* (Cambridge: Cambridge University Press, 1996), 154–201: 165; E.M. Tansey, 'The Physiological Tradition', in W.F. Bynum and R. Porter (eds), *Companion Encyclopedia of the History of Medicine* (2 vols., London and New York: Routledge, 1993), 120–52: 124.
13. L. Brockliss, *Calvet's Web. Enlightenment and the Republic of Letters in Eighteenth-Century France* (Oxford: Oxford University Press, 2002), 147–9.
14. A.H. Maehle, *Kritik und Verteidigung des Tierversuchs: Die Anfänge der Diskussion im 17. und 18. Jahrhundert* (Stuttgart: Steiner, 1992); A.H. Maehle and U. Tröhler, 'Animal Experimentation from Antiquity to the End of the Eighteenth Century: Attitudes and Arguments', in N.A. Rupke (ed.), *Vivisection in Historical Perspective* (London and New York: Routledge, 1987), 14–47.
15. *De Partibus*, 114.
16. The most extensive and still indispensable biography is L. Hirzel's account, prefixed to his edition of *Albrecht von Hallers Gedichte* (Frauenfeld: Huber, 1882). It does not, however, meet modern critical standards; see E. Hintzsche, 'Einige Kritische Bemerkungen zur Bio- und Ergographie Albrecht von Hallers', *Gesnerus*, 16 (1959), 1–15. A good description of Haller's scientific work is E. Hintzsche's entry in the *Dictionary of Scientific Biography*, vol. 6 (New York: Scribner, 1972), 61–67. The best analysis of his

thoughts as a man of science and religion is R. Toellner, *Albrecht von Haller. Über die Einheit im Denken des letzten Universalgelehrten* (Wiesbaden: Steiner, 1971). For a comprehensive primary and secondary bibliography see H. Steinke and C. Profos (eds), *Bibliographia Halleriana: Verzeichnis der Schriften von und über Albrecht von Haller* (Basel: Schwabe, 2004).

17. The letters and summaries of the correspondence are recorded in U. Boschung *et al.* (eds), *Repertorium zu Albrecht von Hallers Korrespondenz, 1724–1777* (2 vols., Basel: Schwabe, 2002). Haller's correspondence is one of the main sources I have used for this study.
18. The library was bought by the Emperor to complement the collections of the newly founded public library of Milan. The stock of Haller's library is recorded in M.T. Monti (ed.), *Catalogo del Fondo Haller della Biblioteca Nazionale Braidense di Milano* (13 vols., Milan: Angeli, 1983–94).

PART I
THE EVOLUTION OF A NEW CONCEPT

1

Theories of Animal Motion before 1750

General textbooks of medical history often characterise Haller's concept of irritability as a restriction of Francis Glisson's theory of general irritability to the muscular fibre. Driven by the necessity of brevity and comprehensibility, they portray a picture of change within continuity that allows the reader to connect the different parts of the story. Haller's theory, then, appears to be 'rather in line with Glisson's hypothesis.'[1] As both authors used the very same term 'irritability', this connection seems to make sense.[2] But in fact these two concepts have much less in common than a first brief look might suggest. Between Glisson's fully developed exposition in 1677 and Haller's treatise, many different theories of muscular contraction were proposed that are much closer to Haller's view than Glisson ever was. It is the aim of this first chapter to describe the concepts which evolved over these seventy-five years. The two figures best known to historians of medicine are Baglivi and Boerhaave, who presented their mature views in the first quarter of the new century. These are the men who, in the two most authoritative analyses of Haller's physiology, mark the path leading to the new theory of irritability.[3] The history of ideas tends to link the concepts of the 'great' figures and 'original' thoughts, and in our case, the Baglivi–Boerhaave–Haller link certainly is of importance. Nevertheless, it is one of the central tenets of this study, that the history of irritability – and the history of science in general – should be seen as a gradual development of concepts, as an ongoing process of confrontation with thoughts but also with observations, experiments and details into which all participants are integrated. This means, that we have to pay more attention to so-called 'smaller' figures shaping or indicating these ongoing changes. This means, too, that scientists and natural philosophers have their own history of thoughts which often reflects the concerns of others and which we cannot reduce to one highlighted famous statement. This is particularly the case for Boerhaave and for Haller himself. It will thus become apparent that there were several figures of the 1730s and 1740s who to Haller, at least in some respects, were even more important than systematisers such as Baglivi and Boerhaave. Haller, with his indefatigable appetite for medical news and books, was well aware of the currents and trends in the physiology of his time and in his publications was particularly keen to show his familiarity with recent literature. He was as much a consumer of new treatises as he was a pupil of Boerhaave and a reader of Baglivi.

This approach will not only help us to delineate the development of Haller's own thoughts, as presented in Chapter 3, but should give us a broader picture of the situation in which these thoughts emerged. The inclusion of 'smaller' authors is not an attempt to reduce them to forerunners of Haller, quite the contrary, it seeks to depict the variety of ideas of which Haller's was only one. Nevertheless, it should be stressed that this chapter does not aim at a general history of physiological thought. It focuses on authors who developed at least partially *dynamic* views of movement, ie. views that – as Haller's – stood in contrast to a strict Cartesian mechanism. In accordance with Karl Rothschuh we may say that such concepts do not consider the account of structures and motions of corpuscles as sufficient explanations of life and vital actions.[4] They suppose that vital activities are directed either by hidden forces acting as innate or superadded principles of the body or by conscious or unconscious forces of the soul. Such notions Rothschuh called 'biodynamic' or 'psychodynamic', and it will be the *biodynamic* concepts – such as Haller's – that attract our attention.

Although this book is concerned with the debate on irritability and sensibility, ie. with motion *and* sensation, with muscles *and* nerves, this first chapter deals almost exclusively with questions of muscles and motion. There are two reasons for this. First, mechanistic physiology between Descartes and the middle of the eighteenth century was predominantly interested in animal motion. The nerve, of course, played an important role in the machine model, but the attention it received was mainly due to the part it assumed in muscular movement.[5] It was generally considered as a passive transport structure conveying impulses to and from the brain and as such, was not a prime object to be studied on its own. This would change fundamentally in the course of the eighteenth century, and especially in its second half. Whereas in 1700 life was equated to motion, with heart and muscle as its organs, in 1800 life was envisaged as sensibility, a quality inherent in the nerve and the nervous system. Sergio Moravia has aptly entitled this shift as a change 'from homme machine to homme sensible'.[6] Secondly, as Chapter 3 will show, Haller's own interests in nerves and muscles arose, too, mainly from his early preoccupation with animal motion, especially the movement of the heart, the mechanics of respiration, and the role of the nerves in muscular movement. Nerves and sensibility did not occupy any considerable position in his research agenda before 1750.

The heritage of the seventeenth century

The circulation of the blood, discovered by William Harvey (1578–1657) in 1628, was used as a paradigm for mechanical physiology of the seventeenth century. The notion of movement as the basis and expression of life – a commonplace in medical theory since Aristotle – acquired a new significance, and the heart as central moving force seemed to maintain the reigning position that the Greek philosopher had bestowed on it with his dictum 'primum vivens ultimum moriens'. Yet Harvey considered the heart to be a muscle and thereby challenged his own praise of its uniqueness. The muscular structure of the heart then was confirmed especially by Nicolaus Steno (1663) and Richard Lower (1669), and was generally accepted at the turn of the century.[7] Questions about the movement of the heart and the motion of animals in general now had one common focus: the contraction of the muscle. Mechanical explanations prevailed, and with a considerable effort physicians and natural philosophers tried to demonstrate how the muscular machine works on a macro- and microscopic level. Probably the most famous of these attempts was Giovanni Borelli's *De Motu Animalium* (1680–1) with its highly sophisticated mechanical and static analysis of the force of the muscles. He pointed out that the muscles had to be activated through the nerves and their fluid, a notion which was supported almost unanimously until the mid-eighteenth century. How exactly this would happen was a matter of controversy. Steno, for one, did not venture on this subject, but many accepted an iatrochemical theory similar to that of William Croone (1664), Thomas Willis (1670) and Borelli, who attributed the inflation of the muscle to an effervescence arising from the mixture of the nervous fluid with the blood.

This account could at least partly explain how the rational soul as the acknowledged ultimate cause of voluntary motion could affect the muscles. The animal spirits flowing in the nerve tubes were conceived either in a Cartesian manner as consisting of entirely material, very fine corpuscles, or as a kind of refined matter somehow in between materiality and immateriality, or composed of both qualities.[8] Of course, these concepts could not solve the mind–body problem, but the idea of a subtle fluid made its gnawing presence more bearable. The fluid was produced by the brain (a gland) through rarefaction of the blood. The new and original theory of Giorgio Baglivi, first presented in 1702, furnished further evidence for the existence and importance of the spirits. He had observed the rhythmic movement of the dura mater (the dense and outermost envelope of brain and spinal cord), which he described as the motor, even the 'heart' of the circulation of the nervous juice.[9] This discovery of another mechanical tubular system was welcomed by leading theorists such as Friedrich

Hoffmann, but equally was rejected by eminent figures such as Boerhaave, and remained an undecided problem until it was rejected on the basis of experiments in the 1750s.

Involuntary muscular movement proved to be a much more controversial issue. From a Cartesian point of view, these motions could not be governed by the soul, which was responsible only for conscious movements. It was supposed that – as in voluntary movement – the contraction should be due to an influx of animal spirits. This seemed to be confirmed by the experiments of Lower (1669), who observed quiver, palpitation, and ultimate stoppage of the heart after the ligature of the cardiac nerves.[10] Furthermore, Borelli (1680–1) delivered a mechanical interpretation of the oscillatory contractions of the heart: the channels of the nerves would be of such a structure as to make the juice pour into the muscle not in a continuous flow, but in drops. Yet such mechanisms were – even for Borelli – insufficient to account for the ultimate cause of the heartbeat. He stated that the unpleasant accumulation of blood in the heart of the preformed embryo would be perceived by the 'sentient faculty' (*facultas sensitiva*) of the soul through the nerves, which would then prompt the ventricle to contract. The circulation, once set in motion, would continue and the motion would become habitual, unconscious, very much as the violinist's swift play of long practised runs.[11] Thus, involuntary movements were ultimately governed by the soul, truly an animist and not a mechanist idea. As Thomas Fuchs suggests, it seems as though Borelli was somehow forced into such a position because his Cartesianism did not allow the thought of active matter.[12] The initial movements could only be caused by the rational soul, ie. by will, acting through brain and nerves. This was also in accordance with his notion that passions affecting the motions of the heart are caused by the soul. If the heart was controlled by the soul in these instances, this must also be the case in the normal state. Again, he embraced an idea quite opposite to Descartes's view of passions as products of humoral and nervous movements. The case of Borelli is illustrative because it corrects our often too rigid conception of what mechanism would have been in the late-seventeenth and early-eighteenth century. Borelli, one of the *mechanist* standard authorities in this period, conveyed beliefs that came to lie at the core of the theory of the main *animist* authority, Stahl.[13] This is not to say that these tenets undermined Borelli's main framework, nor that they were cardinal to many other physicians, but they argue for a current of non-mechanist thought running along or underneath the essentially mechanist outlook of physiology.

The other, and more evident, non-mechanist tradition was essentially rooted in Aristotelian vitalism and Galen's notion of specific corporeal faculties. It was transferred into the Cartesian area especially by Harvey, Glisson and Baglivi.[14] Galen identified several natural faculties and subfaculties which he used in his description of physiological processes.[15] The gall bladder, for instance, if irritated by a certain quantity or quality of gall, contracted thanks to its *facultas expellens.* Much the same happened in the case of the stomach, intestines, urinary bladder and also the heart, when troubled with matter accumulating inside. Galen's opinion as to whether nerves were required for these reactions, was ambiguous; in some places he said yes, in others no. Certainly the soul did not take part in these bodily functions, but neither were they mere mechanical motions. For the case of the heart he stated that 'the power of pulsation has its origin in the heart itself... by the high virtue of some special element in its nature.'[16]

Such a notion of vital faculties is not very far from Harvey's views, especially those exposed in his later embryological work. Harvey agreed with the physician from Pergamon that there was an innate faculty in the heart reacting to the irritation. The contraction proceeded 'by means of [the heart's] own proper fibres, as by the instrument destined for that use.'[17] No nerves were required for this action. This was suggested by embryological observations to which he was drawn in his Aristotelian search for the primary moving principle. What he saw was the rhythmic pulsation of a spot of blood before the appearance of any nervous structure, and – in contrast to Aristotle – even prior to the appearance of the heart. Even before the pulsation of the blood there was the blood itself. Application of heat in some cases could bring this blood to contraction. Harvey concluded that this was only possible as long as there was some innate heat (*calor vitalis* or *innatus*) in the blood.[18] The presence of this vital principle was a precondition for the contraction. In another passage Harvey specified the duties and abilities of such a principle:

> Lastly, [in the blood] there is an inherent mind, foresight and understanding not only in the vegetative part of the soul but existing even before that soul itself, and procuring and disposing and ordering all things immediately from the very beginning for the being and well-being of the chick, and fashioning its shape and likeness to its parents.[19]

Although this statement seems to make quite clear that there was a principle of development in the blood that preceded the appearance of the soul, other remarks suggest that its power was still due to the Aristotelian vegetative soul.[20] What is certain, though, is that the contraction of the blood and the changing of its amplitude and frequency upon irritation were manifestations of the soul as author of all movement and sensation.[21] Harvey, therefore, was essentially a monist. He believed, that 'the soul cannot be separated from spirit and innate heat.'

Spirit and heat were innate in the blood, which 'is the common bond between the body and the soul, and with it, as a vehicle, the soul penetrates into all the parts of the whole body.'[22] When the blood,

> is taken away, the soul is presently gone, so that the blood would seem to differ nothing from the soul, or at least, would be considered the substance whose act is the soul. For such is the soul, that it is not altogether a body, nor yet wholly without a body; it comes partly from without, and is partly born at home.[23]

Two elements in Harvey's embryology are of importance to us. First, within a generally monist concept there is still a notion of vital innate first principles in some way independent of the soul. And secondly, as the changing movement of the early blood drops and later of the heart show, there is an unconscious perception, independent of nerves and brain. This may also be observed in other cases such as the stomach which is able to distinguish between healthy and noxious food.[24] Both these elements are more clearly developed in the works of Glisson.

Francis Glisson (1597–1677) was an early and important exponent of Harveian physiology and especially a strong supporter of the new theory of circulation.[25] Of particular interest to us is his theory of irritability, which he developed in several writings.[26] We shall only look at his most mature ideas as presented in the last years of his life, when he first laid the more systematic foundations of his thoughts (1672), and then proceeded with the physiological demonstrations in his *Tractatus de Ventriculo et Intestinis* (1677). What distinguished Glisson from Harvey was, above all, the idea of the fibre as the main constitutive element of the solid parts of the body.[27] This idea was mainly developed in the period immediately after Harvey, and was especially fostered by the microscopic investigations of Leeuwenhoek and others. But no proper fibrillar theory had been established before Glisson. According to this author, the body was not made up entirely of fibres, but the other elements were of less importance. The fibre was the morphological unit not only of the muscle, but of most of the parts of the body, and irritability was its quality. Without this quality the fibre would be in constant rest or unchanging movement. Glisson realised, as Harvey, that reaction upon irritation implied a kind of perception. But he distinguished more neatly than Harvey had done between the different faculties. In fact, he described three distinct elements of the process which takes place within the fibre: the *natural perception* of the stimulus, the *appetitus* thereby awakened to act towards a desirable end, and the *motus* in accordance to this end. In the case of the heart, for instance, the muscular fibres perceived the stimulus due to the accumulation of blood in the auricles, tried to expel the provoking matter and therefore contracted. All this did not involve consciousness, but it did in other instances, when the stimulus was transferred to the nerves. This tripartite notion of irritability corresponds, as Roger French has pointed out, to the three faculties of the Aristotelian vegetative soul.[28] But it is not the soul itself which is the ultimate source of life, as in later animist concepts, 'it is matter which contains the roots of life.'[29] Natural perception as the basic principle of life inherent in the fibre, preceded the presence of the soul. The soul might modify life, but it could not give life. Glisson proposed a monistic concept – quite similar to Harvey's – in which both matter and soul are active principles shaping the vital actions.

The third important figure of the seventeenth century to present a dynamic concept of involuntary muscular movement, was Giorgio Baglivi (1668–1707).[30] We have already seen that he had identified two different systems of motion, the circulation of blood with the heart, and the propagation of nervous fluid with the dura mater as its motor. This notion he merged with an original, comprehensive concept of fibres. He ventured a new division of the whole body into membranous fibres on one side, developed through the solidification of nervous juice and depending from the dura mater, and carneous fibres on the other side, emerging from the condensation of blood and governed by the heart. The membranous parts were nerves, vessels, intestines, and glands, the carneous muscles, heart, tendons, and bones. The different functions of these parts were due to their particular fibrillar structure, confirmed by microscopic observation. This was a crucial step. Baglivi – as Haller – created a theory in which different types of living fibres and their structure account for the essential phenomena of movement and sensation. According to Baglivi, the membranous fibres were the instruments of sensation, the carneous, the *fibrae motrices*, those of voluntary movement. Blood and nervous juice triggered the contraction of the muscle but the motive power resided in the fibres themselves. This conclusion he drew from his experiments, in which he could see heart and muscles continuing to contract when irritated after death or cut into small pieces. These observations were not new and were reported even by great authorities such as Virgil, Galen, Vesalius, Francis Bacon and Harvey, and especially in the second half of the seventeenth century by Robert Boyle (1663), Robert Hooke (1664), Nils Stensen (1667), Johannes Bohn (1668), Richard Lower (1669), Francis Glisson (1677), Johann Jakob Wepfer (1679), Johann Conrad Peyer (1682), and others. Haller noted, that such observations 'seem rather to have been dropped accidentally than wrote on purpose.'[31] It was especially Baglivi who stressed their importance. He did not consider them as peripheral phenomena. For him, they were important arguments for his new concept. There was a realm of autonomy independent of the nerves. All the parts of the body were in continuous movement, their apparent rest was an illusion which was due to the activity of the antagonist muscles. Microscopic observation showed that the fibres were in constant oscillation, they possessed a vital tonus. A deviation from the normal degree of tonus – spasm or atony – was the pathological basis of many diseases.

In recounting these different concepts of vital properties we should not forget that the general framework of physiology at the turn of the eighteenth century was mechanistic. Harvey's vital notions were obscured by the reception of his discovery as the paradigm of mechanism. Glisson was an author rarely mentioned in literature on muscular physiology before Haller, and Baglivi's theories, although often printed and widely read, were controversial. This, again, should not lead to the conclusion that mechanism necessarily was Cartesianism. Borelli may serve as a good example of how foreign elements may be introduced into a strict mechanical concept, and the opening decades of the eighteenth century furnish us with yet another range of modifications and appropriations.

Boerhaave and the Leiden school

Boerhaave

Herman Boerhaave (1668–1738) is generally considered the most important representative of a mechanical system of medicine in the eighteenth century.[32] Mechanics, for him, was Newtonian mechanics. As one of the earliest and most important promoters of Newtonian physics on the continent, he considered the new approach and new laws in physics also as important to medical research. Newton's natural philosophy exerted a considerable impact upon the methods as well as the concepts of eighteenth-century physiology. His authority is particularly manifest in English works concerned with questions of muscular and animal motion.[33] One of these English authors seems to have been of singular importance for Boerhaave: Archibald Pitcairne (1653–1713).[34] Pitcairne, a well-known physician from Edinburgh, aimed at the construction of a system of medicine in accordance with the new laws of physics. Boerhaave was particularly confronted with this new system because, in 1692, Pitcairne was appointed professor of medicine in Leiden.

In his inaugural lecture, Pitcairne adapted the Newtonian dictum to medicine and stated that 'Enquiries after Physical Causes as are generally proposed by the Philosophers, are entirely useless and unnecessary to Physicians.'[35] The observation of the actions and reactions of the corporeal bodies should furnish us with the data necessary to discover the forces and establish the laws that direct their movement. If these laws were known, it was of no advantage to the physician to know their causes, ie. the 'nature' of things. The physician should follow the method of the astronomer. Since all bodies, whatever their magnitude or minuteness, suffered from the same effects of motion and change, the 'Laws and Properties of the Fluids and Canals of Human Bodies may be defined, after we shall either have made more Observations, or compared and methodised those that have been already made.'[36] Given contemporary physiological knowledge it should not surprise us that Pitcairne could not deliver a comprehensive system in which these laws and properties were described. His Leiden lectures from 1692–93 – posthumously published in 1717 – are therefore rather a statement of his general mechanical beliefs than accurate, mathematical descriptions of physiological processes. In his view, the body was basically a hydraulic machine, 'composed of Canals of divers kinds, conveying different sorts of fluids.'[37] The most important fluid was, of course, blood, and 'Life itself is either this Circulation, or this the measure of it.'[38] Circulation originated in the motion of the heart, thanks to the power acting in the ducts of its fibres. But how this motion originally started in the embryo could not be explained, ultimately it had to be derived from God himself.[39] Pitcairne also did not deliver any new explanations of muscular contraction; he used a rather simple model of influx of animal spirits and of reflux for the transmission of sensations.[40]

Pitcairne's inaugural lecture delivered in Leiden in 1692 may be considered as the first rigorous statement of English Newtonian medicine. But it may also be regarded as the starting point of Dutch mechanistic, but non-Cartesian physiology. Although Pitcairne had already left the Netherlands in 1693, his voice seems to have left a deep impact in Leiden. His appeal to study the human body according to the laws of mechanics and to follow the method of the astronomers was taken up in a public lecture in 1698 by the philosopher and experimental physicist Burchard de Volder (1643–1709), and – of more importance to us – by Herman Boerhaave in his *Oration on the Usefulness of the Mechanical Method in Medicine*, held in 1703.[41] For Boerhaave, too, life was essentially motion, and particularly the circulation of blood. He illustrated this with the example of a person who had fainted:

> We see, then, a dead man; but in what sense dead? In this body all solid and liquid parts which suffice for life and health are present – the only thing which is lacking is the motion which causes the humours to circulate. And when eventually the nerves of this patient are roused to activity, by whatever means you will, so that the matter which sets the heart in motion resumes its course, then at once happy life returns, the sad spectacle of death is

> banished... What ferment, effervescence, what aggressive salt, oil, or spirit is created or destroyed in such a situation? Nothing is added or taken away, except motion; yet life itself was lost and has been restored.[42]

As to the origin of this motion, Boerhaave, although not quite explicit on the subject, gave a double answer. Changes, alterations and, in the above-mentioned example, the restoration of motion were caused by the soul. Originally animal movement went back to the first motions of the liquids in the embryo, which were themselves derived from the parents and, ultimately, from God.[43] The body itself was neither able to generate nor to alter motion, but somehow it was able – under normal circumstances – to maintain its permanency, to preserve life. These bodily conditions, which guaranteed the continuation of life, had to be studied. Every effect therein, Boerhaave said 'is wholly corporeal, and so, subject to mechanical laws.' He asked:

> Does it matter whether the prime cause of a change is mechanical or not, when, firstly, it is granted to the mechanistic physician, without going into this, to perceive its effect – which is corporeal – and to scrutinise and guide it, and, secondly, when this suffices for his aim?[44]

Whoever had investigated the marvellous structure and motion of the heart 'would search for the ground of life's permanency nowhere but in the mechanical capacity itself of the internal parts.'[45]

This is why Boerhaave stated, 'only the mechanicians are entitled to claim that they are dealing with their proper subject; and there exists nothing in the whole of the body, in as far as it is solid, which does not belong to this.'[46] They could lead us from the known to the unknown. Reporting the many facts that demonstrate the parallels between the macroscopic and microscopic world, Boerhaave stated:

> It follows from this that either nothing can be deduced in a scientific manner from all these facts, or that we must acknowledge the preeminence of mechanics with regard to the investigation, or even the governing of the human body.
>
> But who would believe or maintain that nothing true, nothing certain, nothing useful can be deduced from so many clearly observed facts, when they have been correctly pondered one by one, or with most judicious reasoning comprehended and compared with one another?[47]

If the unobservable processes in the microscopic world were ever to be explained, they had to be done so by mechanical laws. Because chemistry was not able to furnish these laws, Boerhaave in his early years had some reservations as to the general validity of this science: 'Chemistry, then, is able to bring data to light and to define their attendant conditions; but it will never provide us with rules by the use of which one may logically deduce anything from these data.'[48] Chemistry was simply not the right tool to explain the functions of the body. Boerhaave, although a promoter of observation and experiment, believed in full accordance with the reigning rationalism that only the 'logical' science of mechanism could be a basis of 'sound reasoning'.[49]

Boerhaave's belief in the analogy of macro- and microcosm led to his formulation of a new fibre theory.[50] The ultimate structural unit of the body was the fibre, which combined into membranes, which then built the smallest of all vessels, which in turn configured larger vessels.[51] Lastly, the whole body consisted of vessels. He was led to this conclusion mainly by the microscopic observations of his friend, Frederik Ruysch, who detected very small vessels in places where nobody had expected them. Furthermore, he used the investigations of Leeuwenhoek to claim that the blood corpuscles consisted of six serous globulae or thirty-six spherulae. Small vessels would admit only one corpuscle, the so-called serous arteries only a single globule, and the lymphatic vessels only a spherula.[52] The most minute vessel he equated somehow with the nerve, which transported an even more subtle fluid, the animal spirit. He said that these vessels would be 'very similar' to the smallest nerve fibrils[53] and that 'we call the nerve the last and most subtle vessel.'[54] Thus, although he did not clearly identify the nerve with the most minute vessel, he stated that 'the whole solid mass of our body is totally built up of mere nerves as its elements.'[55] He was also drawn to this conclusion by the observation that there was barely a part of the body which did not feel or move.[56] Movement was, of course, dependent upon the influx of nervous juice into the muscular fibres, which were nothing more than tubular expansions of the smallest last nerves, filled with animal spirit.[57]

Boerhaave presented these basic tenets at different places with somewhat different accents, but they remained basically unchanged throughout his career. In this system the fibres have no specific biological qualities, it is their arrangement in larger structures which defines their actions, and these structures are analysable in mechanical terms. The uniformity of the structural unit and the analogy of the macro- and microcosmic world led Boerhaave to stress in his early oration the unlimited validity of mechanical laws in all parts of the body.

Twelve years later, in his *Discourse on the Achievement of Certainty in Physics*, given in 1715, Boerhaave advanced initial reservations as to our knowledge of processes within the microscopic realm. He started his argumentation with the discussion of gravitation. We could not, he said, explain its origin and nature by mechanical principles, but its effects could be excellently elucidated by them. It was their effects, evident to our senses, and the examination of these that led to the discovery of the laws of this universal property. The very same approach should be pursued in the analysis of attraction. Some people proclaimed that at last the first principle causing effects in nature was revealed and that its laws could be disclosed. But Newton himself, Boerhaave said, was much more cautious. He concluded from continuous observation and experiments that there was a principle which acted upon bodies placed at a certain distance and made them rush together. But attraction was only the name of 'an unknown cause that creates an apparently spontaneous motion through which bodies are brought into mutual contact.' It also manifested itself differently between different bodies, strongly between some, less between others, or even not at all:

> Attraction, then, does not display itself beyond the point where it affirms the boundaries of its domain through its material effects. We should therefore establish experimentally whether attraction really prevails between certain bodies before we are entitled to explain effects by assuming this concept. And there will be found to exist as many distinct species of attractions as there are kinds of different bodies – each of which, however, is regulated by its own laws.[58]

Boerhaave mentioned gravitation, magnetism, electricity, and the efficacy of salts as probable modes of attraction. This is, of course, a typical Newtonian stance, derived from the *Opticks*. But unlike some optimistic Newtonians he emphasised that we first had to establish whether these different forces really act in all different kinds of bodies. Boerhaave presented the general framework in which these assertions are placed in the following discussion of the wonders of the seed. Apart from these forces of attraction which might be explained mechanically, there was a 'seminal principle' that 'brings the scattered elements together in the structure of the seed.'[59] No man could discern the way this power worked, and it surely could not be explained by blind collisions of atoms. Yet it was from this principle alone that we tried to explain all the phenomena that arose in the natural world. There were very many different kinds of animals and plants and yet all specimens of the same kind were perfectly similar. These observations led Boerhaave to his preformationist belief:

> Let it suffice to draw from them the conclusion that the creation, the nature and the powers of single bodies presuppose that similar beings, from which they have sprung, already existed before; that these beings do not, therefore, depend on any universal principles, but that each single species acknowledges principles particularly its own; that, therefore, these are as various as there are multiple forms of bodies, that is, infinite in number; so

> that they can never be learnt, except in so far as they may be discovered one by one, through a true study of nature.[60]

Whether certain forces really acted within an animal or a plant depended upon its structure as originally given by God. Boerhaave certainly would not deny that gravitation acts upon all bodies as the universality of this force had been experimentally confirmed. He clearly would not deny, either, that there are certain principles acting in many different types of bodies. Furthermore, he would not deny that these principles would act according to mechanical laws. But whether a specific principle would act in a specific body had to be proved by the study of nature, ie. by observation and experiment. Thus, Boerhaave did not withdraw his mechanistic credentials, put forward in the oration of 1703, but he emphasised the singularity of all bodies and was much more cautious as to the application of all different kinds of principles in realms we could not observe. Boerhaave's early a-priori mechanism had developed into Newtonianism as an analytical instrument or working hypothesis.[61] This development – as Rina Knoeff has shown – was paralleled by an increasing interest in chemistry that, in Boerhaave's eyes, did not search for general laws of nature but 'sticks to the consideration of the sensible powers peculiarly found in each body.'[62]

These particularities were stressed even more in his late *Discourse on Servitude as the Physician's Glory*, delivered in 1730. The main topic of this oration was nature, which the physician should follow as a guide in his research as well as in healing. Boerhaave described the ideal curriculum of a medical student in these terms:

> Learn the general laws of Nature from the mathematicians who explain the general properties of bodies, and the forces arising from these properties, in terms of mechanical, hydrostatic, and hydraulic science. Then you must get to know the single faculties [proprias facultates], proper to each definite body; these are demonstrated in a manner both ingenious through physical and chemical experiments, by men who never grudge exertion when they may expect recognition of their merits. Observe the act of creation through which God constructs the human body from insignificant dust, as is set out by the eminent professor who occupies this Chair [of anatomy] so successfully.[63]

Boerhaave repeated the credentials already evoked in his earlier oration. The general mechanical laws as the basis of medical investigation were not questioned, but each specific body in addition was governed by specific forces, now called faculties, which had to be demonstrated experimentally. Ultimately, the whole structure of the body was due to God's act of creation. The use of the term 'faculty' (*facultas*) is significant; it seems to suggest that these actions were not totally reducible to mechanical laws. In a similar manner Boerhaave talked of the effects of medicines, which were,

> ...completely dependent on the particular qualities (*particulare indole*) with which the Creator has endowed human nature, and they [the medicines] can never be understood in any other way from whatever cause... They are only stirred to action through the activity of life (*actio vitae*).[64]

The whole oration is guided by a qualitative rather than a quantitative approach to the body, which is seen not as mere passive matter, but as a reactive instrument. Medicine, said Boerhaave, is 'the science of those reactions which the body, as created by God, elicits in itself as well as in other bodies; and of the reactions to which it may itself be stimulated by other bodies.'[65] If God was the creator of the body, nature was its operator. All parts had been created together, and when you studied a single part of the body you realised that it was linked to all the others. Reminiscent of the early oration of 1703, Boerhaave used the example of a drowned person. There was no movement and no life in that person. But inflation of air into the lungs would regenerate breathing, this would activate the heart, the circulation set in again, the brain was supplied with blood, and nerves, muscles, and intestines resumed their functions. However, whereas in his earlier lecture he stated that the 'bodily conditions' would guarantee the preservation of life, he now claimed that it was, 'Nature alone who operates simultaneously throughout all parts, continually regenerating these motions once they have been started by her own power, as long as the organic structure remains safe and sound [*quamdiu compagi sua modo constat integritas*].'[66] Nature had replaced the 'mechanical capacity' of the body as the operator of the machine. Haller said of Boerhaave that, 'that what he called *nature*, and which was the object of his respect, was probably not that far from an archeus.'[67] Although this may be an overstatement – probably due to the fact that, in his praise for nature, the Leiden professor sought even the support of van Helmont and Paracelsus – it tells us at least that not everybody considered Boerhaave a rigid mechanist.

The development of Boerhaave's ideas has been treated in somewhat more detail because it has been – until the recent book of Rina Knoeff – almost entirely neglected.[68] This might be due to the fact that accounts of his physiology are often based on his early oration of 1703 – a classical document of mechanistic medicine – and on his *Institutiones Medicae*, in which modifications are much less visible than in the orations.[69] In fact, the reader of the *Institutions* gets rather a picture of Boerhaave as a rigid mechanist. But we should not forget that the spread of Boerhaavian ideas occurred not only through his textbooks but also through his orations and lectures, in which he could stress certain aspects and advance hypotheses more easily. For instance, in his lectures on nervous diseases, delivered in the early 1730s, he ventured the *enormôn* or *impetum faciens*, a part of the soul located in the sensorium commune, as the prime mover of the animal spirits and thereby of the body as a whole.[70] In other lectures of a slightly earlier period (1725–32), and edited by Haller, we find a further interesting statement, namely that there were some voluntary muscles that, from mere habit, could act in an unconscious manner (*nobis non conscius*). Boerhaave used the classical Stahlian example of the wandering philosopher who does not realise that he is walking.[71] Now, this does not make Boerhaave a Stahlian, but it shows that he often – and especially in his later career – expressed thoughts that were not in exact correspondence with a strict mechanism.

The textbook, as mentioned above, was more rigorous, and here his general notions of muscular motion remained virtually unchanged. It has already been said that he thought the muscular fibres to be the 'last' nerves, and that almost all parts of the body would move or feel. As to the general mechanism of contraction, this could only be explained by the influx of nervous juice, caused by an impulse from the brain. How the soul would act on the brain could not be discovered and should not concern the physician. As muscular contraction was stronger than the simple contraction due to elasticity, he called it not an innate but a superadded movement. The muscles for voluntary movement would receive their nerves from the medulla cerebri, and those for involuntary movement from the cerebellum. This was proved by experiment. The alternate contraction of the heart he explained by an ingenious construction, widely adopted by his contemporaries. The systolic expulsion of blood in the aorta would lead to a compression of the nearby cardiac nerve, thereby interrupting the transfer of nervous juice and causing the diastolic relaxation of the heart. The only truly important addition made to his *Institutiones Medicae* is a remark on the heart to be found in his last edition of 1734:

> There is an amazing and occult tendency in the heart to repeat the systoles and diastoles, alternatively, also after death and even in the excised heart, and actually in single parts of the dissected heart.[72]

This is the passage on which Haller based his statement that Boerhaave had 'acknowledged an active force in the heart, and a latent principle of motion in the pieces of it which are cut...'.[73] Although this is not exactly what his teacher had written, Haller's exposition of Boerhaave's thought is, in fact, accurate. In his public but unpublished lectures on the heart, held in 1735–7, Boerhaave actually postulated the existence of an active, innate force of the heart.[74] In his earlier, published works, he had always identified three causes of the heart's action: the influx of venous blood stimulating the heart to contraction, the arterial blood in the coronaries, and the animal spirits flowing through the nerves. Now, on 23 November 1736, reconsidering the descriptions of Harvey, Boyle, Wepfer and others on the post-mortem movements of the heart, he argued that,

> there is also a fourth [cause] to be taken in consideration, that is a property of the heart itself. I will call it its irritability, and by this name I understand its innate aptitude due to its fabric, thanks to which it easily and promptly contracts upon the application of these [three] or other stimulating causes.[75]

Boerhaave did not enlarge upon this topic in his later lectures on the heart. Nevertheless, his statement must be taken as a further sign of the continuous development of his conception of living matter away from a strict mechanist model. It is possible, though, that Boerhaave did not evolve the idea of irritability as a fourth, innate cause in the heart on his own but adopted it from his pupils who, as we will see, had already made similar statements in 1727 and in 1736, some months before his lecture.

The Leiden school

The essential characteristic of Boerhaave's pupils in the Netherlands is that they actually made the step that their teacher only contemplated, viz to openly declare innate bodily faculties responsible for movements that are independent of the soul.[76] The five physiologists of greatest importance to us graduated from Leiden at different periods in Boerhaave's career; Jan de Gorter (1689–1762) in 1712, Bernard Siegfried Albinus (1697–1770) in 1718, Hieronymus David Gaub (1705–80) in 1725, Frederick Winter (1712–60) in 1736, and Abraham Kaau Boerhaave (1715–58) in 1738. The step they took was considerable, but it was not necessarily a step *against* Boerhaave. The presentation of new concepts is often described as a reaction against the old one, and so, too, the search for innate or vital forces has been seen as a reaction against the prevailing Boerhaavian mechanism.[77] But – as the preceding pages should have shown – such a view is somewhat misleading. It is not a coincidence that the new notions should have been fostered by pupils who were especially close to the master and aware of the development of his thought.[78] De Gorter in 1724 was proposed by Boerhaave as professor of medicine in Harderwijk, and it was equally at the request of their teacher that Albinus, in 1719, and Gaub, in 1731, were called to Leiden as lecturers in anatomy and chemistry, respectively.[79] It was Gaub who, in a letter to Haller, emphasised that Boerhaave, in his lectures on nervous diseases, which Gaub had attended and which Boerhaave's nephew Kaau wanted to edit, had shown much knowledge of the property of irritability.[80] The investigation of such properties in the Netherlands was therefore at least as much a development of Boerhaavian thought as it was a reaction against its more reductionist elements. The following discussion is devoted to these investigations as they are documented in the lectures and publications of the 1730s and 1740s of Boerhaave's five pupils.

Until the early 1730s, Jan de Gorter explained muscular contraction in quite a reductionist mechanistic manner. In his *Medicinae compendium* from 1731, he ingeniously imagined that the ultimate nervous tubes and the nervous fibrils would build a latticed network, and that the influx of nervous juice would compress the muscular fibres at the intersections, thereby transforming the fibres into shortened chains of vesicles, the microscopic equivalent to visible contraction.[81] De Gorter's explanatory approach reminds us very much of Boerhaave's smart mechanical description of the alternate contraction of the heart. Nevertheless, even then, de Gorter presented his model as a provisional and uncertain hypothesis. As he stated in later years, his practical experience increasingly made him aware of the fact that there was a hidden something, acting according to the distinct laws of life, which neither chemistry nor mechanics nor hydraulics could explain.[82] In his *Exercitatio Medica de Motu Vitali* from 1734 he argued that, as generally acknowledged, animal motion would be caused by the flowing nervous spirits. But the small amount of this fluid could never explain the strong contractions of the heart or other organs. This vital motion, also called the vital principle, responsible for life, would not be based on the organic structure and was not dependent on the soul, its existence could only be deduced from its actions.[83] It was able to produce from a small cause an immense motion, and followed laws that we could not compare with the laws of mechanics without falling into abstruse error.[84] The vital principle belonged to a category distinct from soul and body, and Gorter did not provide any further clues as to the interaction of these spheres. But he proposed – again in a Boerhaavian manner – a model of the mechanism of vital motions as far as it concerned the structure of the body. This explained the muscle's potentiality for movement (*potentia movendi*) although it could not account for the exquisite force exerted in contraction. It was thanks to the particular, elastic structure of the muscular fibres that the slow and steady influx of animal spirits could result in alternate contractions characteristic of vital motions. This mechanism would continue to function in intestines even after their excision from the body thanks to the remaining fluid in the nerves.[85]

Some fourteen years later, in 1748, Gorter added that there was no part of the body except the nails, hair and epidermis that was not agitated by vital motion, and that its absence led to corruption and death of the parts concerned. From this ubiquitous motion he now distinguished a *motus viventium particularis* responsible for the specific actions of the organs, such as preparation, secretion and excretions of humours. Its extinction did not lead to death but rendered the organ useless. The lack of the particular motion of the muscles, the *motus animalis*, was equivalent to paralysis. Again this motion could not be explained by simple mechanics; its operations and laws had to be detected by experimental investigation.[86]

Bernard Siegfried Albinus is a figure whose importance in the history of physiology is very difficult to assess.[87] For some fifty years (1721–70) he was professor of anatomy and surgery at Leiden and was acknowledged as one of the leading anatomists of his time. He also gained a certain reputation as a physiologist, although he never published any substantial work on this subject. This reputation rested on his lectures on 'human nature', which he delivered from the mid-1720s privately and after 1745 officially as professor of anatomy.[88] Although we do not know how many students attended his early, private lectures – as Haller did in 1725–7 – we may assume that for many of them, they served as an addition or even a corrective to the views of his colleague Boerhaave. Albinus was basically rooted in Boerhaavian mechanism, but already in 1727 he doubted the overall validity of the mechanical model. Like Boerhaave, he identified three causes of the heart's action – venous blood in the heart, arterial blood in the coronaries, and the animal spirits of the nerves – but he put more emphasis on the irritation of the heart and reported his own experiments on smaller animals and on the chicken embryo in which he could activate the heart through artificial stimulation. It was the distension of the muscular fibre which acted as stimulus, and this mechanism could be demonstrated also in the aorta, the pharynx, and indeed in all muscles. With more stress and precision than Boerhaave, he argued that, besides the accepted three causes, one had to postulate a fourth cause. He had observed that the heart of the eel and some other animals continues to contract for hours when extracted from the body. In these cases there was no venous blood in the cave of the heart, no blood in the coronaries, no animal spirits running in the nerves. Albinus concluded that the motion must be due to 'a certain hitherto unknown cause which adheres to the heart itself.'[89] It seems that for Albinus it was there, in the heart itself, that the immediate cause, the *causa proxima* of the movement had to be sought. This was still more or less in line with Boerhaave's statement on the innate force of the heart. But Albinus increasingly presented ideas that went beyond Boerhaave's restrained suggestions.[90] Experiments on living animals furnished him with further insight into the cause of the heartbeat. In his 1741–2 lectures he stated that the languid heart could be reactivated through the injection of lukewarm water or inflation of air. The action of the heart was therefore due neither to a cohesive force nor to a force of the nerves or arteries, it was caused by distension, by irritation and might even be prompted after its excision by pricking. The muscle did not act without an irritating cause, in voluntary motion this was the will, in involuntary motion there was some mechanical cause. But the contraction itself was due to a faculty of the muscular fibre: 'There is a force in the fibres themselves, a life which makes them contract.'[91] The fibres were in continuous oscillation, in constant movement as an expression of the *vis vitalis* or *enormôn*. They had a tendency to contract, which was held in balance by the opposing muscles. The oscillation was due to the perpetual motion of the smallest solid particles or atoms as ultimate products of food. These particles were endowed with the vital force and present in every solid and fluid part of the body. They were a kind of subtle matter attached to inert matter, and when stimulated or irritated, they swell. In the case of the muscular fibre they caused contraction. The most irritable structures of the body were the cerebral medulla and the nerves as they consisted mostly of this subtle matter.

Albinus also used his concept of vital force or *enormôn* to explain the processes of development, nutrition and growth and linked it to questions of regimen and humoral complexion. But it is difficult to say whether he really intended to provide a consistent and complete physiological system. He was a very cautious author and – as Haller observed – in his printed works often advanced a specific notion only to question it subsequently.[92] In his lectures he presented many original ideas and discoveries but he never published them.[93] This led to some unpleasant quarrels as he accused several of his pupils of plagiarism, and the partially edited lecture notes suggest that often his claims were not without foundation. But Albinus's main contribution to physiology should not be seen in specific discoveries but in his – at least partially – experimental argumentation, and especially in his encouragement to many students to venture beyond Boerhaavian mechanism.

The most important of these students for us is Frederick Winter, who had already treated the subject of muscular motion in his inaugural dissertation in 1736, written *sine praeside* and dedicated to Albinus.[94] There he showed experimentally that the ligature of the vessels or the nerve led to a paralysis of the limb in question, which proved that the flow of blood and nervous juice were prerequisites for muscular contraction. But at the same time Albinus's observation, that the muscular fibres of a beheaded animal still contracted, served him as a demonstration of a 'separately and distinctly acting faculty of contraction' inherent in the fibres.[95] And in a manner similar to Albinus – and half a year before Boerhaave – he stated that a languid heart could be reactivated through injection of blood or water thanks to its intrinsic irritability.[96] The chylus would have the same function in the intestines. These two somewhat contradictory aspects of contraction – dependence on nerves and vessels on the one hand and intrinsic capacity on the other – kept Winter thinking for a long time. He presented some further remarks in an oration held in 1746, but the most detailed presentation of his thoughts is to be found in the dissertation of his pupil Johannes Lups, entitled *De Irritabilitate* and published in 1748.[97] Lups stated that a muscle left at rest would not move, it had to be stimulated in order to contract. Voluntary muscles were stimulated by the will, involuntary muscles by an irritation, eg. the water in the urinary bladder or the blood in the heart. As the movement of the excised heart and other muscles showed, there was a faculty of movement inherent to muscles which was independent of the nerves and which Lups named irritability. What this force, also called the vital principle, would be and in what manner it would inhere in the muscles he could not say.[98] Again, the case was not that simple. Lups reported the famous experiments of Lower, in which the ligature of the cardiac nerves led to quiver, palpitation, and ultimate stoppage of the heart. Lups concluded that the nerves alone were not the direct cause of the contraction but that they were necessary for its rhythmic action. Some pages later he added that in the heart and other muscles of vital motion, irritability would be the dominant cause of action but that the influx of nervous juice would support contraction, 'because when irritability is not strong enough, the nervous liquid has to help.'[99] This is very much what Winter said in his unpublished lectures in 1752, where he stated that the sole irritability of the heart would not suffice for life but had to be assisted by the nervous liquid as a source of energy from the brain.[100]

But Lups – the mouthpiece of Winter – did not restrict the domain of irritability to the muscles. He objected to Boerhaave's idea that the nerves as instruments of sensation and movement would pervade the whole body. The placenta, for instance, was devoid of nerves but still irritable. Irritability, the faculty of movement, was independent of nerves. Lups reported the opinion of Winter, that this faculty had its origin in the dura mater, and went on to argue that all the solid parts of the body originated from this structure and therefore were irritable. He showed that the bones were a product of the periosteum which in itself derived from the dura mater, and used the ossification of the muscles as an indication of the identical nature and origin of bones and muscles. Thus he ended up in a particular merging of Albinian and Baglivian concepts, using the former's idea of an independent faculty of movement spread all over the body and the latter's stress on the importance of the cerebral membranes. Lups also used his model to explain the states of health and disease, a temptation barely any author writing about irritation and irritability could resist. According to his views, a well-balanced distribution of irritability in the body was equal to health, but the increase or decrease of motion or inflow of blood would be the cause of disease.[101]

The last two students of Boerhaave – and of Albinus – to be mentioned here, are Gaub and Kaau Boerhaave. They presented less detailed accounts of animal motion than Winter and Lups, and will not be treated in any detail. Roughly speaking, they both in the mid-1740s ventured the existence of a vital principle, the *enormôn*, pervading all the parts of the body that could be activated through irritation and which, in its local presence and independence of the soul, was responsible for specific actions and motions.[102]

Boerhaave's Dutch pupils developed quite a range of different notions of animal motion. This was possible because they were aware of their master's own changing views and did not feel bound to remain within a strict mechanical framework. But even more this was due to a particular climate, which encouraged them to embark on active scientific investigations. There were several specific factors which fostered the growth of science and the interest in natural history in the Netherlands.[103] The long period of fighting for independence resulted in a decentralised organisation of government and the foundation of local universities with flexible structures and without scholastic traditions. Politicians and the clergy were open to questions of science and education, and many members of the regent class supported scientific research. The high standard and status of craftsmanship fostered interest in the practical aspects of science and the philologist tradition stimulated attention to detail. Natural theology, or physico-theology as it was called in the Netherlands and the German countries, was a dominating philosophical current, which supported investigations into nature. No other university but Leiden could boast an uninterrupted tradition of animal experimentation since the middle of the seventeenth century, witnessing its height in the 1660s and 1670s with illustrious figures such as Regnier de Graaf (1641–73) and Jan Swammerdam (1637–80), and continuing with Charles Drélincourt (1635–94), Anton Nuck (1650–92), and their pupil Boerhaave.[104] This tradition led to a continuous scrutiny of established positions. It was with experimental observations, too, that Boerhaave's pupils argued in favour of their respective theories, and – with the exception of Gaub – these were observations they had made on their own. Despite their different positions, it has to be stressed that they shared fundamental beliefs about animal motion. As already mentioned, they held innate bodily faculties as responsible for movement and made them independent of the soul. As their experimental procedure consisted mainly of stimulation, they stressed the fact that this faculty would be especially active when irritated. As heirs of the Boerhaavian fibre theory, they tended to emphasise the uniformity of the elementary bodily structures and conceived this force as acting in all parts of the body. Last but not least, they used the terms 'vital principle' (Gorter, Winter), *enormôn* (Kaau, Gaub) or 'vital force' (Albinus) in order to designate this principle. Taken together, these authors form an important school of thought that should not be neglected in our aim to describe the evolution of the new concept of irritability and sensibility.

Notes

1. R. Porter, *The Greatest Benefit to Mankind: A Medical History of Humanity from Antiquity to the Present* (London: Fontana, 1997), 250 (on Haller), see also 222 (on Glisson). Similarly the older classical textbooks of F.H. Garrison, *An Introduction to the History of Medicine* (4th edn, Philadelphia and London: Saunders, 1929), 318 and C. Singer and E.A. Underwood, *A Short History of Medicine* (2nd edn, Oxford: Clarendon Press, 1962), 153.
2. For a general history of the terms 'irritability' and 'irritation', see H.-J. Möller, *Die Begriffe 'Reizbarkeit' und 'Reiz': Konstanz und Wandel ihres Bedeutungsgehaltes sowie Problematik ihrer Exakten Definition* (Stuttgart: Fischer, 1975).
3. Duchesneau establishes the link Boerhaave–Baglivi–Haller; Monti in her first chapter, mainly based on Grmek, draws the line Glisson–Baglivi–Haller. See F. Duchesneau, *La Physiologie des Lumières: Empirisme, Modèles, Théories* (The Hague: Nijhoff, 1982); M.T. Monti, *Congettura ed Esperienza nella Fisiologia di Haller: La Riforma dell' Anatomia Animata e il Sistema della Generazione* (Florence: Olschki, 1990); M. Grmek, 'La Notion de la Fibre Vivante Chez les Médecins de l'École Iatrophysique', *Clio Medica*, 5 (1970), 297–318.
4. See K.E. Rothschuh, *Konzepte der Medizin in Vergangenheit und Gegenwart* (Stuttgart: Hippokrates, 1978), Chapter 10.
5. For a general history of neurology see J.D. Spillane, *The Doctrine of the Nerves: Chapters in the History of Neurology* (Oxford: Oxford University Press, 1981).
6. S. Moravia, 'From Homme Machine to Homme Sensible: Changing Eighteenth-Century Models of Man's Image', *Journal of the History of Ideas*, 39 (1978), 45–60.
7. See E. Bastholm, *The History of Muscle Physiology from the Natural Philosophers to Albrecht von Haller* (Copenhagen: Munksgaard, 1950).
8. See K.E. Rothschuh, 'Vom Spiritus Animalis zum Nervenaktionsstrom', *Ciba-Zeitschrift*, 8 (1958), 2950–80; R. French, 'Ether and Physiology', in G.N. Cantor and M.S. Hodge (eds), *Conceptions of Ether: Studies in the History of Ether Theories 1740–1900* (Cambridge: Cambridge University

Press, 1981), 110–34; E. Clarke, 'The Doctrine of the Hollow Nerve in the Seventeenth and Eighteenth Century', in L.G. Stevenson and R. Malthauf (eds), *Medicine, Science, and Culture: Historical Essays in Honor of Owsei Temkin* (Baltimore: Johns Hopkins University Press, 1968), 123–41.

9. The definitive form of his concept was published in the *Opera Omnia* of 1704.
10. On the history of heart physiology, see K.E. Rothschuh, 'Geschichtliches zur Lehre der Automatie, Unterhaltung und Regelung der Herztätigkeit', *Gesnerus*, 27 (1970), 1–19; R. French, 'Sauvages, Whytt and the Motion of the Heart: Aspects of Eighteenth Century Animism', *Clio Medica*, 7 (1972), 35–54; T. Fuchs, *The Mechanization of the Heart: Harvey and Descartes*, transl. from the German by M. Grene (Rochester: University of Rochester Press, 2001).
11. *De Motu Animalium*: editio altera, correctior & emendatior (Leiden: Gaesbeek, 1685), ii, 115–16; the English translation is *On the Movement of Animals*, transl. by P. Maquet (Berlin: Springer, 1989), 285–6.
12. Fuchs, *op. cit.* (note 10), 179–80.
13. Haller pointed out that Stahl's theory, which deduces all motion from the soul, would be found already in Swammerdam, Borelli and Perrault, see *Elementa*, i, 481. On Stahl see J. Geyer-Kordesch, *Pietismus, Medizin und Aufklärung in Preußen im 18. Jahrhundert: Das Leben und Werk Georg Ernst Stahls* (Tübingen: Niemeyer, 2000). On Perrault's animism see R. Rey, *Naissance et Développement du Vitalisme en France de la Deuxième Moitié du 18e Siècle à la Fin du Premier Empire* (Oxford: Voltaire Foundation, 2000), 38–42.
14. This paragraph is especially indebted to O. Temkin, 'The Classical Roots of Glisson's Doctrine of Irritation', *Bulletin of the History of Medicine*, 38 (1964), 297–328 and W. Pagel, 'Harvey and Glisson on Irritability: With a Note on van Helmont', *Bulletin of the History of Medicine*, 38 (1967), 497–514. Temkin stresses the Galenic roots, Pagel lays more emphasis on the Aristotelian parlance in Harvey and Glisson.
15. The main accounts are to be found in 'On the Natural Faculties' and 'On the Use of Parts'.
16. *Galen on Anatomical Procedures: De Anatomicis Administrationibus*, transl. of the surviving books with introduction and notes by C. Singer (Oxford: Oxford University Press, 1956), Book VII, Chapter 8, 184.
17. *Exercitationes de Generatione Animalium: Quibus Accedunt quaedam de Partu: de Membranis atque Humoribus Uteri & de Conceptione* (London: Du-Gardian, 1651), exercitatio 51, 302; the best English translation is *Disputations Touching the Generation of Animals*, transl. with introduction and notes by G. Whitteridge (Oxford: Blackwell, 1981), 242.

18. Harvey, *Disputations* (note 17), exercitatio 51, 241.
19. *Ibid.*, 295 (exercitatio 57).
20. *Ibid.*, 381 (exercitatio 71).
21. *Ibid.*, 99 and 240 (exercitationes 17 and 51).
22. *Ibid.*, 243, 248–9 (exercitationes 51 and 52).
23. *Ibid.*, 382 (exercitatio 71).
24. *Ibid.*, 296–9 (exercitatio 57).
25. R.G. Frank, *Harvey and the Oxford Physiologists* (Berkeley: University of California Press, 1980), 22–3; R. French, *William Harvey's Natural Philosophy* (Cambridge: Cambridge University Press, 1994), Chapter 10.
26. Glisson's concept of irritability has received considerable attention and will not be treated here in detail. Substantial analyses have been presented by several authors: G.H. Meyer, 'Glisson's Irritabilitäts- und Sensibilitätslehre', *Archiv für die Gesamte Medicin*, 5 (1843), 1–17; Temkin, *op. cit.* (note 14); Pagel, *op. cit.* (note 14); T.M. Brown, 'The Mechanical Philosophy and the Animal Oeconomy: A Study in the Development of English Physiology in the Seventeenth and Early Eigteenth Century' (Princeton Univ. D.Phil. thesis, 1968); T.M. Brown, 'From Mechanism to Vitalism in Eighteenth-Century English Physiology', *Journal of the History of Biology*, 7 (1974), 179–216.
27. For a history of fibrillar theories see A. Berg, 'Die Lehre von der Faser als Form und Funktionselement des Organismus', *Virchows Archiv für Pathologische Anatomie und Physiologie*, 309 (1942), 333–460; L.J. Rather, 'Some Relations Between Eighteenth-Century Fiber Theory and Nineteenth-Century Cell Theory', *Clio Medica*, 4 (1969), 191–202; Grmek, *op. cit.* (note 3).
28. R. French, 'Ether and Physiology', in G.N. Cantor and M.S. Hodge (eds), *Conceptions of Ether: Studies in the History of Ether Theories 1740–1900* (Cambridge: Cambridge University Press, 1981), 110–34: 114–5.
29. '[...] materiam vitae radicem in se continere.' F. Glisson, *Tractatus de Natura Substantiae Energetica seu de Vitae Naturae Eiusque Primis Facultatibus* (London: Flesher, 1672), ad lectorem, § 8.
30. Baglivi's concept of fibre is treated by Grmek, *op. cit.* (note 3); M. Grmek, 'Il de fibra motrice e morbosa di Giorgio Baglivi', *Medicina nei Secoli*, 12 (2000), 19–27; Duchesneau, *op. cit.* (note 3), 116–26.
31. *Irritable Parts*, 64.
32. On Boerhaave's physiology see L.S. King, *The Medical World of the Eighteenth Century* (Chicago: University of Chicago Press, 1958); L.S. King, 'Introduction', Albrecht von Haller, *First Lines of Physiology* (Reprint New York and London: Johnson, 1966), IX–LXXII; B.P.M. Schulte, 'The Neurophysiology of Herman Boerhaave', in K.E. Rothschuh (ed.), *Von*

Boerhaave bis Berger: Die Entwicklung der Kontinentalen Physiologie im 18. und 19. Jahrhundert (Stuttgart: Hippokrates, 1964), 5–13; Rather, *op. cit.* (note 27); L. Premuda, 'Beobachtungen und Kritische Bemerkungen über die methodologische Grundlage von Herman Boerhaave', *Boerhaave and His Time* (Leiden: Brill, 1970), 40–59; G.A. Lindeboom, 'Boerhaave's Concept of the Basic Structure of the Body', *Clio Medica*, 5 (1970), 203–8; Duchesneau, *op. cit.* (note 3); J.Wright, 'Boerhaave on Minds, Human Beings and Mental Diseases', *Studies in Eighteenth-Century Culture*, 20 (1990), 289–302; A. Cunningham, 'Medicine to Calm the Mind: Boerhaave's Medical System, and why it was Adopted in Edinburgh', in id. and R. French (eds), *The Medical Enlightenment of the Eighteenth Century* (Cambridge: Cambridge University Press, 1990), 40–66; K. Wellman, *La Mettrie: Medicine, Philosophy, and Enlightenment* (Durham: Duke University Press, 1992), Chapter 3; R. Knoeff, *Herman Boerhaave (1668–1738): Calvinist Chemist and Physician* (Amsterdam: Koninklijke Nederlandse Akademie van Wetenschappen, 2002).

33. See Brown, *Mechanical* (note 26); Brown, *Mechanism* (note 26); T.M. Brown, 'The College of Physicians and the Acceptance of Iatromechanism in England, 1665–1695', *Bulletin of the History of Medicine*, 44 (1970), 12–30; R. Schofield, *Mechanism and Materialism: British Natural Philosophy in an Age of Reason* (Princeton: Princeton University Press, 1970); A. Guerrini, 'Isaac Newton, George Cheyne and the "Principia Medicinae" ', in R. French and A. Wear (eds), *The Medical Revolution of the Seventeenth Century* (Cambridge: Cambridge University Press, 1989), 222–45.
34. For Pitcairne see A. Guerrini, 'Archibald Pitcairne and Newtonian Medicine', *Medical History*, 31 (1987), 70–83; L.S. King, *The Philosophy of Medicine. The Early Eighteenth Century* (Cambridge, Mass.: Harvard University Press, 1978), 109–18.
35. A. Pitcairne, 'Oration Proving the Profession of Physic Free from the Tyranny of any Sect and Philosophers', in *The Works... Wherein are Discovered the True Foundation and Principles of the Art of Physic* (London: Curll *et al.*, 1715), 10. The lecture was originally published in Latin in 1692.
36. Pitcairne, *ibid.*, 15.
37. I used the English translation, A. Pitcairne, *The Philosphical and Mathematical Elements of Physick: In Two Books* (London: Bell and Osborn, 1718), 338.
38. *Ibid.*, 8.
39. Pitcairne, 'Dissertation upon the Circulation of Blood in Born Animals and Embryons', in *The Works..., op. cit.* (note 35), 164–7.
40. Pitcairne, *op. cit.* (note 37), 58–9.
41. I have not read the oration of de Volder; for a summary see A.M. Luyendijk-

Elshout, 'Oeconomia Animalis, Pores and Particles: The Rise and Fall of the Mechanical Philosophical School of Theodoor Craanen (1621/1690)', in T.H. Lunsing Scheeurler and G.H.M. Posthumus Meyjes (eds), *Leiden University in the Seventeenth Century: An Exchange of Learning* (Leiden: Brill, 1975), 294–307: 304–5. It was Volder who introduced experimental physics at Leiden, and Boerhaave was one of his students. See E.G. Ruestow, *Physics at Seventeenth and Eighteenth-Century Leiden: Philosophy and the New Science in the University* (The Hague: Nijhoff, 1973), 91–115.

42. *Boerhaave's Orations*, transl. with Introductions and Notes by E. Kegel-Brinkgreve and A.M. Luyendijk-Elshout (Leiden: Brill, 1983), 107 (oration of 1703). For all of Boerhaave's orations I have adopted this reliable translation.
43. *Ibid.*, 108.
44. *Ibid.*, 114.
45. *Ibid.*, 108.
46. *Ibid.*, 102.
47. *Ibid.*, 97.
48. *Ibid.*, 105.
49. L.S. King ('Rationalism in Early Eighteenth Century Medicine', *Journal of the History of Medicine*, 18 (1963), 257–71) stresses Boerhaave's rationalism whereas H.J. Cook, in 'Boerhaave and the Flight from Reason in Medicine', *Bulletin of the History of Medicine*, 74 (2000), 221–40, emphasises his anti-rationalist tendencies. Cook, in my view, underestimates the rationalist and systematic character of Boerhaave's medicine. What Boerhaave demanded was the use of both, experience and reason. Although he claimed that medicine and scientific investigation had to be founded on observation and experiment, he argued that the conclusions inferred by the use of 'strict principles of reason' ('severa lege ratiocinii') were 'certainly as secure as those, which appear to the senses'. *Praelectiones*, i, § 20, 24.
50. See Berg, *op. cit.* (note 27); King, *op. cit.* (note 32); Rather, *op. cit.* (note 27); Lindeboom, *op. cit.* (note 32).
51. There is some disagreement between scholars over whether the fibre (Rather, Lindeboom), the vessel (King), or even the nervous fibre (Berg) would be the most simple structural unit. As Rather and King have shown, we can certainly rule out the nerve.
52. For this part of the concept I have followed entirely Lindeboom, *op. cit.* (note 32).
53. '...vas minimum ex arteria ultimo exortum evadat simillimum fibrillae minimae nervi.' H. Boerhaave, *Institutiones Medicae in Usus Annuae Exercitationis Domesticos* (5th edn, Leiden: Theodor Haak *et al.*, 1734), § 302 (identical in the 1713 edition).

54. '...nervum enim vocamus vasorum id, quod ultimum est, et tenuissimum.' *Praelectiones*, iii (1741), 643–4.
55. '...omnem totius nostri corporis solidam massam meris modo nervis, ut elementis suis, absolute constructam esse....' Boerhaave, *op. cit.* (note 53), § 440 (identical in the 1713 edition).
56. '... vix ullam esse particulam corporis, quae non sentiat, vel non se commoveat; credemus fere, omnes partes solidas corporis contexta esse fibris nervosis atque iis constare.' *Ibid.*, §301 (identical in the 1713 edition).
57. '... has fibras esse nervi ultimi, involucris orbati, expansionem tenuissimam, intus cavam, figurae ut musculus, plenam spiritu ....' *Ibid.*, § 395 (identical in the edition of 1713).
58. *Boerhaave's Orations* (note 42), 163 (oration of 1715).
59. *Ibid.*, 165.
60. *Ibid.*, 168.
61. Duchesneau, *op. cit.* (note 3), 115.
62. H. Boerhaave, *A New Method of Chemistry*, transl. by P. Shaw (2nd edn, 2 vols, London: Longman, 1741), i, 174. Knoeff, *op. cit.* (note 32).
63. *Boerhaave's Orations* (note 42), 262 (oration of 1730).
64. *Ibid.*, 257.
65. *Ibid.*, 256.
66. *Ibid.*, 254.
67. A. von Haller, 'Physiologie', *Encyclopédie, suppl.*, iv (1777), 344–65: 354.
68. Knoeff, *op. cit.* (note 32) describes Boerhaave's development from a firm mechanist to a supporter of non-mechanical powers especially in his changing attitude towards chemistry and relates it to his Calvinism. I have not been able to fully evaluate and integrate all her arguments as I read her book only after this chapter had been written.
69. The first edition of the *Institutiones Medicae* (1708) is much shorter than the second (1713); in the later editions of 1727 and 1734 there are only rather short, but often quite important, additions. A critical edition of this work – often called 'the most important medical book of the eighteenth century' – would indeed be of great value.
70. These lectures were published in 1761, based on notes of pupils. The new edition is based on Boerhaave's own notes: *Hermanni Boerhaave Praelectiones de Morbis Nervorum 1730–35*, B.P.M. Schulte (ed.), (Leiden: Brill, 1959).
71. *Praelectiones*, iii (1741), § 401; interestingly, La Mettrie in his edition of Boerhaave (a translation of Boerhaave's lecture and most of Haller's remarks including some of La Mettrie's own thoughts) did not adopt this passage.
72. 'Mirifica, & occulta, est in corde fabricato proclivitas in reciprocandas systoles & diastoles, vices, etiam a morte, imo & in corde exsecto, denique & in segmentis cordis dissecti.' Boerhaave, *op. cit.* (note 53), § 187.

73. *Irritable Parts*, 63.
74. It is possible that Boerhaave's interest in muscular action was enlivened through working on the posthumous papers of Jan Swammerdam which he acquired in 1727 and edited in 1737–38. Of special interest are the experiments on heart and muscles of frogs; see J. Swammerdam, *Bybel der Natuure..., Hier by Komt een Voorreeden, waar i het Leven van de Auteur Beschreben is door Herman Boerhaave... De Latynsche Overzetting heft Bezorgt Hieronimus David Gaubius* (2 vols., Leiden: Severinus, 1737–38), ii, 835–61.
75. 'Consideratis jam causis tribus, quae cordi applicatae, faciunt illos motus alterne repetitos, quarta tandem etiam consideranda venit, quae est porprietas ipsius cordis. Vocabo hanc irritabilitatem ejus, atque eo quidem nomine intelligam aptitudinem ei innatam ex fabrica, qua facile, & prompte, se contrahit, ad applicationem illarum causarum, vel et ad alias stimulantes.' *Praelectiones Publicae de Corde*, 1735–37, ff. 60–1 (lecture of 23 November 1736). University Library, Leiden, MS. Leningrad XIII. 32 (microfilm of the original manuscript in the Military Academy in St. Petersburg). A catalogue of all the manuscripts of Boerhaave which, surprisingly, have never been used in the assessment of his physiological work, is given by E. Cohen, 'Katalog der Wiedergefundenen Manuskripte und Briefwechsel von Herman Boerhaave', *Verhandelingen der Koninklijke Nederlandes Akademie van Wetenschappen, Afdeling Natuurkunde*, 2nde Sectie, xl/2 (1941), 1–45.
76. For the physiological concepts of Boerhaave's Dutch pupils see A.M. Luyendijk-Elshout, 'Mechanisme contre Vitalisme; De School van Herman Boerhaaven en de Beginselen van het Leven', *Gewina*, 5 (1982), 16–26 and G. van der Waa, '*De Irritabilitate': Een Onderzoek naar de Betekenis van het Irritabiliteitsbegrip in de Geschiedenis van de Achttiende-Eeuwse Nederlandse Fysiologie* (Rotterdam: Erasmus, 1992); on Albinus see H. Punt, *Bernard Siegfried Albinus (1697–1770) on 'Human Nature': Anatomical and Physiological Ideas in Eighteenth Century Leiden* (Amsterdam: Israël, 1983); J.K. van der Korst, 'Anatomia Actuosa et Apta: De Mechanicistische 'Proto'-fysiologie van B.S. Albinus', *Gewina*, 16 (1993), 308–23. See also H. Beukers, 'Boerhaavianism in the Netherlands', *Journal of the Japan-Netherlands Institute*, 1 (1989), 116–29.
77. See eg. R. French, 'Sickness and the Soul: Stahl, Hoffmann and Sauvages on Pathology', in A. Cunningham and R. French (eds), *The Medical Enlightenment of the Eighteenth Century* (Cambridge: Cambridge University Press, 1990), 88–110; F. Huisman, 'Medicine and Health Care in the Netherlands, 1500–1800', in K. van Berkel, A. van Helden and L. Palm (eds), *A History of Science in the Netherlands: Survey, Themes and Reference* (Leiden: Brill, 1999), 239–78: 264–7.
78. There were not that many other pupils of Boerhaave who actually published

on physiology or who were engaged in physiological research. Some of them developed dynamic concepts (most notably Haller and Robert Whytt), some did not (Alexander Monro and Johann Friedrich Schreiber). Two of the most famous pupils, Gerard van Swieten and Anton de Haen, stuck to a mechanistic interpretation, but they were concerned with practical medicine rather than physiology.

79. Gorter consequently baptised one of his sons with the name Hermanus Boerhaave, in 1732; see G.A. Lindeboom, *Herman Boerhaave: The Man and his Work* (London: Methuen, 1968), 43.
80. See his letter to Haller, 14 November 1752, in *Epistolae*, iii, 348.
81. I have not consulted this work myself and rely on Berg, *op. cit.* (note 27), 412–16, Bastholm, *op. cit.* (note 7), 230–2; Rather, *op. cit.* (note 27), 193–4.
82. J. de Gorter, *Exercitatio Medica Quinta de Actione Viventium Particulari* (Amsterdam: Gerrevink, 1748), praefatio.
83. '...hic motus neque *mentis actione*, neque ex *organi structura* continuetur....' J. de Gorter, *Exercitationes Medicae Quatuor. I. De Motu Vitali. II. De Somno et Vigilia. III. De Fame. IV. De Siti* (Amsterdam: Ratelband 1737), 6. This book contains the treatise of 1734 and three further 'exercitationes'.
84. *Ibid.*, 2–8.
85. *Ibid.*, 21–4, 44.
86. Gorter, *op. cit.* (note 82), 1–2, 7–8.
87. Besides Punt, *op. cit.* (note 76), very few authors have attributed any significance to him at all. Korst, *op. cit.* (note 76), argues that Punt overstates the importance of Albinus's vitalist statements. According to Korst, they are embedded in a general mechanist vocabulary, and Albinus has to be considered a strict mechanist. In my view, however, they announce significant changes in the conception of living matter.
88. On the content of the lectures see Punt, *op. cit.* (note 76), who also edits important paragraphs of lecture notes taken by students.
89. 'Erit a causa adhuc quaedam ignota quae in ipso corde haeret.' A. Haller, 'Excepta de Usu Partium harumque Fabrica Subtili, ex Ore Clar. Viri D. Bernhard Siegfried Albini', 1727 (Biblioteca Nazionale Braidense, Milano, AD.XI.6), 43.
90. The following description has to rely not only on the lecture notes (esp. 138–57) but also on the analysis presented by Punt, *op. cit.* (note 76), as he edited only rather short excerpts of the lectures.
91. Punt, *op. cit.* (note 76), 141.
92. *GGA*, 1759, 1315.
93. There is a posthumous edition of Albinus's lectures edited by his brother Friedrich Bernhard in 1775, which, understandably, did not attract great attention at this time.

94. F. Winter, *Dissertatio Medica Inauguralis de Motu Musculorum* (M.D. thesis, Leiden: Luzac, 1736).
95. '... inesse singulis fibris seorsim consideratis, facultatem separatim & distincte se contrahendi.' *Ibid.*, 11.
96. '... inest tamen & cordi ipsius irritabilitas quaedam faciens, ut non diu post mortem, utcunque ad contractionem adhuc stimulari posset.' *Ibid.*, 43.
97. F. Winter, *Oratio Inauguralis de Certitudine in Medicina Practica Dicta a. d. XXVI. Aprilis MDCCXLVI* (Franeker: Coulon, 1746), 80–95; J. Lups, *Dissertatio Physiologico-medica Inauguralis de Irritabilitate* (M.D. thesis, Leiden: Luzac, 1748). Waa, *op. cit.* (note 76) has shown that the dissertation of Lups basically related the content of Winter's lectures on the topic; this is what Lups himself declared also, 28.
98. Lups, *op. cit.* (note 97), 17.
99. *Ibid.*, 17.
100. Winter's lectures on irritability, February–September 1752, edited by Waa, *op. cit.* (note 76), 203–34: 212.
101. Lups, *op. cit.* (note 97), 28.
102. A. Kaau-Boerhaave, *Impetum Faciens Dictum Hippocrati per Corpus Consentiens Philologice et Physiologice Illustratum Observationibus et Experimentis passim Firmatum* (Leiden: Luchtmans, 1745); H.D. Gaub, *Sermo Academicus de Regimine Mentis quod Medicorum Est* (Leiden: Van der Aa, 1747). See also L.J. Rather, *Mind and Body in Eighteenth-Century Medicine: A Study Based on Jerome Gaub's De Regime Mentis* (London: The Wellcome Medical Historical Library, 1965); Punt, *op. cit.* (note 76), 124–6 (for Kaau-Boerhaave).
103. W.D. Hackmann, 'The Growth of Science in the Netherlands in the Seventeenth and Early Eighteenth Centuries', in M. Crosland (ed.), *The Emergence of Science in Western Europe* (London: Macmillan, 1975), 89–109; H.J. Cook, 'The New Philosophy in the Low Countries', in R. Porter and M. Teich (eds), *The Scientific Revolution in National Context* (Cambridge: Cambridge University Press, 1992), 115–49.
104. G.A. Lindeboom, 'Dog and Frog: Physiological Experiments at Leiden During the Seventeenth Century', in T.H. Lunsing Scheeurler and G.H.M. Posthumus Meyjes (eds), *Leiden University in the Seventeenth Century: An Exchange of Learning* (Leiden: Brill, 1975), 278–93. Boerhaave definitely performed experiments in his early years as tutor but it is difficult to say to what extent he continued to do so after his appointment as professor; see Lindeboom, *op. cit.* (note 79), 36.

2

Experimentation in the Göttingen Laboratory

There are two main components of scientific activity.[1] The practical side consists in the physical investigation and may involve observation, preparation, experiment, and measurement. The theoretical side comprises the analysis of the results thus achieved but also of those displayed in other writings, and includes different modes of more or less reasonable argumentation which may be based on presumptions, interpretations, deductions and inductions, calculations, inferences, hypotheses, etc. In the last two decades it has become a commonplace in the historiography of science to argue for the intricate intertwining of both these parts.[2] The focus of attention has shifted from the study of theories to that of the investigative process in the laboratory. But whereas the research procedures in the physical and chemical sciences have been studied in great detail, those in the medical and biological sciences have been much less examined.[3] This is particularly true for the eighteenth century, and it has to do with the relatively small amount of experiments performed in this period and the limited number of laboratory notebooks preserved.[4] The experimental procedures of several eminent eighteenth-century authors – notably Abraham Trembley (1710–84), Charles Bonnet (1729–93) and Lazzaro Spallanzani (1729–99) – have, however, been studied to some extent.[5] As to Haller, the interplay between theoretical reflection and experimental practice in his embryological research has marvellously been demonstrated by Maria Teresa Monti.[6] In this chapter, I examine his investigations into irritability and sensibility, which confronted him with a different set of methodological problems. Together with the discussion of various attitudes towards animal experimentation in Chapter 4, I hope to contribute to the understanding of eighteenth-century experimental biology. Furthermore, I want to stress the central role the experimental approach played in Haller's whole physiology. As we have seen in the previous chapter, the performance of animal experiments was not new at all. Experiments were carried out by many physicians in the mid-seventeenth century and also by the Dutch pupils of Boerhaave, but never before were they performed in such a systematic manner and considered as a far-reaching program in order to establish a new basis of physiology. The description of the specific 'milieu' of research Haller created in Göttingen is important in order to understand not only how he achieved his results but also to understand the position from which he argued. In the first section a general outline of the conditions of the Göttingen 'laboratory' will be given.[7] It will be indispensable to repeat several of the characteristics already revealed by Renato Mazzolini.[8] The two following sections of this chapter will deal more specifically with the question of irritability and sensibility and will examine the experiments performed by Haller and his pupil Johann Georg Zimmermann, who achieved results quite different from those of his teacher and presented them several months before him. Many of these experiments they carried out jointly, although Haller had already started with certain investigations before the arrival of his pupil, and continued his extensive research after Zimmermann's departure from Göttingen. Both of them made notes of their experiments which are preserved almost in their entirety.[9] This allows a partial reconstruction of the sequence of their theoretical and experimental inquiries. The last section deals with an old but important argument which has recently been strengthened by a new analysis of Andrew Cunningham.[10] Cunningham argues that the physiologists before the nineteenth century – including Haller – considered vivisection not as part of the theoretical science of physiology but of the mechanical art of anatomy. And their experimentation on living animals – including Haller's investigations into irritability and sensibility – were not cases of experimental physiology but of experimental anatomy, as they started from anatomical structures and properties instead of physiological. The question raised by Cunningham will serve as a means to re-approach from a new angle the experimental process described in the first sections of this chapter.

The Göttingen experimental community

Haller's fame attracted many students, but his teaching seems not to have been suited for beginners. As some of his pupils said, his demonstrations of anatomy did not reach the necessary degree of comprehensibility and were often devoted to special subjects with which he was occupied at that time. The students had to learn the basics after the lesson from the prosector.[11] Botany he explained unsatisfactorily, too; it was more easily to be mastered with the help of his publications than in his lectures.[12] But this is exactly what Haller had proposed in his inaugural lecture *On the Methodic Study of Botany without a Teacher* (1736), a rather curious title for a professor taking up an appointment.[13] Haller was well aware of the dilemma. On the occasion of the foundation of the Göttingen Academy of Science (1751), he commented with some regret that the professors were bound to teach the whole range of their subject and to comply with the interests of the students instead of being allowed to pursue their own. As they had to dedicate their whole time to instruction they could not contribute to the advancement of science. A scientific academy, on the other hand, was only created for 'invention' (*ad inveniendum*).[14] This separation of research in the academy and teaching at the university was common at that time and Haller, as the main designer of the program in Göttingen, did not fundamentally change this. But since the heads of the different sections of the academy at the same time were professors at the university, there was a close bond between the two institutions. As these heads regularly had to produce essays, necessarily inventive, research came at least partially to be a duty at the university, too.[15]

Haller, for one, found his own way of combining teaching and research. He rather neglected the younger and more idle students, but took all the more care of the advanced and promising among his pupils. At the outset of his Göttingen years (from 1736), the dissertations written under his direction were, as usual, mostly his own accomplishment, but more and more he demanded independent work from his candidates, and especially in his later years (1746–53) he urged them to undertake difficult and time-consuming investigations. Although a serious analysis of authorship, content and originality of medical dissertations of the eighteenth century is lacking, all the information we have points to the fact that this was extraordinary. The author was often the professor himself, the effort involved smaller, and the performance of experiments rare.[16] The dissertations of Haller's pupils were therefore often considered to be works of their master. But Haller rejected this idea:

> We do have to remark in the first instance, that Mr D. attributes the works of Zinn and Zimmermann to our teacher [Haller], and uses the latter in particular to raise objections against Mr v. H[aller], which concern him much the less, as he was not even presiding over Zimmermann's dissertation and even less was he its author.[17]

It was unusual for students to write the whole thesis themselves; the performance of experiments for inaugural dissertations was even more exceptional. At the University of Edinburgh, for instance, towards the end of the century, despite the esteem for experimental research, it was a rarity when a student not only mentioned experiments of others but actually conducted his own.[18] Only during the French revolution at the *École de Santé* in Paris was an atmosphere created which fostered the experimental approach again; twelve per cent of the dissertations at this school were based on animal experiments.[19] In the light of these facts, Haller expected a great deal from his pupils. This is how he justified his demands:

> Something really can be achieved through the candidates, which we can expect less from the professors, if only they [the candidates] invest some time and proceed with patience and diligence. And that is why I advocated

> restricting the practice of writing dissertations to the aim for which the learned societies are created, that is to invent, to settle obscure controversies and to fully establish the single parts of truth.[20]

Haller placed research as it should be pursued by students at the university on the same level as that conducted at the academy. He stressed the particular character of the Göttingen Academy which would also be of use to the students. A dialogue should develop in which the pupils would not only learn from the discoveries and thoughts of the teachers, but to which they could also contribute with their own insights.[21] It made sense, therefore, that in the meeting of the Academy on 4 November 1752 Haller led a student to perform the experiment which was meant to prove the absence of air in the pleural cavity.[22] In doing so, the student was not, however, entering a research community for the first time, he was simply crossing over from one to another. As Zimmermann, in his biography of Haller, remarked, he had been a member of a scientific community before:

> I have now given an account of the history of the Royal Academy of Sciences in Göttingen. But already a fairly long time before another academy of science under the direction of Mr Haller had been present at this place, and with fiery zeal had served the kingdom of truth. The members of this society were not encouraged by rewards, the fame of their comrades was not their fame, they could not pride themselves on titles or on royal protection; they even sacrificed their own fortune for their love for the sciences and made their endeavours public at their own expense.[23]

This community was formed especially in Haller's later Göttingen years (1746–53), when he involved selected students in his own research program. At this time an average of 600 students attended the university, fifty-to-sixty of whom studied medicine.[24] Ten-to-fifteen of these later studied for their doctorate with Haller. They stayed in Göttingen for an average of two and a half years. Four to five of them were in the final year which was normally needed to accomplish a dissertation, and they must have formed the small circle of diligent students that surrounded Haller.[25] Twelve of thirty-eight doctoral candidates tutored by Haller in this time performed experiments on living animals. In his early years Haller himself had not tried many experiments but in 1746 the controversy on the mechanics of respiration with Georg Erhard Hamberger (1697–1755), professor of medicine at Jena University, intensified.[26] Ten years later, Haller recollected:

> The dispute on respiration made me to repeat [the experiments] and bit by bit the taste for it spread. Several of my students wanted to carry out series

> of experiments in order to enrich their inaugural dissertations. I guided their experimentation.[27]

For this, of course, he needed laboratory animals. Haller wrote to Réaumur in 1752, a month after the delivery of his speech on irritability:

> I make my arrangements in order to continue the research on the eggs of the quadrupeds. One needs a stock of dogs and rabbits, which have their difficulty in a small town where everything astonishes and attracts gapers. The destruction of dogs I have caused has rendered their acquisition difficult. But this will only delay my research.[28]

We do not know whether Haller bred dogs and rabbits himself. At any rate, the students had to buy the animals by themselves and had also to look after them.[29] They conducted their anatomical and physiological studies in the anatomy building situated at the edge of the botanical garden, just next to Haller's own house (Figure 2.1). On the upper floor were five rooms and the experiments were presumably performed in the oblong dissection hall with a view of the garden (Figure 2.2). Several of the students tell us in their dissertations that their investigations have been witnessed by others. The frontispiece to Haller's French edition of his collected experiments conveys a certain, although surely not very accurate idea of how the scene might have presented itself (Figure 2.3).[30] It is not very likely that several experiments were performed at the same time in the same room. Haller himself was often but not always present. He never carried out his own experiments alone; they were always witnessed by at least one student. Sometimes they were performed also in the anatomical theatre in front of a greater number of students.[31] He reports the names of seventeen pupils present at different stages of his work.[32] He named his friend and professor of philosophy, Samuel Christian Hollmann (1696–1787), as the only non-student witness. This picture corresponds closely with what we know about Haller's days in Göttingen. He was not on very good terms with the two other professors of medicine, the tense relations sometimes even worsened to open hostility. His contacts with other teachers were not much better. Zimmermann recorded in his private notes: 'The enemies of Mr Haller have everywhere been of a very great number. ... One would have to go through too long a list of professors in Göttingen in order to mention all those he had at this university.'[33] Haller withdrew into his work, surrounded by his disciples. It is not astonishing that a close relationship developed between teacher and some of his pupils during the completion of the dissertations with their series of experiments, which often lasted over several months. The gratitude expressed by students in their dissertations and letters to Haller was, within the rhetorical shell, genuine. And Haller's accompanying letters, inserted in

Figure 2.1
The botanical garden of Göttingen with the anatomy building (left) and Haller's own house (right). Frontispiece to Haller's Enumeratio Plantarum Horti Regii et Agri Gottingensis *(2nd edn, Göttingen: Vandenhoeck, 1753).*

several of his pupils' dissertations, are signs of his sincere regret at their departure. Moreover, several of the students evidently cultivated quite close friendships among each other.

Although the bonds between some members of this community seem to have been quite strong, we should not paint an overly romantic picture. Some of the animal experiments performed in their laboratory were of a cruelty beyond description and must have provoked reluctance at least in some cases.[34] Furthermore, Haller was a very earnest man and his difficult character must have created tensions. Extremely ambitious himself, he encouraged the ambitions of the students, as he considered this to be one of the main driving forces behind research.[35] The pupils who had decided to write a difficult dissertation under the famous professor seem to have been receptive to this kind of incentive. Zimmermann, for instance, wrote in a letter in 1748: 'I lead in this country the life of a man who wants to continue living after his death.'[36] Despite friendships a certain competition must have reigned, not least due to Haller's own conduct, who conveyed his uneven sympathies. He clearly distinguished between 'superior' and 'inferior'

Figure 2.2
The Göttingen laboratory, the second floor of the anatomy building. The animal experiments were presumably carried out in the oblong room ('Praeparir-Saal'). Ground plan from W. Heider, 1829 (from Thode, Brita, Die Göttinger Anatomie 1733-1828 *(Göttingen Univ. M.D. thesis, 1979), figure 10).*

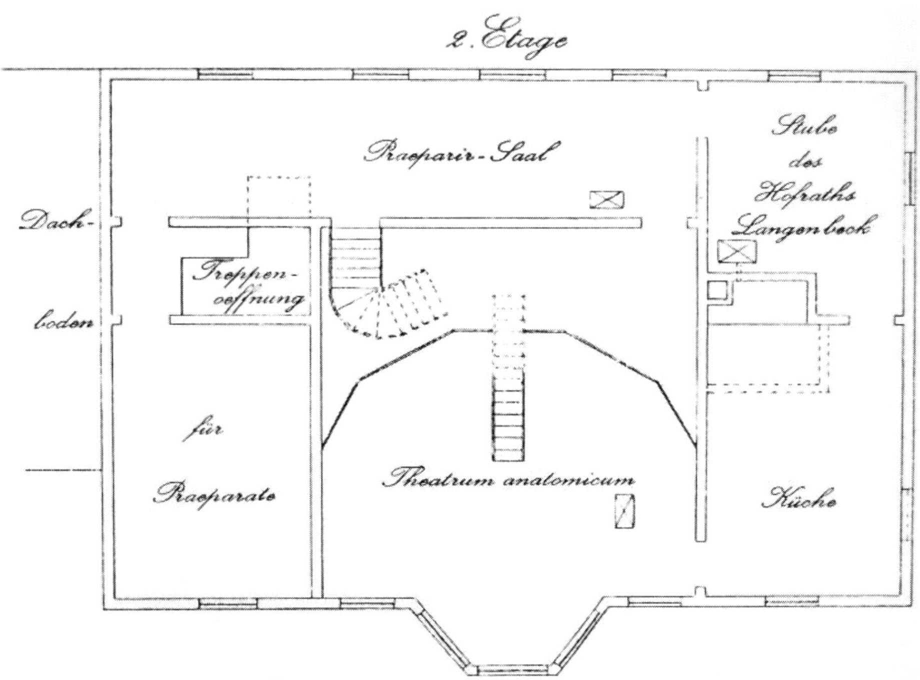

students. His pupil Foelix, for one, he called, 'our disciple of the superior class'.[37]

The common credo of this community was not so much a medical theory which we might call Göttingen or Hallerian physiology but rather the belief in observation and experiment as the basis of research. Haller put forward this conviction in his writings over and over again, and it was repeated by his pupils in their own texts. It is even more significant, though, that several students let their dissertations begin with the description of experiments without any prior methodological explanations. Eight of the twelve dissertations based on animal experiments bear the same type of title: 'Inaugural dissertation consisting of experiments...' (*Dissertatio Inauguralis Sistens Experimenta...*). The experiments are the core and main issue of the dissertation, and usually they are detached from the rest of the text. The authors describe the trials conducted by themselves in the first person. Haller's experiments on the same topic and performed in their presence, on the other hand, they did not usually mention. In the case of Zimmermann we know that Haller carried out many more experiments on irritability during and immediately after the most intensive working phase of his pupil in November and December 1750. Zimmermann must have witnessed these trials but he did not mention them in his dissertation. In only one case he entered in his laboratory notebook: 'The experiments on the other dog... I do not record as they are not mine.'[38] In this community geared towards performance, results and rewards, the experiment was the most precious jewel and the property of the experimenter. Haller always took great care to avoid claiming merits he did not deserve but he vigorously defended what he considered to be his own contribution. This meritocratic principle ruled the small Göttingen experimental community as well.

The topics of the investigations were clearly defined. The dissertations exactly fulfilled what Haller expected from the work of a member of a scientific society:

> Such a member of a learned society does not cover a whole branch of science, he does not present a compendium or, so to speak, a map of an extensive empire, from which, in the limited space, he could bring out only a few cities and none of the market-towns. For such an academic paper one chooses a small area, the hills and streams, market-towns and villages, and almost the single houses of which one undertakes to locate. ...Compelled and at the same time encouraged, [the members] take care of a small part of the true [des Wahren], wherein they do not learn but teach and apply their previously untested forces for the greatest benefit of the sciences.[39]

This deliberate promotion of specialised research was, for Haller, the only way to move on. The single larger or smaller results, though, should not remain isolated but in their totality would help to construct and extend the single branches of science. It was this perpetual process that Haller along with his pupils wanted to set in train in the fields of anatomy and physiology:

> Whenever somebody applied for the medical degree, and to this end was about to draw up a dissertation, I easily convinced him to tackle a difficult part of anatomy to which he had to devote almost two winter terms. This proposal not only gave special credit to the candidate but I could also curtail my own dissections. Indeed, in no other way could one come closer to the perfection of anatomical knowledge than by following this good advice for many years and through whole centuries at a university, which was equipped

Figure 2.3
Performing animal experiments on irritability and sensibility (Frontispiece to the Mémoires*).*

> with all the appropriate facilities, and where, as in Göttingen, this good intention was supported by ambition, particular emulation and public reward.[40]

What Haller achieved with his pupils was only a first step on this long road. His sudden departure from Göttingen and the premature death of several of his pupils possibly prevented the early emergence of a continuous tradition of animal experimentation, which was secured only in the 1790s in Paris.[41] But he clearly had envisaged the general goal of his efforts already in the 1740s: the establishment of a new physiology, based on an experimental foundation. In 1755 he wrote to Charles Bonnet: 'I had decided to repeat everything and to verify absolutely all the physiological experiments, without excluding the most simple ones like those on circulation.'[42] Renato Mazzolini has already pointed out that we have to conceive Haller's single works as parts of this broad research programme.[43] Clearly, it was impossible for him to deliver experimental contributions to all parts of physiology. Nevertheless, he thought it useful to deliver a comprehensive account of the actual state of knowledge. Such general descriptions of different branches of science would help us to see how far one had advanced and what was left to do.[44] His own exposition of physiology he presented in 1757–66 in eight monumental volumes.[45] Inevitably, he was also forced to give accounts of poorly studied areas. But wherever he could, Haller drew upon experiments, preferably conducted by himself or his pupils. They were the raw material he needed for the establishment of his synthesis. It was less the explanations given in the dissertations on which he relied, and much more the observations and experiments themselves which he cited in their hundreds. The interpretation of the phenomena discovered by his pupils he reserved for himself.

The experimental process

After three years of medical studies, Haller's pupil Johann Georg Zimmermann (1728–95) started working on his dissertation in 1750, which would make up his fourth and last year in Göttingen. Like quite a few of his fellow students he was a Swiss, from Brugg in the state of Bern. The presence of fellow countrymen was a certain comfort to Haller who was often plagued by homesickness (*Heimweh*), which in these times was considered to be a typically Swiss malady.[46] Zimmermann was certainly one of the favoured pupils of Haller. After his studies he wrote an extensive biography of his teacher (1755), who procured him the post of a physician in his hometown and, in 1768, that of private physician to the English king in Hanover. He became one of the best known figures in the German countries as a much demanded physician but even more through his popular books on solitude

(1756, revised 1784–5), national pride (1758, revised 1768), experience in medicine (1763–4) and for his published conversations with Frederick the Great (1788).[47] These writings were not to Haller's taste and were partly responsible for the growing alienation between the two.

As a student, Zimmermann lived in Haller's house. He was very diligent and in the evening drank a lot of tea in order to stay awake as long as possible.[48] Haller described him as a pupil 'of extraordinary promise.'[49] It must be considered as a sign of special appreciation that Haller placed the problem of muscular action into his hands. The last chapter outlined the knowledge of this controversial topic as it was available to Zimmermann when he set out to explore it himself.[50] Of course, he was especially familiar with Haller's thoughts, which will be treated in more detail in the next chapter but some of which we must mention briefly here. Similar to Frederick Winter, Haller located a pronounced irritability especially in the heart and the intestines, which explained the movement of these organs even after their detachment from the nerves. Nevertheless, and again in accordance with Winter, the nerves to him seemed to play a certain role, as their ligature resulted in a decrease of the activity of the heart. But he rejected both Winter's theory of the dura mater as origin of all movement and Boerhaave's idea of a sensible and motive fibre as the ultimate structural unit of the body. These concepts, in his view, would result in the attribution of the faculties of sensation and movement to all parts of the body. To such generalised notions he had objected already in his early writings and with more precision in the dissertation of his pupil Johann Gottfried Zinn (1727–19) in 1749 which had proven the insensibility and immobility of the dura mater.[51]

On 15 September 1750 at 2pm, Zimmermann performed his first experiments on two frogs.[52] With a pair of pliers he cut off the head of a living frog and put it on the floor. The frog jumped around three or four times. Half an hour later and again after one and a half hours, it reacted to minute irritations of its upper leg with long leaps. Even at 8.30pm it was still jumping and at 10.20pm still moving. The next morning, though, it was stiff and 'the heart could not be irritated in any manner.' In six further, similar experiments, carried out on 26 September and 19–21 October, Zimmermann managed to reactivate a stopped heart through stimulation. From these sporadic trials he concluded that heart and muscles may be set in motion irrespective of their connection with the brain. But, of course, that is what he already knew before. To us, these early experiments mainly reveal Zimmermann's uncertainty as to what to examine and how to proceed. It seems that Haller was not present at these first trials. Since the series of his pupil Foelix from December 1749 to February 1750, Haller had not performed further experiments of his own. Only on 12 November 1750 did he resume his former inquiries and determined the topic of his new interest: 'a dog, regarding irritability' (*canis ad irritabile*). This notebook entry marks the beginning of Haller's explicit experimental investigation into the subject. And now the quality of the experiments changed. Henceforth, the animals were mostly dogs and cats, the single bodily parts of which were irritated with greater care, both mechanically and chemically and in some cases also by the application of heat or cold.[53] Until 15 April 1751, the day of Zimmermann's last notebook entry, some 200 experiments were carried out, 45 of which were protocolled by the student, 135 by the teacher, and 20 by both of them (see Figure 2.4 and 2.5). The person taking the notes was usually the experimenter himself but very likely the trials were often witnessed by both student and teacher, and sometimes by others, too. From Haller we know that he made the notes immediately after the experiment and in presence of those who had seen it.[54] Whereas Zimmermann completed the greater part of his experimental work after three weeks, in November 1750, Haller carried on with continuous intensity for another year, until November 1751.

On 22 July 1751 Haller wrote to his former pupil Georg Thomas von Asch (1729–1807): 'In two weeks Mr Zimmermann will [publicly] defend [his dissertation on] irritability and, indeed, he will cause a sensation.'[55] Haller was well aware that Zimmermann's dissertation would call for a revision of several traditional assumptions about the movement and sensation of some of the parts of the body.[56] What he did not say in his letter, though, was that already at this stage his own views differed in many respects from those of Zimmermann. His own results he presented some nine months later in his famous speeches at the meetings of the Royal Society in Göttingen. The following discussion tries to explain how it came about that Haller and Zimmermann reached quite different conclusions, even though they worked together in the same laboratory.

The tendon, or when to irritate

In the first experiments performed by Haller on 12 November 1750, the single parts of the body were tested for their capacity for contraction. Zimmermann, who had witnessed these experiments, recorded: 'In the urinary bladder [there was] quite a strong contraction... The uterus definitely could not be provoked to contraction. The stomach showed the same degree of irritability as the appendix.' Suddenly, reporting an experiment which is not mentioned in Haller's notebooks, he recorded: 'The Achilles tendon was insensible [*insensilis*].' Zimmermann's sudden change of terms, from 'irritable' to 'sensible', seems not to have been made on purpose.

Figure 2.4
Experiment on a dog, 25 November 1750. From Haller's protocols (Haller papers, Ms. 30, fol. 121r).

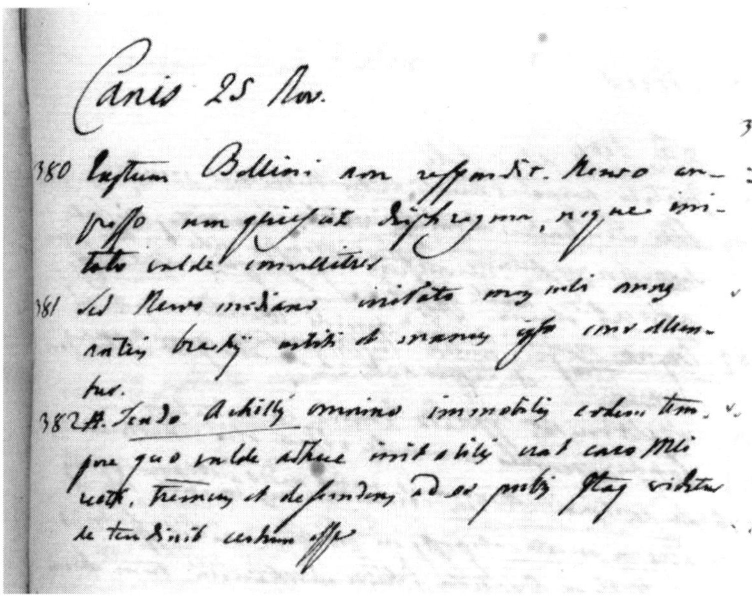

He simply wanted to state that the tendon was not responsive to any irritation. He did not want to exclude any sensation of pain as the animal had already been unconscious for a long time. Quite clearly, he did not have a distinct concept of sensibility. Haller, however, had always associated this term with the transmission of sensations through nerves and their conscious reception in the brain.[57] What Haller had examined was irritability, the capacity of contraction, and there is no evidence that he sought for the sensibility of the tendons. Two days later he noted down that the muscular but not the tendinous part of the diaphragm would contract upon irritation. He reported his findings to his close friend Paul Gottlieb Werlhof (1699–1767), private physician to the King in Hanover, who in his answer from 20 November showed himself astonished at the 'Unirritability of the Tendons.' At the same time he pointed to the distinctions that had to be made:

> I wish the Theory of Irritability put into a clear light, as for the Parts truly irritable, and the Manners, Directions, and Degrees, of the Motions raised: as also, what particular Things irritate particular Parts more, than others: and

Figure 2.5
Experiment on a dog, 25 November 1750. From Zimmermann's protocols (Zimmermann Papers, B1, fol. 235).

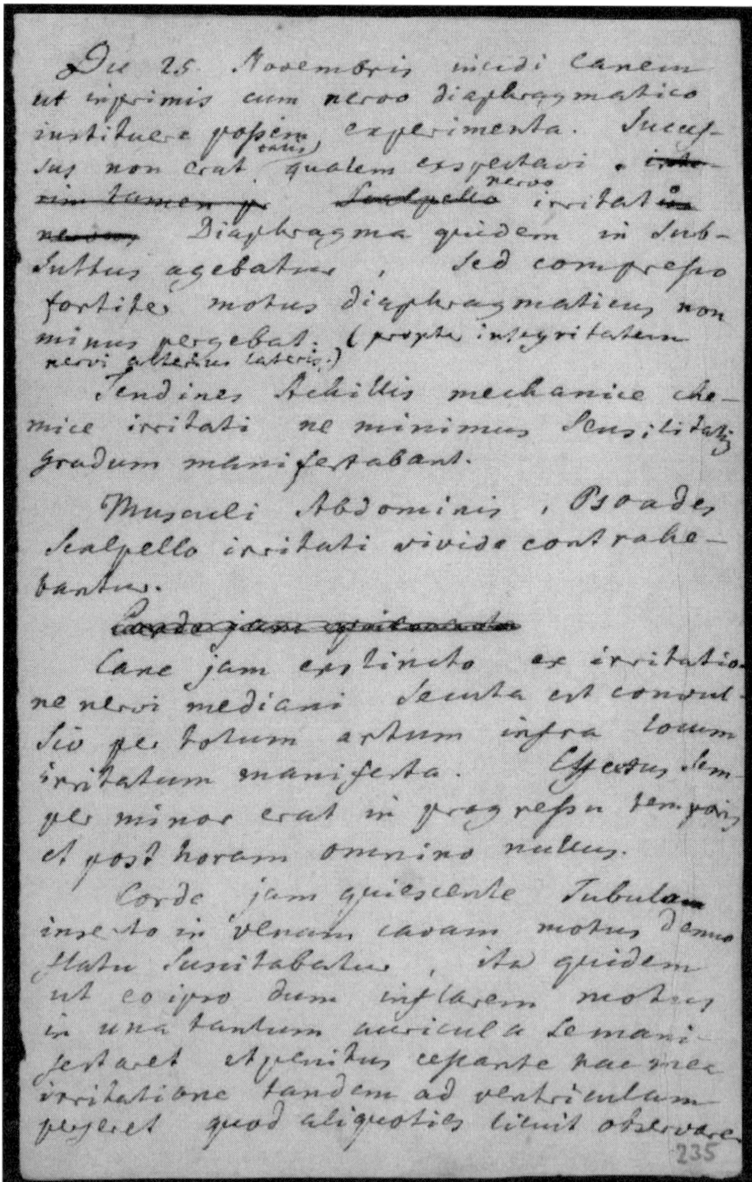
Die 25. Novembris incidi Canem
ut imprimis cum nervo diaphragmatico
instituere possem experimenta. Succes-
sus non erat qualem exspectavi. nervo irritato
Diaphragma quidem in sub-
sultus agebatur, sed compresso
fortiter motus diaphragmaticus non
minus pergebat. (propter integritatem
nervi alterius lateris.)
Tendines Achillis mechanice che-
mice irritati ne minimum Sensilitatis
gradum manifestabant.
Musculi Abdominis, Psoades
Scalpello irritati vivide contrahe-
bantur.
Cane jam exstincto ex irritatio-
ne nervi mediani secuta est convul-
sio per totum artum infra locum
irritatum manifesta. Effectus sem-
per minor erat in progressu temporis
et post horam omnino nullus.
Corde jam quiescente Tubulo
inserto in venam cavam motus demo-
strato suscitabatur, ita quidem
ut eo ipso dum inflarem motus
in una tantum auricula se mani-
festaret. Aperiendo cessante hac mea
irritatione tandem ad ventriculum
pergeret quod aliquoties licuit observare.
235

the Differences of what is only irritated and makes a Reaction, without real feeling, from what is felt by the Soul.[58]

We do not know whether Werlhof's thoughts were a mere echo of Haller's lost letter or were based on his own insight that in the examination of these phenomena one should pay attention to two different things, to movement (reaction) *and* to sensation (feeling). It is certain, though, that from now on Haller explicitly tested the body for its sensibility, too. Thereby he confined himself to testing the sensation of pain, as only a painful stimulus would in all cases provoke a noticeable reaction of the animal. Thus, the animal had to be tortured in full consciousness. On 23 November, in the first experiment he performed after the reception of Werlhof's letter, Haller immediately began with this new approach and recorded: 'I started with the Achilles tendon'. He noticed that the animal showed no sign of pain after its irritation. On the next day he repeated the experiment and wrote down that an irritation would provoke neither a movement nor a sensation. From now on Haller was interested not only in the analysis of one but of two different properties. And Werlhof now spoke of Haller's 'Experiments on sensibility [and] irritability.'[59]

Zimmermann's dissertation, however, remained restricted to irritability. As an assistant of Haller, though, he inevitably became a witness of sensibility tests. But he did not see a fundamental difference between the two qualities. It is true that in his dissertation he furnished, quite similarly to Haller, a definition of irritability as the property of contraction upon irritation, but it is doubtful whether he had such a narrow interpretation in view already at the outset of his investigations. Surely, sensibility for him had something to do with nerves and sensation but not explicitly with conscious perception or, in the case of animal experiments, with sensation of pain. He inclined to use both terms, 'irritability' and 'sensibility', in the sense of Haller, but not exclusively so, and did not aim at a differentiation of these properties. He rather irritated different parts of the body and recorded what had attracted his attention. Mostly this was the presence or absence of a movement or a contraction as these were the phenomena traditionally associated with irritability. Only in the case of some tissues, especially the dura mater, periosteum of the skull (pericranium) and the tendons, did he note in addition whether their irritation would cause pain. As he lacked a clear concept of sensibility or sensation of pain as isolated from irritability, he did not separate these findings from his others. In one single case, on 25 November, he irritated the tendon of an animal which was still fully conscious in the presence of Haller. But three days later he examined three dogs, the brains of which were already severely or entirely damaged, and nevertheless recorded that 'nothing has happened' and that the tendons would 'give no signs of sensibility.' Zimmermann, again, had tested irritability without knowing exactly what he had examined. But although he had performed less precise experiments than Haller and did not develop a distinct concept of sensibility, he still came to the same conclusion as his teacher: the tendons are not sensible.

As we will see in the next chapters, this was indeed an important statement, and it was partly responsible for the attention Zimmermann's dissertation attracted. In fact, Zimmermann had rejected an old doctrine, and that is why his results were called 'paradoxical' by many in the original meaning of the word.

The periosteum, or where to irritate

On 27 November, Zimmermann irritated the pericranium, the periosteum of the skull, for the first time. The dog howled terribly. The next day he repeated the experiment but could not provoke any reactions although he went through the procedure again with great care on different parts of the skull. Zimmermann was uncertain about the facts. He made an additional note in which he recorded the contradiction, but made no attempt to solve it. Eventually, in his dissertation, he published only the first experiment, withheld the following trials and came to the conclusion that the pericranium – and therefore the periosteum in general – would be sensible. This would make sense, he wrote, as one could find many nerves in these membranes. Haller's findings were similarly contradictory. In his first three experiments he found the pericranium to be sensible, in the following five trials insensible. He observed a great amount of nerves, too, but this to him was not an argument *in favour* but *against* the sensibility of the pericranium. He took the view that the apparent sensibility was due to the touch of the nerves themselves which passed through the periosteum but ultimately were responsible for the innervation of the skin and not of the underlying tissue.

How do we account for these differing interpretations? Both authors agreed that the periosteum was a kind of 'connective tissue' (*tela cellulosa*). Their former experiments had shown that structures consisting of connective tissue, such as the dura mater, the tendons, membranes etc., were not sensible. Nevertheless, should the connective tissue in the case of the pericranium be supplied directly with nerves, it had to be sensible. But such a direct innervation could neither be proved nor disproved anatomically or microscopically. The sensibility of the pericranium, therefore, was theoretically possible. Zimmermann, considering the many nerves, argued for this option, but Haller, who in his first experiments had observed sensibility too, subsequently examined the periosteum in different parts of the body. Much more so than Zimmermann's, his view was guided by the concept of elementary structures. One of the main characteristics of his physiology was the determination of specific structures and their specific functions. That is why he tended to isolate the different parts for their examination. In the case in hand this meant that the periosteum should be inspected where it presented itself in its most 'pure' form. For Haller this was the inward part of the tibia.[60] The examination of the 'pure' structure in this place had to provide the answer to the question of its function. In this case, the irritation of the tibia did not provoke any reaction in the animal, and neither did the periosteum appear to be sensible elsewhere. When Haller subsequently repeated the trials on the pericranium, he could not even observe any sensibility in this part anymore.

According to Haller, diverging results had to be attributed to a lack of experimental care, in this case to a kind of 'impurity' of the pericranium because of the nerves passing through it. This approach, of course, tended to deny sensibility to many parts of the body, and from today's view Haller was wrong in some cases – as for instance the periosteum and the large tendons. On the other hand this allowed him to reduce the property of irritability to the muscular, and that of sensation to the nervous fibre. The central role of structural and functional units in Haller's concept becomes obvious in his laboratory notebooks where he often used the terms 'fibre' (*fibra*) or 'flesh' (*caro*), which we do not find in Zimmermann's notes. To put it in slightly exaggerated terms, one might say that Haller thought in terms of (simple) structures and Zimmermann in terms of (compound) organs.

The nerve, or how to irritate

On 14 November 1750 Haller repeated the famous experiment of Lorenzo Bellini (1643–1704): he irritated the phrenic nerve and thereby provoked a contraction of the diaphragm. What did he prove by this? That the movement is caused by some specific action of the nerve? Zimmermann doubted that. He stated that in such an experiment the muscle would rather be activated by a kind of unspecific mechanical pull. If we had to prove that actually the nerve itself was able to induce the movement it had to be touched with chemicals (poisons).[61] This is what he did the very same day. Long after the animal's death he cut out the phrenic nerve, placed it on the table and put on it a few drops of vitriolic oil (sulphuric acid of seventy-five per cent), a chemical irritant widely used in these days. He saw how the distant parts of the nerve moved towards the place of irritation. He carried out a similar trial on 20 November and observed that the nerve miraculously wriggled for several seconds like a worm. It seemed to be proved: the nerve itself is irritable and through its own movement is able to stimulate the muscle. Zimmermann was convinced of his method of irritating detached parts of the body. Two days later he repeated the experiment, now with the vein of a mouse. The result was similar, the vein contracted. Presumably Haller was not present at these trials. But Zimmermann must have informed him about their outcome because the next day Haller tested – as he had already done before – the irritability of the veins. In contrast to Zimmermann, he left the veins in the body and could not observe any movement. The day after, he irritated the tendons with vitriolic oil but could not provoke any contraction and he made one of the few comments to be found in his laboratory notebooks: he noted that we have to be careful not to interpret the (caustic) effects produced by vitriol as a sign of irritability. Even fat contracts upon contact with vitriol. And fat, as a kind of connective tissue for Haller, definitely was not irritable. Zimmermann in this case was less sceptical. At least, he made no critical remarks when he observed the contraction of the fat of the heart on 14 November. However, Zimmermann *and* Haller continued to use vitriolic oil as one of the main irritants in order to test irritability of the different parts of the body. Doubts about the validity of these examinations had been expressed but they remained in the background.

It took half a year until these doubts were raised again. On 6 April 1751, Zimmermann irritated the lungs of a dog with a scalpel, afterwards with vitriol. In the first case he observed a small, in the latter a strong contraction. At this time the intestines – considered to be much more irritable than the lungs – could not be stimulated any more. For Zimmermann this was 'a strong argument' that the lungs themselves were not irritable either and that the observed 'contraction was a chemical effect'. This insight must have challenged his previous view on the irritability of the nerves. Thus, on 15 April he performed a new experiment, which at the same time was his last. He irritated the optic nerve with vitriol and observed, as expected, wriggling movements. After that, he irritated the fat in the orbit but could not induce any contraction. Thus, as the nerves but not the fat reacted to stimulation, Zimmermann concluded that the movement of the nerves would not be due to a chemical effect. But, again, this new experiment stood in contrast to the contraction of the fat of the heart he had observed earlier on. In order to avoid contradiction he assumed that in the earlier case some nerves passing through the fat must have been irritated. Zimmermann, therefore, stuck to the irritability of the nerves. Haller seems to have noticed this. On 4 June he repeated Zimmermann's experiments, yet he placed not only a nerve but also a tendon, an artery, and a vein on the table. On these he put some drops of nitric acid. Nerve and tendon wriggled, and the vessels contracted too. The movement of the tendon served Haller as proof of his assumption that experiments with such caustic acids as vitriol and nitre were not reliable enough to test irritability. In addition, he remarked that such movements of vessels, nerves, and tendons could not be seen when these parts were left in the body. Zimmermann, who probably witnessed Haller's trials, could not or did not want to perform any further experiments, a principal change in his position he seems not to have taken into consideration. Two weeks later he wrote the preface to his dissertation which he submitted two months later.

But Zimmermann did not know what exactly he should do with his experiments on the irritability of the nerves. Certainly, he published them, but at the same time he stressed that he would not attach any great importance to the matter, as tendons too, could be brought to contraction by the application of nitric acid. Nevertheless, without discussing the subject any further and without admitting that he actually had not clearly proved the irritability of the nerves, he still stuck to this notion. It seemed to be obvious; in the nerves 'there is a very marked readiness to movement all the more necessary, because almost all the movements in the animal body depend on them.'[62] Thus, Zimmermann undermined his own definition of irritability as experimentally proved ability to contract upon irritation. He lapsed back into an unspecific use of the term to which he adhered even in a manuscript written several years later:

> There is, not in the simple but in the perfect fibre of the animal, a property, which is different in different parts and which in the vulgar tongue we call irritability. The experiments teach us that this property is peculiar to the muscular as well as the nervous fibre. From the dissection [it is clear that] it manifests itself in the muscles. Through the progression of the impressed motion from the irritated place through the whole length of the nerve to the contracting muscle it is demonstrated in the nerves.[63]

Whereas irritability in the muscles is proved experimentally, in the nerves it is demonstrated by inference. Properly speaking, Zimmermann in his dissertation did not explicitly state the irritability of the nerves but he gave the impression that irritability would somehow depend on them. Thereby, the vague or absent distinction between irritability and sensibility, already present from the outset, became even more obscure. He explained: 'This much has to be remarked at the beginning: occasionally I take sensibility for irritability and vice versa, since I have found by experiments that irritability is the greater the more nerves there are, and that is the reason why sometimes one is taken for the other.'[64]

Haller had no doubts about how to interpret the effect of chemicals upon nerves. Not only had he observed their caustic power before, but he was aware of the fact that in the nerve–muscle units he had repeatedly irritated the nerve mechanically, and this had always been followed by a movement of the muscle but not of the nerve. The idea that the nerve could cause movement without moving itself was not foreign to him. Since the early days of his medical career he had refused the concept of vibrating nerves as supported by Newton and others. Their softness and lack of tension argued against such an opinion. Haller, therefore, favoured, quite traditionally, the nervous juice as transmitter of irritation.

New concepts

It was crucial to realise that the nerves themselves are not irritable. Only if one had recognised this fact could one grasp the meaning of Haller's principal distinction between irritable and sensible structures not only on an anatomical but also on a conceptual level. Haller's pupil Johann Gottfried Zinn resumed his experiments in September 1751 in Berlin and reaffirmed the results he had presented in his dissertation in 1749, ie. that the dura mater was neither moving nor sensible. Thus, he distinguished between the phenomena of sensation and movement, but that did not lead him to a conceptual distinction of two different properties. He inferred: 'Ubi nulla irritabilitas, ibi nulla sensibilitas.'[65] Only a month later, after having performed several experiments on the nerve–muscle unit and after having taken the view that the nerve itself does not move, he reached the conclusion that sensibility could not be taken for irritability and that the muscular fibre alone was irritable.[66] This restriction was fundamental to Haller's concept. Irritability for him was always a property clearly restricted to a specific structure and specific action. He always rejected the concept of irritability as a governing vital force. In the *Elementa* he discussed irritability in detail, but only in the chapters on muscles and nerves; it did not gain any importance in the presentation of other physiological processes.

It was of some importance to the emerging debate that Zimmermann had not distinguished clearly between irritability and sensibility. The fusion of the two properties as presented by Zimmermann would continue to be the main characteristic of all the opponents of Haller. In many authors this led to the postulation of one single vital force, as in Zimmermann, too:

> I have found that irritability especially is marked in the parts the movement of which seems to be most necessary for life, and that it is the less present the less the parts are destined for extensive movements. It is clear therefore that the whole life depends upon this property of animal bodies and that probably nothing else is required.[67]

Some pages later he formulated an even more concise statement: 'Irritabilitatem si tollas, tolleres vitam.'[68] Besides the important discovery of the insensibility of the tendons it was especially this proclamation of a vital principle which drew applause and aroused criticism as well. As several of Zimmermann's manuscripts from the late 1750s or early 1760s show, this was the topic which actually was of interest to him.[69] There he repeatedly cited Diderot's remark from the *Pensées sur l'Interpretation de la Nature* (1754), that there would be two classes of philosophers: some of them have many instruments but few ideas, others have many ideas but few instruments. But one should use the scalpel as well as philosophy. This was what Zimmermann attempted already in his dissertation when he compared irritability with Leibniz's *vis activa.* He continued to develop this idea. A friend wrote to him that he, Zimmermann, tried to 'demonstrate experimentally the powers of nature for which Leibniz had given only metaphysical proofs.'[70] With a similar kind of perspective, in his speech for the defence of his thesis, he had used irritability as a starting point for the description of the temperaments.[71] He extended these studies too. But in none of these did he complete his ambitious projects in a manner which would have allowed their presentation to the public. Only when he moved away even further from the dissection table did he finally publish books which would establish his reputation as a 'philosophical physician', something for which he had longed.

Observation and experiment as a practice

The differences between teacher and pupil were largely limited to the points discussed above. Zimmermann, in contrast to Haller, considered the vessels and nerves to be irritable and the periosteum as sensible. They both agreed upon the insensibility of the tendon and dura mater and upon the irritability of the different organs of the body. Of more importance, though, were the different consequences ensuing from these results. Whereas Zimmermann considered irritability to be a newly detected vital principle, Haller restricted its sphere of action to the contraction of the muscular fibre. Although both had worked in the same laboratory and often together, quite distinct results and even more different concepts had emerged. How did *they* explain this fact?

Zimmermann did not furnish an explicit explanation why his results would differ from those of Haller. In his *Treatise on Experience in Physic* (1763–4), though, he talked generally about the observation of phenomena and addressed the debate on the sensibility of tendons and periosteum. At this time he coined the expression 'spirit of observation' (*Beobachtungsgeist*), which a good physician should have:

> The physician who sees all the circumstances of a disease; he who sees only a part of them, and another who sees none of them, or rather, sees only his own prejudices, must necessarily be of different opinions; and yet they will

all swear by their experience. It is in this manner, that the most opposite opinions are proved. The insensibility of the tendons and periosteum, has been disputed from Moscow, even to Ragusa: one party said the tendons are insensible, we have it from experience; the other said the tendons are sensible, Mr Haller is a Lutheran. Everybody had made his experiments. A man defends, even to his latest moments, that which he thinks he has seen, without asking himself whether he was capable of seeing. ...The appeal to his own experience befits only a person who has the spirit of observation [*Beobachtungsgeist*] in all its strength, who uses it with the utmost precision and who, free from prejudices and passions, sees nature in nature [*Natur in der Natur*].[72]

In order to detect the pure truth, an excellent 'spirit of observation' is needed, hinging on the sensitivity, attention and imagination of the physician. This process requires the senses as well as the powers of reason and is perfected through continuous repetition. The phenomenon does not present itself, it can only be discovered through a combination of observation and reflection. Of course, any sort of prejudice has to be done away with.[73] The lack of this 'spirit of observation' was Zimmermann's explanation of how and why Haller's opponents came to produce diverging (and wrong) results. But, surely, he could not thereby account for the differences between himself and his teacher, because for him there was no doubt: Haller as well as he himself had this 'spirit of observation'.

Haller commented on the problem of poor observation too:

There are two classes of *savants*; there are some who observe, often without writing; and there are others who write without observing. One could neither increase the first of these classes too much nor perhaps diminish the latter too much. A third class is even worse, it is the one that observes badly.[74]

With this remark, though, he in no way argued for a specific talent of observation the lack of which would cause false results. He described his own observations on sensibility as simple sensory perceptions:

I have thought with my master Boerhaave the dura mater to be sensible, the wounds of the tendons to be very dangerous, and the periost to be susceptible of intense pain. It is neither reflection nor reading which has dispelled these prejudices, it is the reiterated testimony of my senses.[75]

Haller clearly distinguished observation from reflection. Whereas for Zimmermann the interlocking of these two faculties was necessary to get a pure idea, for Haller it was the naked observation *without* reflection that would guarantee truth. In his review of Jean Senebier's *Art d'Observer* (1775) he stated that attention and industry rather than genius would be required for observation.[76] There is no 'spirit of observation' needed for the assessment of the results on display, but what is required is an 'art of invention' in order to *design* the experiments leading to these results. Talking about different methods of anatomical and physiological investigations, Haller declared:

> In all these [manoeuvres] a certain art of invention [*ars inveniendi*] is needed which you cannot describe in short and which nature has conceded only to few mortals. It is necessary to embark on the work without prejudice, and not in order to see what a classical author has described but with the intention of seeing what nature has produced.[77]

Thanks to his art of invention the researcher develops a variety of possibilities for conducting an investigation and realises what kind of factors are of importance. For Haller, the experimenter's qualification depends on his manual–reflective ability, on his design of the trials; for Zimmermann it relies on his visual–reflective capability to penetrate and fathom the visible results of experimentation. Haller's esteem for the experimental practice was remarkable for his time, when the general view was that observation required genius and reason whereas the performance of experiments was only a question of skill.[78] In Haller's view the 'art of invention' would, like Zimmermann's 'spirit of observation', only be granted to few persons, but the experiments it had devised could be repeated.[79] Haller issued exact instructions in the technique of experimentation. As to the test of sensibility, for instance, the following points had to be considered: one should not inflict too large wounds but the part to be investigated should be wholly uncovered. The animal had to lie in a position as little painful as possible and to calm down. The single parts might be irritated in different manners; especially in the case of chemicals one had to be extremely careful that they did not eat through to other tissues. It had further to be noticed whether the reaction of the animal was really due to the irritation. Even the approach of the experimenter who inflicted such pains on the animal might provoke its cries. Moreover, the individual animals were different, some were very quiet and others cried constantly without having been touched. Many experiments should not be performed on the same animal as the older wounds might still induce some reactions. But the experiments themselves had to be repeated as often as possible.[80] This point was of particular importance to Haller because only through repetition could the truth be distilled from its distorting contamination: 'after a certain number of verifications only those results remain that arise from the nature of things.'[81] Of course, all these trials had to be conducted without any prejudice or any intention to verify or reject a certain idea.

Whoever would consider these instructions was welcome to repeat Haller's experiments:

> It is with these conditions that I invite everybody of the profession [*metier*] to verify my experiments. They are equitable, they can only prevent error, without mixing it with the events. I am totally persuaded that by submission to these natural precautions one will never see anything contrary to what I have seen.[82]

But whoever would not respect these conditions, 'everybody, in a word, who strongly wanted to see differently to how I have seen, would achieve this easily and would succeed to impose [this view] on himself and to impose it on others.'[83] This was the group of the 'poor' observers Haller had mentioned. They were not lacking any 'spirit of observation' but clearly defined conditions under which to perform their research. Inventive minds might push forward research with a greater pace, but Haller's general picture of a 'scientist' was not that of an original inventor but that of a methodologically scrupulous, diligent and patient worker.[84]

In accordance with his own instructions, Haller accused Zimmermann of having neglected the caustic effect of chemicals. In his *Mémoires* (1756–60) he collected a group of experiments which Zimmermann had conducted with vitriol, and he concluded that these experiments would not prove any irritability.[85] Zimmermann's experiment on the sensibility of the pericranium he considered to be sound as it seemed to have been performed with due care. As this contradicted his own experiments, he stated – in contrast to his treatise of 1752 – that the sensibility of the pericranium was still controversial and that further trials would have to be made.[86] With that, the problem for him was sorted out: the differences between his own results and those of Zimmermann were only due to the technical shortcomings of some of the experiments of his pupil.

Zimmermann and Haller both stressed an aspect of experimentation which is of importance. First, there is barely any observation on which several witnesses would agree at first sight.[87] On 8 June 1751 Haller irritated nerve and artery with vitriol and recorded in his notebook: 'Non irritabilis nervus et arteria.' We do not know what Haller had seen; was there no movement at all to be seen with the naked eye or had the nerve or artery trembled very little, the artery perhaps a little bit more than the nerve? Whatever Haller had seen, to him this had not been a contraction in the sense of irritability. It was Haller's particular skill (or 'spirit') of observation, an ability developed by his own set of experiences and thoughts, which led him to see this.

Secondly, the development and performance of experiments is a manual–reflective process.[88] Haller's and Zimmermann's mental and manual preoccupations with tendons, pericranium and nerves were deeply intertwined. In the case of the examination of the tendon it was of importance whether the experimenter realised what kind of property he was studying; this determined whether he would take the state of consciousness of the animal into account. In the case of the periosteum it was crucial that Haller examined it not only on the skull but also at other places; this, and his tendency to think in structures rather than organs, made him consider the pericranium as a 'contaminated' special case. Being aware of that, he produced different results in his second set of experiments. In the case of the nerves it was significant what kind of experiments were performed: either mechanical and chemical irritations of the isolated nerve or of the nerve–muscle unit, or the chemical irritation of other structures. Different kinds of reflection pointed to which experiment had to be considered more reliable than another.

The three examples of the examination of tendons, periosteum and nerves provide only a sketchy description of some of the main twists in the winding process that went on in the heads of the experimenters and at the dissection table for several months. But they illustrate that the form of interaction between mental and manual preoccupation with a problem varies from person to person. Marino Buscaglia speaks of a 'pattern' of experimental procedures that is characteristic of a scientist.[89] Obviously, there is a temptation to look for these patterns in the methodological statements made by the scientists. One could argue that Zimmermann relied somewhat more on his 'spirit of observation' and drew his conclusions – as in the case of the irritability of the nerves and sensibility of the pericranium – somewhat faster. Haller, instead, used his 'art of invention' and examined the periosteum of the whole body before he made a decision on the sensibility of the pericranium. This, to some extent, seems to be true. But these are only certain tendencies, which we should not overrate. They are not least due to the fact that Zimmermann simply had less time and could perform fewer experiments than his teacher. We cannot merely infer from the methodological remarks of authors their own method of research. The differences between Haller and Zimmermann were relatively small in comparison to those between other experimenters. They were both members of the same laboratory and shared a lot of technical know-how and theoretical assumptions. But the dynamic of the experimental process transformed some smaller variations into larger divergences, which, in the end, included rather far-reaching differences in interpretation.

These differences were only possible because the Göttingen research community was not built on a common single theoretical model or working towards its creation, but was performing experiments and thereby continuously producing new phenomena which then could serve as raw material for the conception of such a model.[90] This, of course, does not mean that all this happened in a space void of theories. Our example seems rather to confirm the view of Ian Hacking that the intersection of theoretical assumptions of scientists is usually not a clear-cut theory that may be handed over to others but is more a collection of notions.[91] This was the case even *within* the Göttingen research laboratory. The experiments did not serve to decide between already existing well-defined theories. The case of Zinn is especially illuminating. He did not have a concept of sensibility as different from irritability when he irritated the nerve. Only the observation that the nerve is not irritable allowed him to develop such a concept. Nor did Haller and Zimmermann have a specific hypothesis on irritability they wanted to test. They had, however, a wide variety of physiological notions, some of which they shared and some not. A great deal of contemporary knowledge about nerves and muscles they both took for granted. But whereas Haller tended to isolate simple structures and to identify their specific function, Zimmermann was thinking in compound structures and looking for general functions explaining the fundamentals of life. The more their research advanced, the more certain facts emerged that could, for both of them, serve as elements of a coherent concept or a theory. Haller seems to have got his concept straight only in September 1751, shortly before Zimmermann submitted his dissertation. It is true, he had already gained some important insights which departed from the views of his student, but we do not know to what extent he communicated them to Zimmermann. However, it remains doubtful whether this would have had any crucial influence on Zimmermann's own concept as the pupil stuck to his own interpretation even ten years later.

From experimental anatomy to experimental physiology

Having described some aspects of the experimental process we now have to ask Andrew Cunningham's question whether Haller had performed experimental anatomy or experimental physiology. Instead of starting from a physiological phenomenon and seeking its explanation in the organism – which was the procedure of François Magendie, Claude Bernard and modern nineteenth-century physiology – Haller started, according to Cunningham, 'from anatomy to establish structure and properties of the parts, and only then ha[d] recourse to vivisectional experiment to clarify function.'[92] The success of modern experimental physiology led us to regard every case of vivisection as an early instance of experimental physiology. Haller and his contemporaries, however, considered their experiments as an anatomical investigation, a manual art, whereas physiology was not an experimental but a purely theoretical discipline and therefore – according to the notions of their time – a branch of science. Anatomy delivered the facts and physiology the interpretation.

Haller on the method of vivisection

In my view, Cunningham's thesis is essentially right as to the time before Haller but it is only partially correct as far as Haller and some of his contemporaries like Felice Fontana and Lazzaro Spallanzani are concerned. I will first reply to the claim that Haller considered vivisection as an anatomical method. Cunningham has pointed to the fact that many authors talked of 'anatomical experiments' and that Haller himself entitled two of his treatises, *De Respiratione Experimenta Anatomica* (1746) and *De Sanguinis Motu Experimenta Anatomica* (1754) although they were mainly based on vivisection. This is, however, only one half of the story; in his later works, Haller mostly used the term 'physiological experiments'. In his articles for the *Encyclopédie*, the expression *expériences anatomiques* usually denotes anatomical preparations such as wax injections. Instances of vivisection are mostly called *expériences physiologiques* and they are not reported in the lengthy article *Anatomie* but in the equally voluminous entry *Physiologie*.[93] And at the end of this article Haller says:

> I have given a skeleton of the history of physiology. I have admitted only the ancients, and from the moderns those who have performed experiments and original research; I have omitted those who have only collected or reasoned.[94]

The experiments do not provide the skeleton on the top of which a new discipline – physiology – is constructed, they are the skeleton of physiology itself. Those who added to the erection of this skeleton were good physiologists. Stephen Hales was, 'without being physician, unquestioningly one of the first physiologists. He performed very many and very difficult experiments on living animals.'[95] By contrast, Descartes' works were 'physiological novels [*romans physiologiques*]', not, however, because of their lack of theoretical coherence but because they are 'hypotheses founded neither on the structure of the human body nor on [physiological] phenomena', *viz* they are not tied to the experimentally constructed skeleton of physiology.[96] Assumptions are not to be rejected as such, Haller argued in his essay on hypotheses (1750), but he demanded a clear distinction between ascertained facts and probable speculations.

> We can assume what we wish, if we openly remind the reader concerning the assumed things, that our probable conjecture is still greatly, moderately, or only slightly removed from the truth, when we admit that in order to be convinced, we are still lacking some unperformed experiment, or measurement, or [knowledge of] the structure of still undetermined parts. Can anyone complain if small change is declared to be small change, and its value is established no higher than the value of the silver in it? He alone is deceived who accepts it for pure silver.[97]

Hypotheses, Haller said, were useful as they connected the single parts of our knowledge that would otherwise remain fragments. They led to novelties and truth, and no discoverer, not even Newton, could do without them. Most importantly, they posed questions which would not have occurred to us and which called for experimental testing.[98] Hypotheses, therefore, were intimately linked with experimentation: they were based on experimentally verified facts and they asked for further experimentation. In this concept it is the experiment that constitutes truth whereas the hypotheses provide only various grades of incomplete certainty.[99] And experimental results could, in Haller's view, provide physiological knowledge devoid of any theoretical speculation. His separation of irrtability and sensibility was just such a case:

> I have admitted no hypothesis, and I often wonder that they write about my system, who I have only said that those parts feel or are moved that I have seen feeling or moving... There could be an error in the experiment but there is not even a shadow of a hypothesis.[100]

Factual knowledge without any hypothesis, that was Haller's ultimate ideal of physiology; actual physiology, however, consisted of a skeleton of experimentally verified facts linked by various hypotheses based on these facts. In his 1753 treatise on irritability, the description of nerves as sensible and muscles as irritable was – in Haller's view – just an experimentally verified physiological fact; the only hypothesis he admitted was the notion that irritability seemed to reside in the Gluten of the fibre (see Chapter 3). It is in this sense that we have to read the frontispiece of the *Mémoires* (Figure 2.3). All six men are performing animal experiments and the sitting man is not – as Cunningham has mistakenly maintained – writing.[101] The picture does not illustrate a separation between experimenting anatomists and the writing physiologist. All the persons shown are physiologists and they produce physiological facts.

In a similar manner, as hypothesis was opposed to experience, so was Haller's notion of theory. He never called his concept of irritability and sensibility a theory and he never used the term 'theory' in a title of his many physiological publications. Of course, he considered physiology as the theoretical discipline of medicine. But it was *not only* a theoretical science – as Cunningham argues – and all theoretical reasoning enclosed the danger of leading astray from the facts and thus from truth. In the *Elementa*, Haller therefore continuously referred to experimental results, thus securing the skeleton of physiology. Chapters, for which only little experimental knowledge was obtainable, he considered unsatisfactory. To Giambattista Morgagni he reported that the second volume on the blood, its movement, and the secretion of the humours would be less appealing to him as he was 'compelled to add far too much gained from mere theory and hypotheses.'[102] In a letter to Bonnet he explained how he composed his physiological textbook: 'Generally, I do not consider the physiological subjects but by subjects [i.e. each subject separately]. I therefore collect all the facts, foreign or seen by myself, and let myself guide by their competition.'[103] The facts themselves, in Haller's view, constitute physiology. Of course, the process implies also interpretation and theoretical reasoning, but the main characteristic which distinguishes Haller's physiological works from many of his contemporaries is his restraint in the formulation of hypotheses and his attempt to tie them to an experimentally secured basis of knowledge, thus transforming theoretical into experimental physiology. That is why authors like Condorcet or Vicq d'Azyr called him the 'creator of physiology'.[104] That is also why Haller called his physiology *anatomia animata*.[105] This expression stresses the intimate correlation between structure and function and points out that physiological accounts have always to be based on the anatomical body. The *Elementa* is not only a narration of the physiological functions of the living body, it equally contains extensive anatomical descriptions.[106] To emphasise this physical basis vivisections could also – although with less precision – be called 'anatomical experiments'. The living body, however, is 'infinitely more animated' than anatomical plates and more than a simple machine.[107] Physiology cannot, therefore, be a mere theoretical explanation of anatomical facts. Physiology is not 'theoretical anatomy', it is 'animated anatomy'. The main method of physiology thus is the study of the animated, the living body. Vivisection is the method proper to physiology, it is a physiological method.

Haller's notion of physiological science is portrayed in the frontispiece to the second volume of the *Elementa* (Figure 2.6, overleaf). We see, in the background, a skeleton, representing anatomy upon which physiology is built. At its left, hanging in an upright position, we spot a goose or a large bird, presumably signifying comparative anatomy which, according to Haller, furnishes many insights into the function of the organs. In the foreground, there are two muses of equal standing, both assisted by two putti. The left group is performing vivisection and thus stands for experimental physiology. The muse on the right sits in front of a library and is contemplating and writing, her putti do geometry and astronomy. This part stands for the act of composing a physiological text in the manner Haller described it to Bonnet, ie. by comparing and synthesising the knowledge obtained from anatomical, experimental, literary, mathematical and other study into a more or less coherent narration. One might call this the theoretical part of physiology, although Haller might have had his problems with this expression. The illustration shows that experimentation is not a slave to theory, it is equally important and respectable.[108] As we have seen, Haller had a high esteem for experimental practice. It was not a simple art which could be performed by anyone. One needed an 'art of invention', a special gift, in order to design revealing experiments. Haller praised the experimental works of Trembley, Bonnet and Réaumur whenever he could. 'Not everyone', he said, 'is able to see as Mr Bonnet. This demands attention, time and accuracy [in a degree] which even a financier is barely prepared to devote to the care of his fortune.'[109] Haller opposed this laborious process of experimental investigation to the laziness and vanity which was often the reason for the construction of a new comprehensive theory. But God's intentions made no allowances for our convenience and glory. Mankind had repeatedly been humiliated by the painstaking work of researchers like Trembley who repudiated its theoretical beliefs.[110] Not the theorists but the practitioners of science enlarged our knowledge:

> Better telescopes, rounder glass lenses, more accurate units of measurement, syringes and scalpels, have done more for the augmentation of the kingdom of the sciences, than have the creative spirit of Descartes, the father of order Aristotle, or the erudite Gassendi.[111]

The construction of instruments and the performance of experiments were the methods which created new knowledge and thus helped to uncover God's laws of nature. In this sense, vivisection was a noble, demanding, and an important, if not the most important part of physiology.[112]

On Haller's method of vivisection

Having argued that Haller regarded experimentation on living animals as a physiological method we still have to ask whether his experiments may be called physiological in our modern sense. Cunningham maintained that Magendie's experimental approach was new and entirely opposed to that of Haller. Instead of investigating the structure of an organ in order to deduce function, and then conduct experiments on the living animal to test his deductions, Magendie began with the experimental animal and removed organs in order to see what function was thereby lost or left over. As an

Figure 2.6
The Muses of physiology (Frontispiece to the Elementa, *Vol. 2).*

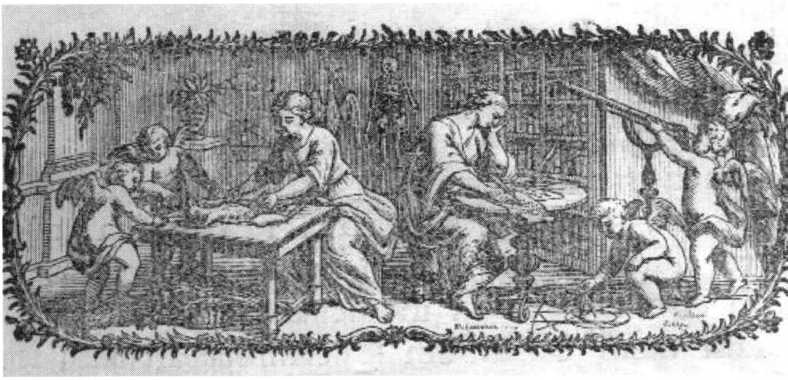

example, Cunningham reported the experiments on the role of the stomach in vomiting. In a first experiment Magendie rendered the diaphragm immovable, in a second the abdominal muscles, and in a third he replaced the stomach by a pig's bladder. In each case he tried to induce vomiting thus successively removing each of the three possible candidates. After repetition and combination of these experiments in several animals he came to the conclusion that not the stomach but both the diaphragm and the abdominal muscles could produce vomiting and usually co-operated in their action.[113] Haller's assertion that the stomach was the essential cause of vomiting, Magendie concluded, was therefore wrong.

Magendie's account of what Haller would have said is, however, incomplete, and so is Cunningham's interpretation of Haller's methodology. In the *Elementa*, Haller discussed vomiting at length.[114] He considered, as Magendie said, the stomach as the necessary agent of vomiting but – based on experimental evidence – he also argued similarly to Magendie that the diaphragm and the abdominal muscles were the main forces acting in vomiting. In contrast to Magendie, however, he and other authors had observed the motion of the stomach in vomiting, also when the phrenic nerve was cut, the diaphragm or the abdominal muscle dissected and even when the stomach was taken out of the body.[115] Haller's technique of removing possible candidates was similar to Magendie's approach, although he made different observations. Additionally, he did not replace the stomach by a pig's bladder. But the general approach which Cunningham considers as typical of modern experimental physiology was one of Haller's favourite methods. Haller used it throughout his experimental enquiries and it was the reason why he supported the study of comparative anatomy:

> If functions are carried out in animals deprived of a certain class of parts, then these parts are not the unique and necessary causes of this function. If animals without head and nerves are irritable, then irritability can be carried out without nerves. A thousand examples of the use of this single rule could be given....[116]

I will briefly refer to some major instances of vivisection where Haller used this rule as a guiding principle. He performed animal experiments from the early 1730s but it was only the dispute with Georg Erhard Hamberger (1697–1755) in the 1740s that caused Haller to carry them out in a systematic manner.[117] The controversy centred on the mechanics of respiration, and Hamberger maintained, among other things, that because of their geometrical position, the external parts of the intercostal muscles elevated the ribs and thus were responsible for inspiration, whereas the internal parts lowered the ribs and thus caused expiration. Haller contested these facts on experimental grounds.[118] He first used human corpses as models and studied the motion of the chest by adding weights to the ribs. These experiments showed that the external as well as the internal parts of the intercostal muscle had to be responsible for inspiration. As Hamberger stuck to his own notion, Haller performed, half a year later, experiments on living animals. He laid bare the chest of a living animal and observed its motion. But only the diaphragm contracted and ensured respiration, the ribs did not move. Haller therefore perforated the diaphragm, thus opening the lungs and causing respiratory stress in order to see whether other muscles would help in respiration. In inspiration the ribs were elevated and the intercostal space narrowed due to the visible contraction of the intercostal muscles. As this narrowing and elevation could be due to both parts of the intercostal muscles Haller then removed the external part of several muscles and observed that in all cases the ribs were still elevated and the intercostal space narrowed. Both the external and the internal parts, he concluded, were thus responsible for inspiration.[119]

The controversy on respiration shows Haller's shift between anatomical and physiological experimentation. He first observed the anatomical structure and its (supposed) movement in order to deduce its function. Then, with presumed ignorance of this function, he performed experiments on living animals and destroyed other structures in order to see what function the structure in question carried out. The mixture of both modes of enquiry are visible in most of Haller's later important investigations. They are based, however, somewhat less on an anatomical and more explicitly on a physiological approach than his earlier research on respiration. We will look at three of them: they concern the action of the heart, the motion of the brain, and irritability and sensibility. These inquiries were mainly performed in Haller's late Göttingen period (1749–53). After his return to Switzerland in 1753, Haller had no laboratory at his disposal and thus only rarely had the opportunity to experiment with larger animals. His later investigations on the microcirculation of the blood or on embryology were conducted mainly on frogs and chickens and were more of an observational and less of an experimental, interventionist character.

The search for the cause of the motion of the heart had always been a difficult problem of physiology, and Haller himself had performed experiments on this topic from the 1730s. It was, however, not until his research on irritability and sensibility, that he came to design the experiments he considered decisive. He had long observed that the motion of the right auricle and ventricle generally survived that of their left counterpart. This happened to be the case also when he isolated the right heart by ligating its afferent veins (superior and inferior caval veins) and its efferent artery (pulmonal artery). In a experiment, Haller ligated again the caval veins but opened the pulmonal artery and pressed and thus evacuated the right ventricle. At the same time, he ligated the efferent artery of the left heart, the aorta. He could now observe that the right auricle stopped first and then the right ventricle while the left ventricle – which due to the ligature of the aorta was still filled with blood – continued beating. When he opened the ligature of one of the caval veins or when he inflated air in the right auricle, the right heart resumed its action. For Haller, these experiments proved that blood or another irritating agent was necessary to induce the heart to action.[120] Whether nerves were a necessary agent, too, Haller did not yet know. At this moment in June 1751, shortly before completing his paper on irritability and sensibility, he had not yet a clear idea of the role of the nerves in muscular movement. In 1751 and 1752, however, he made further experiments on the role of the nerves in the motion of the heart.[121] He cut the vagal nerve, on each, and on both sides, and still the heart continued beating. He cut the sympathetic nerve, and the heart's action did not stop. He cut the whole spinal cord, and the heart nevertheless contracted. Disconnected from all nerves, the heart was still moving. Ultimately, Haller removed it from the body, and could even now observe its motion. Based on all these experiments, Haller stated with conviction that neither the soul nor the nerves but only the blood was a necessary agent in the motion of the heart.

In another set of experiments, mainly performed in 1751 and 1752, Haller investigated the motion of the brain.[122] It had long been known that upon trepanation of the skull one could observe a motion of the brain. This observation had received special importance in the concept of Giorgio Baglivi, who assumed an active motion of the dura mater, which would be responsible for the propagation of the animal spirits into the nerves. Haller rejected Baglivi's theory and considered, like his master Boerhaave, the visible motion as due to the pulsation of arteries. In 1744, however, a new explanation was presented by Johann Daniel Schlichting who detected that the motion of the brain happened synchronously with respiration. When Haller started with systematic experimentation on this matter in 1751, he first observed only the much faster pulsation of the vessels he had seen earlier. Only when he detached the dura mater from the skull could he see – not without surprise, as he said – dura mater and the brain descending in inspiration and ascending in expiration.[123] The motion of the brain persisted when he removed the dura mater; it was, therefore, not due to the membrane and it did not happen in the natural state when dura mater and brain were fixed to the skull. As the motion increased in intensified respiration it had certainly to be linked to the motion of the lungs, but in what manner was yet unclear. Haller then opened the thorax and observed that the caval and jugular veins swelled and ascended in expiration, just as the brain did. For him, this was proof enough that the swelling of the veins was the missing link between the lung and the brain. But he did not – as a strict experimental physiologist would have done – ligate the jugular veins in order to see whether this would stop the brain from moving. In this instance, Haller relied on his anatomico–mathematical knowledge that presupposed the transmission of pressure from the vessel to the brain. As to the mechanism by which respiration affected venous blood pressure, Haller argued on two levels. He compressed the thorax of an animal, in which the brain showed no movement. This caused a swelling of the veins and an elevation of the brain which, Haller argued, was due to an increased thoracic pressure. Such a heavy compression of the thorax was, however, only visible in cases of forced exspiration but not in the normal state. He concluded, however, from older experiments what happened in these instances: in animals, which do not breathe any more but the hearts of which still contract to some extent, the blood is accumulated in the right heart and the caval veins. Upon inflation of the lungs, the heart is reactivated and the congested blood is thrown into the lungs. In inspiration, therefore, blood enters the lungs more easily and thus diminishes its accumulation and the pressure in the veins.

Haller's experiments on irritability and sensibility originated in the search for the explanation of the phenomenon of irritability, ie. the visible contraction upon irritation. Haller's question was not – as Cunningham has suggested – what properties muscles and nerves had but which organs and structures were responsible for irritability. His first notebook entry reads: 'canis ad irritabile', not 'canis ad musculos'. Just as Magendie wanted to determine which structure was a necessary agent in vomiting, Haller wanted to identify the necessary agent in irritability. The difference lies in the fact that Magendie had only to test a selected group of candidates, whereas Haller had to examine many different parts of the body. He thus did not examine all the organs with the same precision, but in many instances he proceeded roughly in the same manner as Magendie. Haller irritated the parts of the body, placed a ligature around the nerves with which they are furnished, irritated the nerve above and below the ligature, and removed the parts from the body. The reaction of the structures upon this variation of experiments made him deduce which parts are irritable and that irritability does not depend on nerves. The last step in the development of Haller's concept – the reduction of irritability to the muscular fibre – however, is based on anatomical reasoning and not on physiological experimentation. The visible contraction of fat and tendons upon chemical irritation had – according to his definition – to be considered as a case of irritability. Only his conviction that fat and tendon were made of connective tissue and thus of a structure unable of active motion led him to conclude that only the muscular fibre is irritable. Two other elements of the final procedure of defining irritability – the reduction of the irritability of entire organs to their muscular structure and the separation of irritability from elasticity – are equally based on anatomical inference: the details of Haller's argumentation are given in the next chapter.

As to the research on sensibility, the procedures in achieving the different results were opposite to those in the case of irritability. The main statement, the reduction of sensibility to the nerve, was based not on anatomical inference but on physiological experimentation. The irritation of a nerve above and below a ligature or of organs with or detached from nerves were as simple as certain proofs of the nerve as a necessary agent of sense and pain perception. The procedure of isolating organs, thus eliminating the neighbouring structures as possible candidates producing pain, was in itself a physiological method, too. But the question how far a certain part had to be isolated in order to be tested was an anatomical one. Whether the sensory nerves surrounding the structure in question were part of this structure or not, was decided on anatomical assumptions.

To sum up, Haller's work contains numerous instances of experimental physiology in which he started from a physiological phenomenon and seeked its explanation in the organism by vivisection. But many of his experiments were also guided by his knowledge of the anatomical structure and the presupposed function deducted therefrom. Haller thus represents a phase of transition from anatomical to physiological experimentation. He certainly was not the only researcher of his time to perform experimental physiology. Just think of the work of Lazzaro Spallanzani, and especially his experiments on digestion.[124] To what extent they were followed by others has yet to be discovered. It is certain, though, that experimental physiology was not established as a prime method of biological research until the nineteenth century. As Chapter 4 will show, the controversy on irritability and sensibility was of some importance in this development.

Notes

1. This chapter is a revised and enlarged version of my article 'Divergierende Resultate eines Forschungslabors des 18. Jahrhunderts: Zimmermann und Haller über Irritabilität', *Cardanus*, 1 (2000), 93–115.
2. Two authors who have particularly enlightened and stressed this aspect are Ian Hacking and David Gooding. See I. Hacking, *Representing and Intervening: Introductory Topics in the Philosophy of Natural Science* (Cambridge: Cambridge University Press, 1983); D. Gooding, *Experiment and the Making of Meaning: Human Agency in Scientific Observation and Experiment* (Dordrecht: Kluwer, 1990).
3. Early and still important works in the field of physical sciences are – besides those mentioned in note 2 – A. Franklin, *The Neglect of Experiment* (Cambridge: Cambridge University Press, 1986); P. Galison, *How Experiments End* (Chicago: University of Chicago Press, 1987); D. Gooding, T. Pinch and S. Schaffer (eds), *The Uses of Experiment: Studies in the Natural Sciences* (Cambridge: Cambridge University Press, 1989).
4. The lack of studies in this period is stressed by M. Grmek, 'La Théorie et la Pratique dans L'Expérimentation Biologique au Temps de Spallanzani', in G. Montalenti and P. Rossi (eds), *Lazzaro Spallanzani e la Biologia del Settecento. Teorie, Esperimenti, Insituzioni Scientifiche* (Florence: Olschki, 1980), 321–44; M. Grmek, 'L'Apport des Correspondants Genevois de Spallanzani à La Méthodologie Expérimentale', *Gesnerus*, 48 (1991), 287–306; M. Buscaglia, 'Pour une Histoire Spécifique de la Méthode en Biologie', *Archives des Sciences et Compte Rendu des Séances de la Société de Physique et d'Histoire Naturelle de Genève*, 47 (1994), 137–54; M. Buscaglia, 'The History of the Experimental Method in the Life Sciences as an Illustration of the Versatility in Interpretation', in J. Montangero *et al.* (eds), *Conceptions of Time over Change* (Cahiers de la Fondation Archives Jean Piaget, 13, 1994), 45–64.
5. M. Buscaglia, 'La Pratique, la Figure et les Mots dans les Mémoires d'Abraham Trembley sur les Polypes (1744) Comme Exemple de Communication Scientifique', in M. Galuzzi *et al.* (eds), *Le Forme Della Comunicazione Scientifica* (Milan: Angeli, 1998), 313–46; M Buscaglie, 'Bonnet dans l'Histoire de la Méthode Expérimentale', in M. Buscaglia *et al.* (eds), *Charles Bonnet. Savant et Philosophe (1720–1793)* (Geneva: Editions

Passé Présent, 1994), 283–313. M.T. Monti, 'Forma della Scrittura e Forme della Comunicazione: Dai Mémoires di Haller al Prodromo di Spallanzani', in Galuzzi *et al.*, *op. cit.* (note 5), 277–311.

6. Monti, *op. cit.* (note 5); Albrecht von Haller, *Commentarius de Formatione Cordis in Pullo.* Edizione critica a cura di Maria Teresa Monti (Basel: Schwabe, 2000).
7. The term *Labor* at this time in the German-speaking area was not yet used but it fits the Institute of Anatomy directed by Haller in Göttingen quite well. In France, *laboratoire* was mainly the place where chemical experiments were performed; see C. Salomon-Bayet, *L'Institution de la Science et l'Expérience du Vivant: Méthode et Expérience à l'Académie Royale des Sciences 1666–1793* (Paris: Flammarion, 1978), 372–80.
8. See R.G. Mazzolini, 'Le Dissertazioni degli Allievi di Albrecht von Haller a Göttingen (1736–53): Una Indagine Bio-bibliografica', *Nuncius,* 2 (1987), 125–94, esp. 179–91.
9. Haller's notebooks are completely preserved (Burgerbibliothek Bern, Haller Papers, Mss. 23–31), only a few of Zimmermann's experiments are lacking (Niedersächsische Landesbibliothek Hannover, *Zimmermann Papers*: MS XLII 1933, 'de Irritabilitate': B1).
10. A. Cunningham, 'The Pen and the Sword: Recovering the Disciplinary Identity of Physiology and Anatomy before 1800', *Studies in History and Philosophy of the Biological and Biomedical Sciences,* 33 (2002), 631–65; 34 (2003), 51–76. His argument is indebted to L.G. Stevenson, 'Anatomical Reasoning in Physiological Thought', in C. McC. Brooks and P.F. Cranefield (eds), *The Historical Development of Physiological Thought* (New York: Hafner, 1959), 27–38. See also W.R. Albury, 'Experiment and Explanation in the Physiology of Bichat and Magendie', *Studies in History of Biology,* 1 (1977), 47-131.
11. These are the judgements of pupils, see J.A.R. Reimarus, *Lebensbeschreibung von ihm selbst Aufgesetzt: Nebst dem Entwurf einer Teleologie zu seinen Vorlesungen Bestimmt* (Hamburg: Campe, 1814), 10; E. Bodemann, *Von und über Albrecht von Haller: Ungedruckte Briefe und Gedichte Hallers sowie Ungedruckte Briefe und Notizen über denselben* (Hanover: Meyer, 1885), 189–90.
12. Reimarus, *op. cit.* (note 11), 13; Bodemann, *op. cit.* (note 11), 201.
13. A. Haller, *De Methodico Studio Botanices absque Praeceptore...* (Göttingen: Vandenhoeck, 1736).
14. 'Oratio Dicta [de Utilitate Societatum Literariarum] cum Die Natali Georgii II. Societas Regia Scientiarum Primum Publice Conveniret', *Commentarii Societatis Regiae Scientiarum Gottingensis,* vol. 1 (Göttingen: Vandenhoeck, 1752), XXXVII–LVI. For the German edition see *Sammlung Kleiner*

Hallerischer Schriften (3 vols., Bern: Emanuel Haller, 1772), ii, 173–206.

15. See R. Toellner, 'Die Verbindung von Lehre und Forschung an der Jungen Georgia Augusta zu Göttingen', *Hippokrates*, 39 (1968), 859–63; O. Sonntag 'Albrecht von Haller on Academies and the Advancement of Science: The Case of Göttingen', *Annals of Science*, 32 (1975), 379–91. For the general background, see J.E. McClellan, *Science Reorganized: Scientific Societies in the Eighteenth Century* (New York: Columbia University Press, 1985).
16. See U. Tröhler and S. Mildner-Mazzei, *Vom Medizinstudenten zum Doktor: Die Göttinger Medizinischen Promotionen im 18. Jahrhundert* (Göttingen: Vandenhoeck & Ruprecht, 1993).
17. Review of a work of Hermann Friedrich Delius, *GGA*, 1752, 457.
18. See L. Rosner, 'Eighteenth-Century Medical Education and the Didactic Model of Experiment', in P. Dear (ed.), *The Literary Structure of Scientific Argument: Historical Studies* (Philadelphia: University of Pennsylvania Press, 1991), 182–94.
19. See R. Rey, 'L'École de Santé de Paris sous la Révolution', *Histoire de l'Éducation*, 57 (1993), 23–57: 41.
20. From Haller's covering letter to the dissertation of A. Swainston, *Dissertatio Inauguralis Medica de Purpura quam... Praeside ... Alberto de Haller...* (Göttingen: Vandenhoeck, 1752).
21. Haller, *op. cit.* (note 14), German edition, 204.
22. See *GGA* 1752, 1127.
23. J.G. Zimmermann, *Das Leben des Herrn von Haller* (Zurich: Heidegger, 1755), 298–9.
24. See Johann Stephan Pütter to Haller, 6 January 1768 (*Haller Papers*) and D. von Maerker, 'Die Zahlen der Studierenden an der Georg-August-Universität in Göttingen 1734/37 bis 1984', *Göttinger Jahrbuch*, 1979, 141–58.
25. These numbers may be gathered from Mazzolini, *op. cit.* (note 8).
26. For details, see Chapter 4.
27. *Mémoires*, i, 4.
28. Letter from 9 June 1752 (*Haller Papers*).
29. See Haller to Réaumur, 23 January 1753 (Archives de l'Académie des Sciences, Paris); J.H. von Brunn, *Dissertatio Inauguralis Medica Sistens Experimenta quaedam circa Ligaturas Nervorum in Vivis Animalibus* (M.D. thesis, Göttingen: Vandenhoeck, 1753), 2.
30. The artist of this etching is not known, neither do we know how he came to depict the scene in this manner.
31. See J.S. Pütter, *Versuch einer Academischen Gelehrten-Geschichte von der Georg-Augustus-Universität zu Göttingen* (Göttingen: Vandenhoeck, 1765), 234.

32. *Mémoires*, i, 103, 106.
33. Bodemann, *op. cit.* (note 11), 216.
34. This is testified in the case of at least one student; see Bodemann, *Haller* (note 11), 201. Haller, too, spoke of 'hated cruelties'.
35. This statement is found at various places in his writings, see eg. Haller, *op. cit.* (note 14), liv.
36. S.A.D. Tissot, *Vie de Zimmermann* (Lausanne: Fischer & Vincent, 1797), 10.
37. Mazzolini, *op. cit.* (note 8), 169.
38. 14 November 1750 (fo. 245r).
39. Haller, *op. cit.* (note 14), German edition, 195–8.
40. *Elementa*, i, preface, IX; a similar statement is to be found in *GGA,* 1771, Zugabe, clxiv.
41. J.E. Lesch, *Science and Medicine in France: The Emergence of Experimental Physiology, 1790–1855* (Cambridge, Mass.: Harvard University Press, 1984). For further details see Chapters 2 and 4.
42. Letter from 15 March 1755, *Corr. Bonnet*, 62–3.
43. R.G. Mazzolini, 'Sugli studi Embriologici di Albrecht von Haller negli Anni 1755–1758', *Annali dell'Istituto Storico Italo-germanico in Trento = Jahrbuch des Italienisch-deutschen Historischen Instituts in Trient*, (1977), 183–242: 205–7.
44. *Elementa*, i: preface, i.
45. The actual work on this textbook he began in 1751 at the latest; see A. von Haller, *Primae Lineae Physiologiae in Usum Praelectionum Academicarum. Auctae et Emendatae* (Göttingen: Vandenhoeck, 1751), praefatio.
46. See U. Boschung, 'Heimweh, die "Schweizer Krankheit": vor 300 Jahren erstmals beschrieben', *Inselbote: Hauszeitschrift für das Personal des Inselspitals Bern*, 2 (June 1988), 22–8.
47. See H. Schramm (ed.), *Johann Georg Zimmermann: Königlich Grossbritannischer Leibarzt (1728–1795)* (Wiesbaden: Harrassowitz, 1998). A good introduction to his writings is *Johann Georg Zimmermann: Mit Skalpell und Federkiel – ein Lesebuch*, A. Langenbacher (ed.), (Bern, Stuttgart, Vienna: Haupt, 1995).
48. Johann Georg Zimmermann, *Von der Erfahrung in der Arzneykunst.* (2 vols, Zurich: Heidegger, 1763–4), ii, 338.
49. Letter to Johannes Gessner, 14 September 1749. *Albrecht von Hallers Briefe an Johannes Gesner (1728–1777)*, H.E. Sigerist (ed.), (Berlin: Weidmann, 1923), 170.
50. Besides his laboratory notebooks, Zimmermann also left some excerpts on irritability but they cannot be dated; some of them definitely were made after 1751. The most frequently mentioned authors in his dissertation are

Baglivi, Boerhaave, Boyle, Haller, Le Cat, Leibniz, Lups, Winter, and Woodward.

51. J.G. Zinn, *Dissertatio Inauguralis Medica Sistens Experimenta quaedam circa Corpus Callosum, Cerebellum, Duram Meningem, in Vivis Animalibus Instituta... Praeside ... D. Alberto de Haller* (M.D. thesis, Göttingen: Vandenhoeck, 1749).
52. It is possible that the only three undated experiments (fo. 222–4) were carried out shortly before. As all the other notes are in chronological order no page references to Zimmermann's – and to Haller's – laboratory notebooks (see note 9) will be given.
53. Neither Zimmermann nor Haller stimulated electrically. A general description although not a thorough analysis of Haller's methods is given by K. Mallepree, *Die Physiologischen Reizmethoden Albrecht von Hallers* (Münster Univ. M.D. thesis, 1966).
54. *Mémoires*, i, 105
55. E.F. Rössler (ed.), *Die Gründung der Universität Göttingen: Entwürfe, Berichte und Briefe der Zeitgenossen* (Göttingen: Vandenhoeck & Rupecht, 1855; reprinted Aalen: Scientia-Verlag, 1987), 337.
56. J.G. Zimmermann, *Dissertatio Physiologica de Irritabilitate* (M.D. thesis, Göttingen: Vandenhoeck, 1751). See also H.-J. Schmallenbach, *Johann Georg Zimmermann und die Irritabilitätslehre* (Münster Univ. M.D. thesis, 1967). This analysis is not based on manuscript sources and differs widely from my own interpretation.
57. Haller, too, used sometimes the adjective 'sensibilis' or the French 'sensible' in the sense of 'responsive', but never – as Zimmermann (see for instance the experiment on the heart of a frog, 15 September 1750) – the noun 'sensibilitas'.
58. This is not a translation, the correspondence was conducted mainly in English. Werlhof's almost 1,600 letters are preserved in Bern (*Haller Papers*), Haller's letters have not survived.
59. Werlhof to Haller, 4 December 1750 (*Haller Papers*).
60. *Mémoires*, i, 138–39.
61. Zimmermann, 'de irritabilitate' (*Zimmermann Papers*), fo. 147 (undated notes).
62. Zimmermann, *op. cit.* (note 56), 36.
63. 'De Viribus Naturae Insitis Dissertatio' (*Zimmermann Papers*, B 6d), fol. 4–5.
64. Zimmermann, *op. cit.* (note 56), 2.
65. Zinn to Haller, 11 September 1751 (Haller papers).
66. Zinn to Haller, 16 October und 23 November 1751. *Epistolae*, iii, 236, 251–2.

67. Zimmermann, *op. cit.* (note 56), 64.
68. *Ibid.*, 70.
69. See his essays: 'Von den Temperamenten' (*Zimmermann Papers*, B 6a); 'De Viribus Naturae Insitis ad Ingeniossissimos et Eruditissimos Italiae Medicos et Philosophos Dissertatio epistolica' (B 6c); 'De Viribus Naturae Insitis Dissertatio' (B 6d); 'De Principio in Homine Impetum Faciente Dissertatio' (B 6e); 'Obs. Medicae et Physicae', with a section on idiosyncrasy (B 19c).
70. Samuel Schmid von Auenstein to Zimmermann, 6 January 1759 (*Zimmermann Papers*, AII 83).
71. 'Oratio de Temperamentis' (*Zimmermann Papers*, B 6b).
72. Zimmermann, *op. cit.* (note 48), i, 174–5; I have only partially adopted the English translation which departs markedly from the original: *A Treatise on Experience in Physic* (2 vols., London: Wilkie, 1778), i, 133–4.
73. Zimmermann's 'spirit of observation' is closer to the immediately perceiving, rather instinctive 'coup d'oeil' than to the palpating naked clinical 'gaze' which emerged towards the end of the century. See M. Foucault, *The Birth of the Clinic: An Archaeology of Medical Perception* (London: Travistock, 1973), Chapter 7.
74. Haller to Bonnet, 5 January 1759. *Corr. Bonnet*, 153.
75. *Mémoires*, i, 105.
76. *GGA*, 1775, 419.
77. *Elementa*, i, preface, iv.
78. See Salomon-Bayet, *op. cit.* (note 7), 337.
79. It has to be noted that Haller's term 'art of invention' does not – as in Ciceronian and also eighteenth-century rhetorics – denote a skill, which can be acquired through careful study and perseverance. Although for Haller, the scientist in general is a diligent technician rather than a genius, his 'art of invention' is, like Zimmermann's 'spirit of observation', a unique gift.
80. *Mémoires*, i, 107–10.
81. *Ibid.*, iv, 25. The same argument is found in Zimmermann, *op. cit.* (note 48), i, 206–9; Engl. edn, *op. cit.* (note 72), i, 156–8.
82. *Mémoires*, i, 111.
83. *Ibid.*, i, 112.
84. O. Sonntag, 'The Mental and Temperamental Qualities of Haller's Scientist', *Physis: Rivista Internazionale di Storia della Scienza*, 19 (1977), 173–84.
85. *Mémoires*, ii, 39–51.
86. *Ibid.*, i, 149.
87. See Hacking, *op. cit.* (note 2), 181.
88. See Gooding, *op. cit.* (note 2), esp. 131–2.
89. Buscaglia, *Histoire Spécifique* (note 4), esp. 146–9.
90. This is the general characterisation of experimentation as given by Gooding,

op. cit. (note 2).

91. See Hacking, *op. cit.* (note 2), 264.
92. Cunningham, *op. cit.* (note 10), 656.
93. A list of Haller's 200 articles is given in H. Steinke and C. Profos (eds), *Bibliographia Halleriana: Verzeichnis der Schriften von und über Albrecht von Haller* (Basel: Schwabe, 2004).
94. Albrecht von Haller, 'Physiologie', *Encyclopédie, Suppl.*, iv (1777), 344–65: 365.
95. *Ibid.*, 357.
96. *Ibid.*, 349.
97. 'The Haller Preface to the German Translation of the Histoire Naturelle (1750)', in J. Lyon and P.R. Sloan (eds), *From Natural History to the History of Nature: Readings from Buffon and his Critics* (Notre Dame, Ind.: University of Notre Dame Press, 1981), 295–310: 306. The original text is to be found in *Allgemeine Historie der Natur nach allen ihren Besondern Theilen Abgehandelt, nebst einer Beschreibung der Naturalienkammer.... Erster Theil* (Hamburg and Leipzig: Grund and Holle, 1750), ix–xxii.
98. Haller, *op. cit.* (note 97), 300–4. Haller advocated the *esprit systèmatique*, the building of theories based on experience, but he refused the *esprit de système* which was solely nourished by conjectures.
99. Simone de Angelis has convincingly shown that Haller's concept of hypothesis is built on that of Willem 'sGravesande which is itself rooted in natural law theory. For 'sGravesande, not only our actions, thoughts and social behaviour (Pufendorf's *socialitas*) are based on the anatomico–physiological constitution of our body but also our understanding of the external world. Not only the physical objects are contingent, our own sense organs and therefore our perceptions are contingent, too. God, however, guarantees the correspondence between external object and idea. Evidence gained through sense perception is not mathematically but morally secured, and 'moral' is not inferior but complementary to mathematical evidence. 'sGravesande transformed Newtonianism in so far as he admitted not only a mathematical but also an empirical understanding of nature. On the level of 'moral' evidence, man is not able to know the essence of things or God's ultimate laws of physics and nature but only more simple laws as deducted from the phenomena he perceives. We have therefore to distinguish various levels of certainty and to allow for the use of hypotheses which can lead from uncertain to more ascertained knowledge. See S. de Angelis, *Von Newton zu Haller: Studien zum Naturbegriff zwischen Empirismus und Deduktiver Methode in der Schweizer Frühaufklärung* (Tübingen: Niemeyer, 2003).
100. 'Hypothesin nullam admisi, & saepe miror de systemate meo scribi, qui

unice sentire eas partes dixi, aut moveri, quas sentire videram & se agitare... Error poterat esse in experimento, hypotheseos neque umbra est.' *Elementa*, viii (1766), v.

101. Cunningham, *op. cit.* (note 10), 656.
102. Letter from 18 September 1758, *Albrecht von Haller, Giambattista Morgagni: Briefwechsel 1745–1768*, E. Hintzsche (ed.), (Bern and Stuttgart: Huber, 1964), 92.
103. 'En general, je ne medite pas les matieres de Physiologie, que par matieres. Je ramasse alors tous les faits, etrangers ou vus par moi meme, et je me laisse guider par leurs concours.' To Bonnet, 4 March 1760; *Corr. Bonnet*, 193.
104. M.J.A.N. Caritat de Condorcet, 'Éloge de M. de Haller', *Histoire de l'Academie Royale des Sciences*, année 1777 (Paris: Impr. Royale, 1780), 127–54: 135–6; *Œuvres de Vicq d'Azyr*, J.-L. Moreau de la Sarthe (ed.), (6 vols., Paris: Duprat-Duverger, 1805), iv, 12.
105. A. Haller, *Primae Lineae Physiologiae in Usum Praelectionum Academicarum* (Göttingen: Vandenhoeck, 1747), prefatio.
106. The anatomy of the heart, for instance, is described on 130 pages: *Elementa*, i (1757), 256–385.
107. *Bibliothèque Raisonnée*, 33 (1744), 44 (from the review of the *Praelectiones*).
108. For Cunningham, *op. cit.* (note 10), 655, the two muses signify the lower art of anatomy and the higher science of physiology. If they were wearing hats, he argues, these would be different and show the difference in status.
109. *Bibliothèque Raisonnée*, 36 (1746), 184 (from the review of Bonnet's *Traité d'Insectologie*).
110. *Ibid.*, 178–83.
111. Haller, *op. cit.* (note 97), 298
112. How much Haller esteemed experimental writings with scant theoretical reasoning is best visible in his *Bibliotheca Anatomica* (2 vols., Zurich: Orell, Gesnner, Fuessli et Socc., 1774–7). Spallanzani's books, for instance, he repeatedly called 'noble works' (ii, 609).
113. F. Magendie, *Elementary Compendium of Physiology: For the Use of Students*. Transl. from the French by E. Milligan (4th edn, Edinburgh: Carfrae, 1831), 286–7.
114. *Elementa*, vi (1764), 281–92.
115. *Ibid.*, 282.
116. A. von Haller, 'Oeconomie Animale', *Encyclopédie, Suppl.*, iv (1777), 104–5: 104.
117. The earliest experiment reported dates 14 April 1731 (*Haller Papers*, Mss. 24, fol. 1r). Some more details on the debate with Hamberger are given in Chapter 4.
118. A. von Haller, 'De Respiratione Experimenta Anatomica quibus Aeris inter

Pulmonem et Pleuram Absentia Demonstratur, et Musculorum Intercostalium Internorum Officium Adseritur', in A. von Haller, *Opuscula Pathologica* (Lausanne: Bousquet, 1755), 163–292.

119. From our modern point of view, Haller was actually wrong. The internal part is responsible for inspiration, the external for expiration, just as Hamberger had said.
120. A. von Haller, 'Experimenta de Cordis Motu a Stimulo Nato: Lecta d. 10. Novembr.', *Commentarii Societatis Regiae Scientiarum Gottingensis*, Vol. 1 (Göttingen: Vandenhoeck, 1752), 263–6. The experiments upon which this essay is based, are recorded in *Mémoires*, i, 362–8. See also *Elementa*, i, 459–506.
121. See especially the dissertation of his pupil J.H. Heinrich Brunn, *Dissertatio Inauguralis Medica Sistens Experimenta quaedam circa Ligaturas Nervorum in Vivis Animalibus* (M.D. thesis, Göttingen: Vandenhoeck, 1753).
122. See *Mémoires*, i, 158–86, *Elementa*, ii, 330–6 and iv, 171–80 and the dissertation of Haller's pupil J.D. Walstorff, *Dissertatio Inauguralis Medica Sistens Experimenta circa Motum Cerebri, Cerebelli, Durae Matris et Venarum in Vivis Animalibus Instituta* (M.D. thesis, Göttingen: Vandenhoeck, 1753).
123. *Mémoires*, ii, 160.
124. L. Spallanzani, *Dissertazioni di Fisica Animale, e Vegetabile* (Modena: Soc. Tipografica, 1780). See especially the edition of his protocols: *I Giornali delle Sperienze e Osservazioni*, C. Castellani (ed.), (6 vols, Florence: Giunti, 1994), Vol. 2 (Il Giornale della Digestione).

3

Haller's Changing Views on Irritability and Sensibility

Haller delivered a first comprehensive account of his new concept in his orations *De Partibus Corporis Humani Sensilibus et Irritabilibus*, held in April/May 1752, and published in 1753.[1] The importance of this treatise was immediately recognised and it was translated within two years into French, English, German, Italian and Swedish. Just as among Haller's contemporaries it is still nowadays considered the classic statement of the new doctrine, and is usually the main and often even the sole source used in discussions of Haller's achievement. But it is only partially suited to giving a real understanding of his thought. Its main purpose was to present the results of his experimental research and to a large extent, therefore, it is a catalogue listing the parts which are irritable, sensible, or not. The central tenets of the theory – only the muscles are irritable, only the nerves are sensible – were presented here, but many aspects and problems only emerged during the following debate and in the *Elementa* (1757–66), where Haller was forced to elucidate the significance of irritability and sensibility for various parts of physiology and for the animal economy in general. The reformulations and adjustments put forward, not only in his physiological handbook, but also in minor treatises, the articles to the Yverdon and the supplement of the Paris *Encyclopédie* and many book reviews, altered the shape of his concept to a considerable extent.[2] The orations of 1752, therefore, stand for only one step, albeit the most important, in the development of Haller's thought. As we will see in the next chapters, the authors more closely involved in the debate were well aware of some of these later changes and even tried to depict them as a renunciation of the original position. Although Haller did not accept such an interpretation, he was acutely aware of the constant evolution of scientific knowledge and considered himself as part of it. Three times in his life he tried to give a comprehensive account of what he saw as the up-to-date knowledge in physiology: 1739–44 with the edition of Boerhaave's *Praelectiones*, 1757–66 in the *Elementa* and again 1777 with a new edition of the *Elementa*, of which only the first volume appeared during his lifetime. When his friend Marcantonio Caldani (1725–1813) reported that Morgagni and his pupils would point to contradictions between his edition of Boerhaave and the *Elementa*, Haller answered that he did not want to deny them. It would not be strange, he said, to revise many things after an interval of twenty to twenty-five years, and the whole effort of his adversaries would be in vain.[3]

It is mainly thanks to Maria Teresa Monti, and to a lesser degree to François Duchesneau, that we have come to be aware of the development of Haller's thought.[4] Duchesneau delivered a very thorough analysis of Haller's doctrines, mainly based on the *Mémoires sur la Nature Sensible et Irritable du Corps Animal* (1756–60) and the *Elementa,* but he neglected both earlier and, especially, later publications. Monti drew upon the whole range of Haller's writings and identified several elements which changed. This chapter in no way seeks to challenge the assessments of these two authors, but it does seek to stress the importance of these changes, not only because they are significant, but also because they help us to understand certain aspects of Haller's position. It is important to realise that there was no such gradual development towards a coherent concept of irritability and sensibility as Haller wanted to make us believe. In 1750, Haller was not one step closer to his later ideas than he was in 1740, and he advanced the very same ideas as other pupils of Boerhaave and Albinus in the Netherlands. He could not resolve any of his contradictions before he started with systematic experimentation in late-1750, but from then on, in a relatively short period, arrived at his classic statement of 1752. It is equally revealing to learn that later he tended to reconnect the properties that he had earlier separated with such an effort, and that these tendencies were not provoked by any further experimental insight on his part, but were on one side a reaction to the many controversial writings his treatise had provoked and on the other side necessary adjustments within a more comprehensive description of motion and sensation.

General assumptions

Before we look more closely at the details of Haller's changing views we should briefly consider some of the general assumptions upon which his early as well as his late physiology were based. Probably the most fundamental conviction he shared with many of his contemporaries was his belief that all the parts of nature are built for a certain purpose. Haller never doubted that heart, muscles, intestines, etc. had the perfect structure needed to assume their functions.[5] Even the deformed organs of so-called 'monsters' were arranged in such a manner as to most easily fulfil their duty.[6] This teleological postulation underlied Haller's physiological argumentation and allowed two modes of deduction. First, from structure to function and second – at least as important – from function to structure. Both these approaches Haller used in his research on irritability and sensibility. Once the correlation muscle–irritability and nerve–sensibility was established on what Haller considered a sufficient experimental basis, the deduction could be drawn. Wherever he detected a muscular fibre he expected to find irritability and wherever he observed irritability he assumed the presence of muscular fibres. The teleological argumentation was based on the assumption that the objects in our world are constructed upon an intelligent design. It was not nature itself that would aim at a certain purpose, it could only act in accordance with certain laws given by God. It was God who had arranged the structures, laws, and functions in nature, and it was man, or more precisely, the 'scientist' who could try to discover them. However, this was only the outward appearance of nature; the 'inner nature', the ultimate laws, structures and purposes were known only to God. This is what Haller expressed in his famous lines of 1730:

> Into inner nature no created mind penetrates,
> He is very fortunate when nature shows its outer shell.[7]

Although we could not know the 'inner nature', the examination of its 'outer shell' revealed how wonderfully adapted all the different parts and organs of the body were to their surroundings, hence how wisely designed God's creation was. This was the other side of the argument: the teleological assumption about the correlation between structure and function served not only as a methodological basis to draw scientific deductions, it was also taken as a physico-theological proof of God's creation.[8] What we could observe, therefore, could never be against religion. As the visible phenomena could not be denied, it was not 'science', however, but religion which was endangered by such a confrontation. Haller said: 'Wanting to oppose religion to visible truths is the most dangerous thing one can undertake against it.'[9]

The separation between creator and creation allowed the envisaging of nature as a machine which runs according to the prescribed laws of mechanics. For Haller, the discovery of a few single principles that could account for complex and far-reaching movements seemed to be the highest goal science could achieve. 'I am persuaded', he said 'that the physicist [*physicus*] who reduces an event to a constant, experimentally confirmed law fulfils his duty. We cannot expect anything more from him than the demonstration of the truth and constancy of the law.'[10] Haller had already hailed Newton in his *Gedanken über Vernunft, Aberglauben, und Unglauben* (1729, published 1732), presumably the first reference to Sir Isaac in German poetry.[11] He never doubted the general validity of these laws, and he agreed with Boerhaave that it was not impossible to apply mathematical laws to the human body, but he stressed that this was very difficult.[12] In order to understand the movements of a machine, you had to know all its different parts. But the vessels and fibres in the human body were incredibly fine and complicated, and even detailed anatomical knowledge would not suffice to determine their motions. It was necessary to dissect living animals and to put them under the microscope in order to see the actions of their organs.[13] To some extent, Haller replaced earlier attempts to calculate the body's movements by their detailed observation and description. He did not thereby dismiss mechanics as the main explanatory model but emphasised that not all these motions could be reduced to mere mechanics. Through all the editions of his physiological textbook, from 1747 to 1771, he stated that the power exerted in muscular contraction would be 'remote from any mechanical proportion'.[14] Haller's famous description of physiology as *animata anatome* reflects this notion.[15] Besides its stress on the intimate correlation between structure and function it conveys the more general idea that the body is 'infinitely more animated' than anatomical plates and more than a simple machine.[16]

In this sense, Haller's physiology was not a radical break with Boerhaavian mechanism but rather the full acknowledgement and elaboration of non-reductionist elements in the late conceptions of his teacher. But there is one important pillar of Boerhaavian physiology with which Haller broke from the outset: the fibre theory. As we have seen in the first chapter, Boerhaave stated that there is barely a part of the body which does not feel or move, and that the whole body is built up of nerves which he also called ultimate vessels. It is true that, as L.J. Rather has remarked, Boerhaave only *called* and did not *identify* the nervous fibres with the *vas ultimum*, and that his readers – including Haller – seem to have overstated this equation.[17] However, the essential feature is that Boerhaave did not clearly *separate* these structures but sometimes considered the body as consisting only of vessels and sometimes as a mere mass of nerves. It was especially this notion of the omnipresence of nerves and the faculties of sensation and movement that Haller opposed. His argument was based on anatomical demonstration, physiological observation, and analogy. There were certain parts in the body, he asserted, that were not furnished with any nerves, such as the membranes of the brain or the placenta. Furthermore, some parts of the body had no nerves and did not move, nor were they sensible, such as the connective tissue which was widely spread throughout the whole body. Lastly, there were animals with fibres quite similar to ours but which had neither nerves nor brain.[18] Thus, from the very beginning of his physiological research, Haller restricted the realms of movement and sensation and denied the anatomical foundation that would attest their ubiquity. Instead, he introduced the idea of the connective tissue (*tela cellulosa*) which had neither of these two qualities. This was not yet the exposition of a new coherent fibre theory – which he would only present in the *Elementa* – but it was a crucial step which enabled him to develop his notion of specific structures and specific functions.

Early notions

There has never been a doubt that Haller's concept of physiology originates from the Boerhaavian model. But the towering figure of the Leiden master has led to neglect of the fact that Haller, equally, was a pupil of Albinus. Certainly, Albinus was mainly praised for his superior anatomical skill and knowledge, which were indeed far above Boerhaave's, but he also seems to have drawn Haller's attention to certain physiological problems that lie at the core of the pupil's later investigations. Haller, indeed, not only attended Albinus's anatomical dissections but, in his last few months of study in spring 1727, he also frequented his private physiological lectures. Haller's lecture notes have survived,[19] and his later statement that he had given up reading these notes because they were barely legible does at least confirm that he consulted them in his earlier days.[20] Most interesting to us (and probably to Haller, too) are Albinus's observations on the motion of the heart. As shown in the first chapter, he identified, like Boerhaave, three causes of the heart's action – venous blood in the heart, arterial blood in the coronaries, and the animal spirits of the nerves – but he put more emphasis on the existence of a fourth, 'hitherto unknown cause which adheres to the heart itself.'[21] Haller, who never had heard Boerhaave talking of this cause in such an explicit manner, fully adopted Albinus's position. In his commentary on Boerhaave's lectures, he noted that we had to admit such a fourth cause although Boerhaave had not mentioned it.[22] And in his late *Encyclopédie* articles he stated that Albinus 'had dealt with the motion of the heart as independent of the heart... and refused the influence on the action of the muscles which is usually attributed to the nerves.'[23] As we will see, Albinus's notions of muscular movement were actually closer to Haller's early views than Boerhaave ever was. It comes, therefore, as no surprise to hear that Haller, still a student but now in Paris, addressed Albinus, and not Boerhaave, in order to obtain information about the exact movement of the heart and its valves.[24] And a more thorough analysis of Haller's lecture notes (which cover the whole range of animal economy but, indeed, are difficult to read) might reveal even more similarities between Albinian and Hallerian physiology. One further similarity, certainly, is the insistence on animal experimentation, which Albinus not only preached but actually carried out.[25] A number of experiments are mentioned in the lecture notes and these seem to have left a certain impact on Haller. Twenty years later when, in his *Homme Machine*, La Mettrie stated that muscles separated from the body would move upon irritation, and that each fibre of animal bodies would move due to a principle inherent to itself, Haller concluded that La Mettrie – who had attended Leiden – must have gathered these facts from his own writings and Albinus's experiments.[26]

Haller's own early physiological notions are well documented in his seven-volume commentary on Boerhaave's lectures on physiology from 1739–44 – to which I shall refer as the *Praelectiones* – and in the first two editions of his own physiological textbook, the *Primae Lineae Physiologiae* from 1747 and 1751. The *Praelectiones* can be described as an attempt to define the actual state of knowledge by means of an analysis of the strengths and the flaws of his teacher's theories. Although envisaged as an impressive demonstration of erudition and experience in physiology, after their completion they served mainly as an indicator of unsolved problems which showed Haller to which parts and areas his own research should be directed.[27] At the same time, the detailed assessment of Boerhaave's statements was a means of distancing himself from his teacher. Such a critical treatment of the great master was not to everyone's taste and, indeed, created a controversy about the Boerhaavian heritage.[28] But the step Haller had taken was irrevocable and the publication of his own physiological textbook in 1747 was only its logical result. Here, and in the second edition of 1751, Haller presented a first, more-or-less coherent formulation of his own views.

In his orations of 1752 Haller referred to these early works in order to trace his prior insights into the question of irritability and thereby created the idea of a continuous progress of his insights into the question of animal motion.[29] But his quotations were carefully selected and if we compare all the important passages in his early works we get quite another picture. The first statement from the *Praelectiones* (1740), with which Haller wanted to demonstrate his own early understanding of the problem, is actually nothing other than what he had heard already in the lectures of Albinus in 1727, ie. that the 'heart is moved by some unknown cause [which] lies concealed in the very structure of the heart itself.'[30] Haller contradicted himself some pages later in the very same volume, where he argued against the notion of an innate faculty of motion. We should distinguish, he said, between the weak, natural contraction of fibres and the much stronger muscular contraction. Furthermore, the miraculous readiness to motion that was observed in the excised parts of some animals was certainly not seen in man. It was experimentally established that in the human body all these strong motions would be totally suppressed by the ligature of the nerves.[31]

Similarly misleading is Haller's recall of what he said in the fourth volume of the *Praelectiones*, in 1743, namely that,

> ...all animal muscular fibres when they were irritated contracted themselves, that this character distinguished them from those of vegetables, and that the perpetual irritation alone was the cause of the continuation of motion in the vital organs, while the animal organs ceased to act.[32]

Haller wanted to convey the idea that already at this stage he had reached the rather 'advanced' notion that only the muscular fibres would contract upon irritation and that vital but not animal motions would be independent of the connection with nerves and brain. But we cannot agree with this interpretation. Haller simply omitted a restriction he had made in the very same paragraph which he quoted, namely that vital actions would indeed come to a halt when they were totally devoid of nervous spirits. He also said that not only the muscular but each animal fibre would contract upon irritation and that this faculty, to be distinguished from elasticity and from muscular motion, would be responsible for post-mortem activities.[33]

The passage he quoted from the first edition from the *Primae Lineae Physiologiae* (1747) is equally ambiguous. It is true, that in § 113 he 'positively ascribed the motion of the heart to the force of a stimulus', and he even added that this contraction would also occur in excised hearts and without any help of the nerves.[34] But Haller seems not yet to have been convinced of such an entirely autonomous action as he affirmed in § 408–9 that after ligature of the nerves no motion similar to the vital contraction would be possible and that the palpitation observed in excised muscles would be due to the nerves.[35] In § 248 he ventured that the exquisite irritability of the heart in the foetus might be attributed to the higher proportion of nerves to be found at the foetal than at the adult stage.[36]

In 1747 Haller was barely any step closer to his later concept of 1752, and struggled with the very same inconsistencies as in 1740. It was clear to him that there was in the body – especially in the muscles, and most prominently in the heart – the ability to move upon irritation, but the exact status of this faculty and its relation to the nervous system was uncertain. His assertions were still contradictory. At one point he affirmed that there were motions that survive even after the destruction of the brain (§ 562) but he also ascertained that there was no motion in the body independent of the brain (§380). His position was very close to those of the Dutch pupils of Boerhaave and Albinus. In 1748 he wrote in his review of Frederick Winter's *Oratio Inauguralis de Certitudine in Medicina Practica* (Franeker 1746):

> We only mention one important truth that is presented in this oration... The true cause of the movements in the fibres of the animals and therefore of all movements of the animal is determined by it. It consists in the irritable [*in dem irritabili*] due to which all these fibres in their natural state and also after

death contract and in which the whole animal life ultimately consists. Mr W[inter] rightly remarks that the great power of this movement comes from the nerves but that it nonetheless originally lies not in the nerves but in the structure of each mobile fibre and that it remains efficient also without the nerves, although less powerfully so.[37]

Haller had to admit that he had himself no more precise notion of irritability than his colleague in the Netherlands. He was well aware of the insufficiency of his knowledge and in the fifth volume of the *Praelectiones* had stated clearly what the major weakness of his work was: he had performed only a few animal experiments.[38] Only from 1746, when the general conditions at the university improved and more students arrived, did he find opportunities to perform more experiments. His systematic investigations into irritability started only in November 1750, but then, within a year, he developed an original concept. This process has been described in more detail in the previous chapter. Now we have to look at his statements of 1751 which already reflect his new investigations but precede the more coherent formulation of 1752. The passage of most importance to us comes from the second edition of the *Primae Lineae Physiologiae* and was presumably written in spring 1751:[39]

The irritated [muscular] fibre contracts, palpitates, alternatively acts and slackens. This irritable nature is totally absent in the cellular tissue [*cellulositas*], and is very weak in the membranes that derive from the cellular tissue as well as in the ligaments and tendons, but is inherent in the muscles, and especially in the heart and the intestines, so that they can be activated for a considerable time even after death. And it is not to be withheld that almost nothing of the irritable nature disappears, especially in the heart and the intestines, when a certain irritable part is excised from the body and the nerves removed and detached from the connection with the brain.[40]

Thus, Haller repeated some of his earlier statements, although in a somewhat bolder manner and without apparent contradictions: irritability was a property that resided predominantly but not exclusively in the muscles and which to a certain, although quite small extent seemed to be dependent on the nerves. Essentially, he had no clear idea of the relationship between irritability and nerves, not least due to the disturbing fact that there is no muscle without nerve.[41] But he made a confident announcement:

This power is new and different from any other hitherto known property of the bodies, as it depends neither on weight nor on attraction nor on elasticity, and resides in the soft fibre and vanishes when it dries up.[42]

Such an unequivocal proclamation of irritability as an independent property was new, and was a prerequisite for the discovery of its exact characteristics. But without a clear description of its extension and its qualities, the affirmation of its specificity looked like a rather rash conclusion. It took Haller another year to fill this gap.

Such worrying inconsistencies and gaps as we have encountered in Haller's early concepts of irritability did not trouble his notion of sensibility. He never doubted that all sensations were caused by the impression of objects on the nerves that transmitted the impetus to the brain, where they were then presented to the soul.[43] The nerves, although called sensible, were only the carrier as Haller could conceive sensation solely as a conscious process, which ultimately had to be located in the brain. He favoured the nervous liquids as the medium of transmission because the softness of the winding nerves seemed to exclude the possibility that they could convey impulses by vibration. Thus, unlike irritability, sensibility as a distinct property on its own was never questioned. Haller's difficulty lay in separating irritability not from the *sensory* but from the *motive* power of the nerves.[44] Given his notion of sensation as a conscious process he was never able to understand how somebody could confound it with the local reaction of the fibre. Sensibility – in contrast to irritability – was not perceived as a field of research surrounded with problems that deserved serious attention. Haller's first irritation of an uncovered heart can be dated back to 1731, and from then on the movements of this organ were always of prime importance.[45] Experimental investigation into sensibility, on the other hand, started only in 1748 and served as an experimental refutation of Baglivi's concept of the dura mater as the origin of sensation. The results, published by Haller's pupil Johann Gottfried Zinn, proved the absence of nerves, the insensibility and immobility of this membrane, and were the first step in the scheme to demarcate the limits of sensibility.[46] But the main conclusion was not yet drawn with clarity. Although Haller maintained that the connective tissue which was void of nerves was also insensible, that every nerve when irritated provoked an intense pain, and that the sensibility of an organ seemed to be in proportion to the amount of nerves it was furnished with, he did not before 1752 positively declare what all these statements seemed to demand, ie. that only the nerve is sensible.[47]

The orations of 1752

The orations *De Partibus Sensilibus et Irritabilibus*, Haller's classic statement of his theory, contain the result of experimental research. Within roughly one year, from the start of his systematic experimentation in November 1750 to the end of 1751, Haller proceeded from single, partly inconsistent and even contradictory statements to a new and coherent concept that could and would be considered as a new theory of animal motion and sensation.[48] Some aspects of the experimental process leading to this theory have been treated in the previous chapter. Here, we are dealing with the question of how Haller merged his experimental results into a consistent framework. Haller himself, in 1762, very briefly described the steps he had taken in order to define the properties and extension of irritability. He first separated it from elasticity, then from the motive force of the nerves (*vis nervosa*), and thereby came to the view that the motion of heart and intestines was solely due to their irritability. Lastly, he reduced the property to the muscular fibre and fixed some further aspects of its operation.[49] The description of the procedure as separation and reduction is essentially correct but, of course, incomplete. The following paragraphs try to add the missing parts to this story.

Both concepts, of irritability and of sensibility, were defined from three different angles, both qualities might be represented as a specific visible reaction, as a functional entity, and as a property of a specific structure. The definition that Haller gives at the beginning of his orations reflects only some of these aspects:

> I call that part of the human body irritable, which becomes shorter upon being touched; very irritable if it contracts upon a slight touch, and the contrary if a violent touch contracts it but little. I call that a sensible part of the human body, which upon being touched transmits the impression of it to the soul; and in brutes, in whom the existence of a soul is not so clear, I call those parts sensible, the Irritation of which occasions evident signs of pain and disquiet in the animal.[50]

Despite its incompleteness and although written after the performance of the experiments, this definition actually represents Haller's situation at the outset of his investigation. It mainly contains the description of the properties as visible reactions that Haller tested both from late-November 1750 onwards. Without any clear concept of its functional separation from elasticity and *vis nervosa* and without reduction to the muscular fibre, in his notebook Haller described as irritable those parts which would contract upon irritation. At this stage, his conceptualisation of irritability existed only on the level of visible reaction; sensibility, however, he perceived on two levels already. As we have seen, Haller had always considered sensibility as the conscious reception of impressions transmitted through the nerves. Thus he already conceived it as a property on its own. In order to make this functional entity coincide with an observable reaction, he had to restrict it to the feeling of pain, which to him was just one of the different modes of sensation induced by the irritation of nerves.[51]

During most of the period of experimentation, Haller was more concerned with the apparently more difficult problem of irritability than with sensibility. He investigated, however, the aspects of the latter quality first. In his view, the experiments clearly demonstrated that many parts of the body, contrary to the general belief, were not sensible, such as the tendons and ligaments.[52] Finally, in April 1751 at the latest, he came to the conclusion – one he had probably never doubted but also never positively drawn – that only parts furnished with nerves, or indeed the nerves themselves, were sensible.[53] And in the orations he specified that sensibility resided in the medullar part of the nerves, which originated in the brain.[54] Having thus clarified this major point of the structural location of sensibility, Haller only incidentally performed further experiments on this subject until February 1752 when, together with his pupil Peter Castell, he started another two-month series of trials to establish with more certainty the grades of sensibility in the different parts of the body.[55] Although this is not clearly stated in the orations, the general result of this investigation seems to be that the degree of sensibility is in proportion to the amount of nerves with which a part is furnished.[56]

With regard to irritability, Haller had more steps to take in order to reach a similar degree of conceptualization. One of the first things he probably became aware of during experimentation was the distinction between irritation and response, a distinction which was often lacking in his earlier writings but which he clearly made in *De Partibus*, although he did not explicitly emphasise it.[57] His experiments – along with his knowledge of the publications of the Leiden school from the 1740s – seem to have supported the assumption, put forward as early as 1743, that there was another motive faculty to be distinguished from elasticity and the *vis nervosa*. Already in the 1751-edition of the *Primae Lineae*, some five months after his first trials, he unambiguously declared irritability as a specific force on its own. But the arguments he could provide were not entirely convincing. He pointed out that irritability, in contrast to elasticity, diminished when the fibre dried up.[58] Thus, we may conclude, the retraction observed in a dry fibre was due to its elasticity. But where is the dividing line between the actions of elasticity and irritability? Haller could not tell whether the movement of a tendon – at this time thought to be slightly irritable – would be due to elasticity or irritability. The latter might be a much stronger force than the former. But were these impressive contractions really due to a specific force? Did not Haller himself declare that this force would, to some extent, diminish when detached from the nerves? Haller had not yet clearly separated irritability from the other forces of contraction. The idea of a specific faculty only became persuasive when it was linked to its location within a specific structure.

Already the early experiments of November 1750 served Haller as proof that the cellular tissue and fat were not irritable. As the text of the *Primae Lineae* of 1751 shows, he was for some time still uncertain about the tendons, but these doubts were finally resolved on 16 November, when he noted: 'Tendo v[ero] non irritabilis.'[59] Of even greater importance was the insight, gained in June of that year, that the nerves were not irritable either. What then, was left to be irritable? The heart, of course, the intestines, the muscles, urinary bladder, and some other organs and structures at various degrees. This is all the evidence we can get from Haller's notebooks, which evidently is far away from his well-known statement in *De Partibus*, viz that only the muscular fibre is irritable. In order to understand why Haller made this inference we have to touch upon some further elements of his fibre theory, such as they were elaborated before the middle of the century. Like Boerhaave, he believed that almost all parts of the body consisted of simple fibres, which in their minuteness were inaccessible to the senses. These fibres would form more compound, still invisible structures, such as small vessels, membranes, nervous, muscular, tendinous, ligamentous and bony fibres, and the cellular tissue. Unlike the other compound structures, the cellular tissue did not have an oblong but a broad, flat shape. It was spread in all parts of the body and, together with the vessels, nerves, muscular, and tendinous fibres built up the whole of the even more composite muscles, glands, and internal organs, which thereby received their firmness and stability.[60] Thus the structures accessible to the experimenter were of a rather complex nature. If Haller wanted to maintain the mechanical belief in the correlation between structure and function, he had to relegate irritability to an unchanging element within such dissimilar parts as heart, intestines, and urinary bladder. Of course, this was the muscular fibre, and, most importantly, *only* the muscular fibre. With this attribution the outlook of his fibre theory changed radically, and Haller was well aware of the fact. He indeed created, as he said, 'a new division of the parts of the human body', distinguishing between the irritable muscle fibre and the sensible nerve or, more precisely, the nervous fibre.[61] What was left over were small vessels, membranes, tendinous and bony fibres, and the cellular tissue, none of them either irritable or sensible, but elastic as all fibres of the animal body. All these structures seemed to be of a very similar nature. In *De Partibus* Haller stated that all membranes are made of 'cellulosity' and later, in the *Elementa*, he confirmed that also the vessels and, presumably, the tendons and bones essentially consisted of cellular tissue.[62] Thus, although there were certain differences between these structures regarding their compactness and elementary particles, they could be considered as a third entity besides the muscular and the nervous fibre.

The essential characteristic of Haller's concept, however, was neither the maintenance of the correlation between structure and function nor the division of the body into irritable and sensible parts but – and this has always been stressed – the postulation of a strict connection between *specific* structure and *specific* function. François Duchesneau has described this as a 'special mechanist hypothesis'; *mechanist* because of the correlation between structure and function, *special* because of both their specificity,[63] and *hypothesis* because the fibres which should account for the property were not experimentally accessible, they were only visible in their more compound formations as nervous and muscular filaments. The role of the fibre in this concept is considerably different to the one it had in the older fibre theories. Although there is still a correlation between fibrillar structure and organic function this relation is unintelligible; the function cannot simply be deduced in a geometrical or mechanical manner from the structure and neither can the exact structure be gathered from the function. The source of both irritability and sensibility lies in the 'intimate fabric' of the fibres beyond the reach of knife and microscope.[64] At this level, the mechanistic model is not openly rejected but the exact mechanism is unknown. Conceptually, this was not a new position; a certain scepticism as to the application of the mechanical model to the micro-structures was genuine to the mechanical philosophy of the seventeenth century.[65] Regarding the problem of muscular motion, however, Haller drew fresh attention to such a mechanically unexplainable gap. For him, muscular power was not, as in many physiological explanations of his time, an elastic force that reacted mechanically upon stimulation; it was obvious to him that the reaction was much stronger than the irritation. Already in 1747 he had stated that muscular contraction would be 'remote from any mechanical proportion'. Thus, it could not just be a transformation of an external impulse but had to be due to a power inherent in the muscle. Haller did not address this issue in *De Partibus*, but he did so in two reviews, written almost at the same time, in which he examined Robert Whytt's important *Essay on Vital Motions* (1751). He saw no problem, he said, in attributing active power to matter and exhibited surprise that an Englishman could consider such a position as unphilosophical.[66] He added in a Newtonian manner that it was not the duty of the physician to inquire into the nature of such powers, but to trace certain phenomena to a constant, experimentally verified law.[67] In a slightly earlier review Haller had called irritability a 'quality essential to the structure of plants and animals' but he specified that this and other classes of power did not 'take part in the essence' of matter as they were not common to all its parts. It was God who had endowed different classes of matter with different powers.[68] Thus, we might argue, power is theoretically detachable from matter. But such a speculation was certainly not of primary interest to Haller. He never presented a precise analysis of the relation between force and matter. What was important to him was the disclosure that a force – call it corporeal or physical – was acting *in* the fibre and that there was no need to suppose another force – physical or immaterial – acting from outside upon the fibre.[69] This is why later he favoured the term *vis insita*, an expression he did not yet use in *De Partibus*. Although the exact mechanism of this power was unknown, and although the mechanistic model on this level of microstructures was suspended, Haller still called irritability a mechanical explanation, thereby stressing that no incorporeal forces – and especially not the soul – would interact in this process.[70]

Haller's attempt to localise irritability in the elementary structures of the fibre has to be seen as a further effort to ground contraction in some kind of material reality and to alienate the property from speculative hidden qualities. He thought that the fibre was made of earth, gluten and other chemical substances. The earthy molecules were arranged in rows and were held together by the glutinous substance which consisted of oil and water. The harder a substance was the more earth and the less gluten it contained.[71] Given these premises, Haller argued that irritability seems to reside in gluten, because this moist substance tends to retract upon irritation, whereas the dry earth never changes its shape by itself. Furthermore, younger animals, which possess more gluten and less earth, are more irritable than older ones.[72] In *De partibus* and later, Haller presented this theory only as a hypothesis, but even then this posed some problems because gluten was present in all, but irritability only in the muscular fibres. However, as François Duchesneau reminds us, irritability is not defined on the level of gluten but emerges only in the muscular fibre. Although gluten is a necessary material constituent, only the more complex structure of this fibre can account for the property.[73]

As we have seen, Haller defined irritability and sensibility in three different ways, as an observational, a functional, and a structural entity. His first approach was observation and experiment. It was this method which led to the dissolution of the unitary fibre theory, whose proponents either had thought all parts to be irritable or sensible (Boerhaave), or had postulated entire systems of movement and sensation (Baglivi). Haller did not, as Baglivi did, simply declare the sensibility of the dura mater, but he tried to prove by experiment whether this was the case, and by anatomical investigation whether there were any nerves which would support such a hypothesis. Still, as Chapter 2 has shown, the case was not as simple as that. Tendons contracted upon chemical irritation and according to Haller's definition therefore should be called irritable. But maceration of tendons had shown that they mainly consisted of cellular fibres, which Haller thought not to be irritable. Equally, the irritation of the pericranium sometimes produced pain and there were nerves creeping on its surface. But as Haller thought the pericranium to be an ordinary membrane he expected it to be insensible. The nerves, therefore, in his view did not actually penetrate the pericranium but were only passing through it. Thus, Haller used his structuro–functional definitions not only as a conceptual means to define the two properties but also as a corrective to the experimental approach. Only the three approaches taken together determined the qualities, appearances and extension of the properties.

In the whole discussion of the conceptual framework of irritability and sensibility, one tends to forget that *De Partibus* is to a great extent a summarised report of the experimental investigation, determining which parts are irritable and which are sensible, and to what degree. Let us first have a look at the irritable parts. The most irritable structure was the heart, followed by the intestines and the diaphragm. Still quite irritable were the muscles themselves, the stomach, and the uterus. The arteries might be irritable although this could not be experimentally confirmed. Definitely non-irritable were the nerves, cellular tissue, fat, tendons, membranes, the lungs, the liver, and other internal organs. Thus, Haller concluded, the vital parts were the most irritable, and required only a weak stimulus to be put in motion, whereas the others were not moved except by the determination of the will or by very strong irritations. This was Haller's solution to an old problem, viz to explain why vital organs such as the heart and intestines constantly move and fulfil their functions. It was only due to their irritability. But how is it to be explained that the heart is much more irritable than other muscles even though all are almost entirely built from the same fibres? Haller's answer is remarkable. He admitted that the question was very difficult and that his explanation was only a hypothesis (which nevertheless he adopted). The intestines, he said, have only a few nerves but these are very much exposed, as we can see from the symptoms provoked by the abrasion of their mucus. This would explain the great irritability of the intestines and might also be the cause of the intense irritability of the heart, which likewise has only a few, rather small nerves. But, he continued, there was not enough anatomical knowledge to decide whether the nerves were exposed in a similar manner in the ventricles of the heart.[74] This explanation at first sight seems to contradict Haller's entire project, which stresses the independence of irritability from the nerves and the distinction between irritability and sensibility. And indeed, Haller did not feel entirely comfortable with it. Later, in the *Elementa*, he proposed the same hypothesis, but added that he would gladly listen to any other explanation.[75] Still, we have to note that

Haller did not say that the nerves were the cause of irritability but only that they would increase its power. This notion he mentioned in passing in two other passages in *De Partibus* and more openly in later writings in many variations.[76] Thus, even in the orations of 1752, where Haller presented the most austere separation of movement and sensation of all his writings, these two processes were not entirely disconnected.

The most sensible parts according to Haller were, of course, the nerves themselves. Also sensible were the skin, muscles, retina and choroidea of the eye, tongue, and genitals. As insensible, he described the cellular tissue, fat, tendons, membranes, periosteum, peritoneum, pericardium, and pleura, dura and pia mater, and the iris of the eye. The degree of sensibility seemed to be in relation to the amount of nerves with which an organ was furnished. Lungs, liver, spleen, and kidneys, therefore, must be sensible to a certain small degree although this cannot be proved experimentally. The same reason Haller used to say that the bones are not sensible, although he had made no adequate experiments. The denial of sensibility to certain parts was a general attack on the reigning unitary fibre theories. More particularly, it was the insensibility of the dura mater and the tendons that opposed accepted beliefs. Haller not only proved the insensibility of the dura mater but also its immobility. The motion of this membrane and of the brain, which can be observed through the opened skull was, he said, solely due to the increase and decrease of the venous blood pressure during respiration. All the parts in the skull moved according to this pressure but there was no motion under normal circumstances when the dura mater adheres to the closed skull.[77] Thus, Haller opposed his experiments to one of the main pillars of the unitary fibre theories, viz their argumentation that all sensibility (Baglivi) or all irritability (Frederick Winter) derives from the dura mater. The denial of the sensibility of the tendons was of less theoretical and more practical relevance. It had been a dogma since antiquity that lesions of tendons provoke intense pain and can cause dreadful accidents with high fever, stupor, cramps, and often death, and Haller had shared this belief with almost all his contemporaries.[78] He explained this general error, into which even the most respectable authors had fallen, through the fact that the ancients used the word 'neuron' equally for nerve, tendon, and ligament. The pain and the accidents, Haller concluded, were actually due to the lesion of the nerve. One should no longer be afraid of the wounds of the tendons, and their treatment with hot oil of turpentine should be dropped.[79]

Considering the importance and range of his investigations, Haller stated with a good deal of confidence that 'the experiments which I have made, are the source of a great many changes, both in the physiology, pathology and surgery, and discover several truths contrary to the opinions generally received.'[80] Nobody would have denied that they were indeed contrary to the general opinions but whether they would cause the changes Haller had expected, had yet to be seen.

Later views

Despite its austerity and mainly descriptive character Haller's treatise was considered by many as a presentation of a new medical system competing with such comprehensive theories as those of Boerhaave, Hoffmann, or Stahl. Haller objected to such an interpretation:

> They continue to impute a system to Mr von Haller, although he presents nothing more than general conclusions drawn from many experiments. Irritability does not derive from the nerves, this is what the experiments show, and besides that he says not a single word.[81]

This statement was, of course, a simplistic one, especially as it was made in 1765, when Haller had said many more words on irritability than in 1752. It also neglected the systematic character already inherent in *De Partibus*, where Haller had spoken of 'a new division of the parts of the human body'. The separation of three different structural and functional components – muscular fibres for movement, nervous for sensation, and connective tissue (*tela cellulosa*) as a supporting framework – in itself could be seen as the core of a new explanatory model of the human economy. Haller himself used the postulation of a motive faculty independent of the brain to refute one of the major physiological systems, Stahlian animism. Still, we have to agree with Haller that this was not tantamount to the construction of a comprehensive new system. Such a system Haller would never present. What he did though, was to integrate his ideas on irritability and sensibility into a more complete concept of animal motion and sensation than that presented in 1752. This integration took place mainly in the *Elementa*, where Haller delivered an ample exposition of the entire physiology. Here, and in the later articles for the *Encyclopédie*, he ventured thoughts that went far beyond his earlier rather descriptive account. Nevertheless, the different processes in physiology were treated as distinct problems and were not subjugated to a well-rounded picture of the animal economy, nor was a single quality such as irritability considered as an all-dominating power. Certainly, the concepts of irritability and sensibility were the most important element of Hallerian physiology, as they were concerned with the essential phenomena of motion and sensation, but there are entire volumes in the *Elementa*, such as those on the blood, on respiration and on digestion in which these notions are of virtually no significance. As Haller explained to Bonnet, he treated physiology subject by subject, 'collect[ed] all the facts, foreign or seen by myself, and let myself guide by their competition.'[82] Haller's physiology is the result of the critical assessment and interpretation of a vast amount of detailed information, gathered from many credible authors but in great part also furnished by his own anatomical, physiological, and especially experimental research. Thus, as a whole and in its single parts, it is constructed in a very deliberate manner, and its tenets are intimately intertwined with the way in which they are gained; its content cannot be separated from its methodological approach.

For the extensive treatment of the different aspects of motion and sensation in the *Elementa*, Haller was in need of a more precise and detailed description of irritability and sensibility, as well as a clarification of the relations between these two and other properties. He could do with the same thing for the controversy his treatise had roused. One important question was the relation between the different forces of contraction: elasticity, irritability, and *vis nervosa*. In 1755 Haller, with more lucidity than before, distinguished between these three qualities.[83] Elasticity was a property of all fibres of the body, and had nothing to do with life. Irritability was present only in the muscular fibre, lasted somewhat longer than life, and was the only force able to move the muscles detached from the human body, or those of animals without any nervous system. *Vis nervosa* indicated the power of the muscles derived from the nerves. It was activated through the irritation of the nerves or the will of the soul and was much stronger than the other two forces. Apart from that, all three forces produced more or less the same effect, viz a contraction of the muscle. Interestingly, the effort to distinguish the three forces resulted instead in them being tied together. Once their domains were located they could again be associated without their specificity being violated. Some years later, in the *Elementa*, Haller described elasticity as a 'kind of dead irritability' and did not exclude the possibility that irritability might be only a 'stronger grade of the dead contraction'.[84] The influx of nervous juice in the *vis nervosa* was nothing other than another, very strong kind of stimulation of the muscle, which was necessary to bring the less irritable voluntary muscles into contraction.[85] In the 1765 edition of his physiological textbook he talked of a *triplex vis* acting in the muscles, and in the same year, in the seventh volume of the *Elementa,* he imagined a *scala motus animalis* with three different levels.[86] Similar statements are to be found in the *Encyclopédie* where Haller presumed that elasticity might serve as a basis for all animal movement, only becoming irritability in the muscle because of the greater aptitude for contraction in these fibres.[87] Lastly, he took a final step and declared: 'In the fibres there are three contractive powers, or, if we wish, three degrees, but very distinct degrees, of the same power; the dead power [*force morte*], irritability, and the nervous movement.'[88] In his tendency to reduce movement to one single power, Haller went as far as he could go without giving up the concept of irritability as a specific property.

In a similar manner to the way in which he tended to unite the different forces of contraction, in his later writings Haller narrowed the gap between the realms of motion and sensation. We have seen that already in *De Partibus* he had argued that the nerves might extend the irritability of the heart. With more frankness and precision he now declared his agreement with Robert Whytt that an increased sensitivity of the muscles would add to their power and to the regularity of the heartbeat.[89] Thus the nerves assumed a third function which was neither conscious perception (sensibility) nor the voluntary contraction of the muscles (*vis nervosa*) but a kind of unconscious, involuntary, modulatory role. They assisted the regularity and intensity of the vital and animal motions. Even more, in the 1770s – making overtures to his opponent Johann August Unzer – Haller asserted that the nerves would add to the necessary integrity of the muscle.[90] This had to be understood as a statement, not only about the function of the muscles but also about their composition. In the *Encyclopédie* Haller maintained that,

> ...it is even almost certain that the medullary pulp [of the nerves] is one of the essential elements of which the muscular fibre ist composed, and it is quite natural that the good condition of the fibre supposes that of the nerve which is one of its parts.[91]

Certainly, Haller did not repeat the Boerhaavian notion that the muscular fibres are the 'last nerves' (ie. nerve-ends) – a notion he had opposed from the outset of his career – but his description of the nerve as a part of the muscular fibre was a step backwards in the direction of the abandoned unitary fibre model. Whereas in his oration of 1752 he had neatly separated the muscular from the nervous fibre, he now tended to bring those two parts together again.

The nerves achieved a certain independence of the brain in another question, too. It had long been known that there are cases of commmunication between certain parts of the body, usually called 'sympathy' or *consensus partium.* For instance, in the case of sneezing, an irritation of the nerves of the nose led to convulsion of several other parts of the body. Many authors had explained this through a direct contact between the nerves, and Haller in the *Primae Lineae* of 1747 mentioned several other explanations such as connections between blood vessels, structural similarities between certain parts, the continuity of membranes or the connection of the nerves in the brain, without favouring any of these hypotheses.[92] In the *Elementa,* he accepted the idea that this would happen through the nerves, but, presumably still under the impression of his own concept of sensibility as a conscious phenomenon, he favoured the notion that the communication would happen in the brain.[93] Ultimately, in the *Encyclopédie*, he ventured that there is a direct communication between the nerves through the ganglia.[94] He thereby highlighted the mutual dependence and solidarity between the organs of the body, a notion for which Haller was presumbly indebted to Whytt and which was typical for vitalistic theories of this time. It should be noted, though, that Haller only briefly touched upon this idea and never considered it an important pillar of his physiology.

There are some passages which suggest that Haller even played with the possibility that irritability might depend on sensibility. The opening sentence of the article *Irritabilité* reads as follows: 'Irritability is entirely different from sensibility; *even if it depended on the latter*, it would still be distinct because it causes invariably a shortening which is never the effect of sensibility.'[95] Irritability remained a property with its own effects but Haller did not exclude it altogether to be subordinate to sensibility, which acquired a more central position in physiology than previously. He simply stated: 'Sensibility is the essential character of the animal. That which feels is an animal, that which does not feel, is not.'[96] But, as sensibility itself is not an observable quality, we have to examine the movement of a being in order to decide whether it is an animal or not. Haller asserted, all animals move, and even more, that 'all animals are irritable and... touched with a power proportional to their sensibility, contract and provide some sign of sensation [*sentiment*] by trying to withdraw from that which causes the sensation.'[97] Thus, irritability effectively seemed to require sensibility, and motion appeared to be an expression of sensation. However, we have to be careful not to overrate these somewhat confusing statements, as we also find passages in which Haller faithfully adheres to his strict independence of the two properties. They should not be taken at their face value but rather as a further indication that in the course of the debate Haller softly, but visibly, tended towards a more unitary view of the body and did not always maintain his early clear-cut notions.

Haller also made several adaptations and alterations of smaller significance (such as the notion of increased sensibility in inflammation), which nevertheless gained considerable attention in the debate and will therefore be treated in the second part of the book, together with other questions such as the use of experiment or the role of religious beliefs. Before we pass on to the debate and its relation to the prevailing medical systems of the eighteenth century, we must try to assess Haller's position within this framework of mechanism, vitalism, and animism. In *De Partibus*, Haller had used the detection of a contractive force independent of nerves and brain to argue most decidedly against an animist position, be it that of the Stahlians in general or of Robert Whytt in particular. The 'Stahlian sect', as he called it, had always been the favoured target of his criticism against medical systems. With the discovery of irritability he thought he had found definitive proof for its refutation. But what kind of alternative did his theory offer, was it a mechanist or a vitalist concept? As we have seen, the orations of 1752 themselves did not prescribe a specific position and were open to both interpretations. What was Haller's own view? Ten years later, in 1763, when Robert Whytt called him a member of the 'mechanist sect', Haller answered that this was not an unmerited attribute.[98] More precisely, he called those who, like himself, stressed that the power exerted in muscular contraction was greater than that of the irritating cause, 'the new mechanists'.[99] But what was a 'new mechanistic' explanation of vital functions for him? In the article *Faculté Vitale*, he wrote:

> Everytime we see effects the mechanical cause of which is unknown to us, we can mark this cause by the name of faculty, just as we call an unknown quantity x. If illuminating experiments or perfected anatomy discovered the mechanism which produces this effect, we would then erase the provisional term [*nom d'attente*], just as we erase the letter which stands for an unknown quantity.[100]

Irritability, the main subject of this article, was also for Haller such a *nom d'attente*, and he explicitly stated that he did not search for its mechanical or physical causes.[101] It would ultimately be describable in physical terms because it was an entirely corporeal faculty which could not contradict the physical laws of nature. But, he said, such a description was far away. In his article *Méchanisme* he stated:

> The mechanism of the movements of the human body is undoubtedly the object of the most fervent wishes of the true physician. If it was known, if we knew the corporeal causes which produce digestion, circulation, and the other animal faculties, then we could determine which remedy could restore the movements to the natural condition, once they are deranged, or at least demonstrate that this restoration is impossible.
>
> Unfortunately we are far away from knowing this mechanism. There is almost only the eye of which we know precisely the function of the organ, the structures of its parts and the manner how each part serves its purpose.
>
> This is the triumph of physiology, unfortunately it is almost the only one.[102]

Physiological processes were physical processes, but in their complexity they were far above the mechanics of simple moving bodies. In this sense, they formed a category of their own. They could not be explained by our rather simple knowledge of mechanics or described mathematically as the course of the rays in the eyes, but had to be investigated by experiments on living bodies. The forces which direct these processes were themselves not reducible to simple mechanics. In particular, the power exerted in muscular contraction was – as Haller had already stated in 1747 – 'remote from any mechanical proportion'.[103] Although further research might ultimately enable us to measure the forces at work, there would always remain a mechanically unexplainable gap: 'We will never know the mechanical source out of which the movements following irritation are born; but we will come closer, we will perhaps be able to measure exactly its effect and to compare this effect with the force of the stimulus...'.[104] This is why Haller could maintain that irritability was 'not subjected to the common laws of movement'.[105] Nevertheless, it was a physical force and there was no need to call for non-corporeal powers as causes of movement. Haller followed Newton in the belief that new motion may be generated just as old motion may be destroyed.[106] This generation was caused by forces such as irritability, gravity, and effervescence and, as a physical process, took place in matter itself. In this sense, Haller could declare that it was matter itself that produced the movement: 'Nothing is less true, to finish this discussion, than the inability [*impuissance*] of matter to produce movement.'[107] With his preference for the term *vis insita* for irritability, Haller wanted to stress this material aspect and wanted to emphasise that the faculty resided *in* the fibre.[108] This was not to be understood too narrowly. The exact connection of this force to the fibre was of no particular importance. Haller explained to Bonnet: 'I do not examine whether this power [*puissance*] resides in the fibres themselves or in an foreign element attached to the fibre.'[109] As we have seen, irritability was not a simple force acting in a simple structure, but rather a complex physiological property that emerged only on the composite level of the muscular fibre. Haller continued to assert that it was God who had instilled the forces in the bodies but this was an early act of creation antedating the birth of the animals themselves.[110] Correspondingly, Haller often called irritability – besides *vis insita* – *vis innata* and thus pointed out the intimate bond between force and matter.

If we have to locate Haller's physiology within eighteenth-century medical systems, it is probably best seen as a non-reductionist mechanism. The use of the term 'mechanism' does not mean that Haller fought against the notion of a vital faculty as a property necessary for life. He described death as a loss of the heart's irritability, his article *Faculté vitale* discussed, in fact, irritability, and, in a few cases, he overtly called it a vital faculty.[111] It is true, he preferred the expression 'innate faculty' (*vis insita*), but not in order to directly oppose it to the other term, rather to stress its material location and, on a more technical level, to be precise, because irritability lasted somewhat longer than life.[112] Furthermore, Haller's later tendency to reduce the differences between properties, which, earlier, he had made such an effort to set apart, may be regarded as a movement from a mechanistic account separating different processes towards a more vitalistic interpretation emphasizing the unity of the body. Although the term 'non-reductionist mechanism' reflects neither this change nor the experimental approach and the specificity of properties, which are both inseparable from Hallerian physiology, it at least takes into account its intermediate position, denotes the background from which it developed, and the 'system' in which Haller had allowed himself to be counted.

The expression 'Newtonian physiology', put forward by Shirley Roe and others, is somewhat more problematic.[113] It makes sense in so far as it reflects the fact that Haller's concept of irritability and sensibility by some physicians was considered as a solid and simple fundament of physiology comparable to that which Newton had laid for physics.[114] Even more, it may be justified with respect to Haller's methodology which, in its experimental approach, its search for natural laws instead of hidden qualities, and its refusal of unfounded hypotheses, certainly followed the Newtonian model. But in his description of the law or, more precisely, the property accounting for it, Haller transformed Newton's concept. As François Duchesneau has shown, the inaccessibility of the cause of motion led him to present an explanation on a level of more complex structures than those of the atoms of attraction.[115] Corresponding to these compound structures, irritability is a more complex and specific property, which – as we have seen – is in some sense 'not subjected to the common laws of movement'. At this point the talk of 'Newtonian physiology' becomes questionable. Even more so since Haller – as many adherents of Newtonian thought – had a notion of force and matter somewhat different from that of Newton, although he had never properly elaborated on the subject. Whereas for Newton forces had no material existence and were closely linked with space, for Haller they were properties of a substance. Haller did not need the concept of ether as an embodiment of forces in order to explain the power acting in the muscle; it truly resides in the fibre itself, and matter is not passive but active.[116] For Haller there was no intermediate level of forces, there were only two clearly separated worlds, the physical, corporeal world to which such forces as gravity and irritability belong and the incorporeal world of the soul and spirits, and only the former was a subject of scientific research.[117] Apart from this Cartesian dualism on the level of substances, Haller established also a separation between body and soul on the level of functions.[118] This happened in two respects. As regards the voluntary movement, Haller drew upon Leibniz's pre-established harmony. Thanks to this harmony, the soul generated movements precisely corresponding to its wishes although it did not know which muscles were needed for the action. How the transition from the incorporeal to the corporeal realm happened, was altogether unknown; in this respect, Haller followed Boerhaave.[119] As regards the involuntary, vital movement, he stressed, with his concept of irritability, the autonomy of the living and created a domain of the body independent of the soul.

Ultimately, all efforts to categorise Haller's physiology with simple labels are unsatisfactory. Calling Haller's concept 'Newtonian physiology' without any further indication as to how this description may be justified might be somewhat misleading, as it overstates the validity of certain analogies. Similarly, the classification as non-reductionist mechanism has its deficiencies, because it directs the attention towards the ontological status of the physiological forces, with which Haller was not particularly concerned. His concept rather focused upon the specificity of properties and structures and – as we will see in Chapter 5 – the main difference between his supporters and opponents was therefore not whether they called irritability and sensibility mechanical or vital powers, but whether they accepted their specificity, the restriction of their realms and the clear distinction to be drawn between them. On the other hand, Haller's physiology is characterised by its experimental approach, which cannot be separated from its content. And this, again, was often – as the next chapter will demonstrate – a matter of disagreement between opponents and supporters.

Notes

1 *De Partibus.* Some aspects of this chapter I have already discussed in 'Haller's Concept of Irritability and Sensibility and its Reception in France', *La Lettre de la Maison Française d'Oxford*, 14 (2001), 37–69.

2. Most of Haller's two hundred articles appeared in the supplements of the Paris *Encyclopédie* and the majority was also published in the Yverdon edition. Some twenty articles – as the entry 'Faculté Vitale' to which I will refer – appeared only in the Swiss edition. A list of the articles and further explanations are given by E. Hintzsche, 'Albrecht von Hallers Tätigkeit als Enzyklopädist', *Clio Medica*, 1 (1966), 235–54. On the Yverdon *Encyclopédie*, a 'Protestant' and religiously more orthodox edition, see K. Hardesty Doig, 'The Encyclopédie of Yverdon', in F.A. Kafker (ed.), *Notable Enyclopedias of the Late Eighteenth Century: Eleven Successors of the Encyclopédie* (Oxford: Voltaire Foundation, 1994), 85–116.

3. Haller to Caldani, 14 March 1769, *Corr. Caldani*, 186.
4. M.T. Monti, *Congettura ed Esperienza nella Fisiologia di Haller: La Riforma dell'Anatomia Animata e il Sistema della Generazione* (Florence: Olschki, 1990); F. Duchesneau, *La Physiologie des Lumières: Empirisme, Modèles, Théories* (The Hague: Nijhoff, 1982).
5. See R. Toellner, *Albrecht von Haller. Ueber die Einheit im Denken des Letzten Universalgelehrten* (Wiesbaden: Steiner, 1971), 146–61; R. Toellner, 'Die Bedeutung des Physico-theologischen Gottesbeweises für die Nachcartesianische Physiologie im 18. Jahrhundert', *Berichte zur Wissenschaftsgeschichte,* 5 (1982), 75–82; M.T. Monti, 'Théologie Physique et Mécanisme dans la Physiologie de Haller', in Ä. Bäumer and M. Büttner (eds), *Science and Religion* (Bochum: Brockmeyer, 1989), 68–79.
6. See M.T. Monti, 'Haller et les Monstres: Objets de Recherche Positive ou Preuves Éclatantes de la Liberté Infinie de Dieu', in E. Fierens *et al.* (eds), *Actes du XXXIIe Congrès International d'Histoire de la Médecine, Anvers, 3–7 Septembre 1990* (Brussels: Societas Belgica Historiae Medicinae, 1991), 181–7; U. Boschung, 'Une Manifestation de la Sagesse Divine: Le Foetus Bicéphale Préparé par Albert Haller en 1735', in C. Mörgeli (ed.), *Les Objets Médicaux et leurs Textes* (Lyon: Fondation Mérieux, 1996), 26–38.
7. 'Ins innre der Natur dringt kein erschaffner Geist, / Zu glücklich, wann sie noch die äußre Schale weist!' From the poem 'Die Falschheit Menschlicher Tugenden', in L. Hirzel (ed.), *Albrecht von Hallers Gedichte* (Frauenfeld: Huber, 1882), 74; for Haller's concept of 'inner' and 'outer' nature see Toellner, *Haller, op. cit.* (note 5), 56–67.
8. Andrew Cunningham has stressed that we should distinguish the teleological from the natural theological argumentation. The first was a methodological tool that allowed scientific deductions, the latter 'a piece of polemic in favour of God.' It is important to realise these differences, most 18th century natural philosophers, however, did not separate the arguments properly. See A. Cunningham, 'The Pen and the Sword: Recovering the Disciplinary Identity of Physiology and Anatomy before 1800', *Studies in History and Philosophy of the Biological and Biomedical Sciences,* 33 (2002), 631–65; 34 (2003), 51–76: 72–4.
9. 'Sichtbaren Wahrheiten die Religion entgegen setzen zu wollen, ist das Gefährlichste, was man wieder sie vornehmen kann.' *GGA,* 1760, 1356.
10. *Relationes de Libris Novis,* 3 (1752), 175 (from Haller's review of Whytt's *Essay*).
11. See S.A. Roe, 'Anatomia Animata: The Newtonian Physiology of Albrecht von Haller', in E. Mendelsohn (ed.), *Transformation and Tradition in the Sciences: Essays in Honor of I. Bernard Cohen* (Cambridge: Cambridge University Press, 1984), 273–300: 288.

12. For an early statement along these lines see W.L. Taube, *Dissertationem Inauguralem de Vera Nervi Intercostalis Origine... Praeside D. Alberto Haller... Publice Defendet...* (M.D. thesis, Göttingen: Vandenhoeck, 1743), 10.
13. See for instance the review of his Boerhaave-edition in the *Bibliothèque Raisonnée*, 33 (1744), 44–6.
14. 'His dispendiis omnibus conputatis, adparet vim, quae a musculis in actione inpenditur, valde magnam esse, & ab omni mechanica ratione remotam ...' *Primae Lineae Physiologiae in Usum Praelectionum Academicarum* (Göttingen: Vandenhoeck, 1747), § 415.
15. 'Erunt, qui objiciant meram me scripsisse anatomen. Sed physiologia est animata anatome.' Haller, *op. cit.* (note 14), prefatio.
16. 'Notre Corps est infiniment plus animé, que les Tableaux de Vulcain.' *Bibliothèque Raisonnée*, 33 (1744), 44.
17. L.J. Rather, 'Some Relations Between Eighteenth-Century Fiber Theory and Nineteenth-Century Cell Theory', *Clio Medica*, 4 (1969), 191–202: 195.
18. *Praelectiones*, ii (1740), § 301 and iii (1741), § 440.
19. A. Haller, 'Excepta de Usu Partium harumque Fabrica Subtili, ex Ore Clar. viri D. Bernhard Siegfried Albini', Biblioteca Nazionale Braidense, Milano, AD.XI.6.
20. A. von Haller, *Ad Viri Illustris Antonii de Haen Difficultates Apologia* (n.p., 1761), 14. The illegibility is due to the great cold in which Haller had to take the notes.
21. 'Erit a causa adhuc quaedam ignota quae in ipso corde haeret.' Haller, *op. cit.* (note 19), 43.
22. *Praelectiones*, iii (1741), § 409.
23. A. von Haller, 'Physiologie', *Encyclopédie, Suppl.*, iv (1777), 356.
24. See Albinus's letters to Haller, 17 January and 20 February 1728, in *Epistolae*, i, 2–3, 10–11.
25. 'Experimenta varia instituenda sunt in solidis et fluidis hominis mortui, animalium mortuum et inprimis vivorum.' Haller, *op. cit.* (note 19), 6.
26. *De Partibus*, 158.
27. Haller actually made a list of problems which had to be resolved either anatomically or experimentally. See *Elementa*, i, preface, IX.
28. See E. Lesky, 'Albrecht von Haller, Gerard van Swieten und Boerhaavens Erbe', *Gesnerus* 15 (1958), 120–40.
29. *De Partibus*, 156–7; similar in the *Elementa*, iv (1762), 462.
30. 'Ergo cor movetur a causa aliqua, quae neque a cerebro neque ab arteriis derivatur, ignota et in ipsa fabrica cordis latente.' *Praelectiones*, ii (1740), § 187. Cf. the already cited passage in the Albinian lecture notes: 'Erit a causa adhuc quaedam ignota quae in ipso corde haeret.'
31. *Praelectiones*, ii (1740), § 285. The same statements are also to be found in

vol. iii (1741), § 401, 409.

32. *Irritable parts*, 64, refering to *Praelectiones*, iv (1743), §586. In this English translation of *De Partibus*, 156, the adjective 'muscular' has been omitted although it appears in the original Latin text.
33. *Praelectiones*, iv (1743), § 600.
34. '[Cor] vi irritabili, & stimulo, quo fibrae in contractionem aguntur, constringitur ... Ita vero penitus insidet cordis fibris impatientia stimuli, ut etiam in emortuo pene viscere diversis locis ...; deinde evulsum cor, & frigens, conpunctum, inflatum, lacessitum, se constringat, & fibrae dissecti cordis orbiculatim se corrugent, nullo nunc nervo, nulla arteria cordi ferente suppetias.' Haller, *op. cit* (note 14), § 113–14.
35. 'Quae vero in musculis animalium nuperrime enectis superest palpitatio, aut a venenis aut ab asperis instrumentis irritabilitas, eam ad nervos, ad motum adhuc habiles, refero. ... Ligato, resecto nervo musculi cujuscunque, languet resolutus musculus, & nulla vi ad motum, vitali similem, revocari potest.' *Ibid.*, § 408–9.
36. *Ibid.*, § 248.
37. *GGA*, 1748, 567.
38. *Praelectiones*, v/1 (1744), praefatio.
39. Haller stated that this passage was written 'post mea experimenta', ie. after he had started to perform experiments. *De Partibus*, 156. The preface of the work is dated 24 April 1751.
40. *Primae Lineae Physiologiae in Usum Praelectionum Academicarum: Auctae et Emendatae* (Göttingen: Vandenhoeck, 1751), § 408.
41. 'Certum est nullum in corpore humano musculum esse, qui nervum non habeat.' *Praelectiones*, iii (1741), § 395.
42. 'Haec vis ab omni alia hactenus cognita proprietate corporum diversa & nova est. Neque enim a pondere, neque ab adtractione, neque ab elatere pendet, cum in molli fibra sedeat, a durescente evanescat.' Haller, *op. cit.* (note 37), § 408.
43. *Praelectiones*, ii (1740), §284, 301; Haller, *op. cit.* (note 14), § 381.
44. Only in very few cases did he not clearly distinguish between irritability and sensibility, eg. Haller, *op. cit.* (note 14), § 248.
45. See Haller's laboratory notebooks. The first experiment he performed on a cat on 29 April 1731 (Haller papers, Ms. 24, fol. 2v and 3r).
46. J.G. Zinn, *Dissertatio Inauguralis Medica Sistens Experimenta quaedam circa Corpus Callosum, Cerebellum, Duram Meningem, in Vivis Animalibus Instituta ... Praeside ... D. Alberto de Haller* (M.D. thesis, Göttingen: Vandenhoeck, 1749).
47. Haller, *op. cit.* (note 14): § 378; *Praelectiones*, ii (1740), § 301.
48. Half a year before his lectures on irritability and sensibility, on 10 November

1751, Haller presented already a first result of his studies to the Royal Society of Göttingen. He declared that the motion of the heart depended solely on internal irritation by the venous blood. 'Experimenta de Cordis Motu a Stimulo Nato: Lecta d. 10 Novembr.', *Commentarii Societatis Regiae Scientiarum Gottingensis*, vol. 1 (Göttingen: Vandenhoeck, 1752), 263–6. The experiments for this paper are also discussed in Chapter 2.

49. *Elementa*, iv (1762), 462; see also Duchesneau, *Physiologie* (note 4), 151–2.
50. *Irritable Parts*, 4–5, the translation of *De Partibus*, 116.
51. Haller, *op. cit.* (note 14), § 422.
52. Haller's first doubts about the sensibility of the tendon arose in May 1748 when a student did not feel any pain upon the irritation of his uncovered tendons. See *Mémoires*, i, 129–30.
53. 'Inveni experiendo, contra quam omnes fere credunt, sensu carere viscera, duram m[atrem], tendines, ligamenta, periosteum, unice sensiles esse nervos, et partes, in quibus ii abundant.' Letter to Bernard, 26 April 1751, 'A. v. Hallers Korrespondenz mit Johann Stephan Bernard', ed. E. Hintzsche, *Clio Medica*, 1 (1966), 324–40: 328.
54. *De Partibus*, 134.
55. Cf. Haller to Bernard, 9 March 1752: 'In eo sum, ut in singulis partibus gradum sensilitatis per experimenta definiam.' Hintzsche, *op. cit.* (note 53), 334.
56. 'Vidimus, quae partes sensiles sunt, *nervos* nempe partesque corporis, quae nervis abundant.' *De Partibus*, 133. The proportionality is plainly declared only in *Elementa*, iv (1762), 293, and later works.
57. This distinction is lacking in Haller, *op. cit.* (note 14), § 113–14, 248, 408 and eg. *Bibliothèque Raisonnée*, 43 (1749), 331 (review of Senac's *Traité de la Structure du Coeur*). It is explicitly made only in *Elementa*, i (1757), 505 and in later writings.
58. See note 40; the same difference is mentioned in *De Partibus*, 152.
59. *Haller Papers*, Mss. 31, fol. 67r.
60. Haller, *op. cit.* (note 14), § 1–23: 1–14; *De Partibus*, 117.
61. *De Partibus*, 114.
62. *Ibid.*, 140; *Elementa*, i (1757), 19–21; iv (1762), 432.
63. 'Hypothèse mécaniste spéciale', Duchesneau, *Physiologie* (note 4), 156; the precision of this term has been stressed by Monti, *op. cit.* (note 4), 110.
64. 'In intima enim fabrica latere hanc utriusque potestatis scaturiginem, & longe ultra scalpelli, aut microscopii vim poni, ego quidem persuadeor.' *De Partibus*, 115.
65. Haller's concept corresponds, to some extent, to the corpuscular model of Boyle who argued that complex structures would have properties that could not be described in mechanical terms (it has to be noted that there is a

disagreement between modern scholars whether Boyle allowed for unmechanical explanations or not). Haller read most of Boyle's works but did not mention him in his various expositions of his concept of irritability. He was mainly interested in Boyle's experimental and not in his philosophical work. In 1746, for instance, Haller noted that he read Boyle's *The Origin of Forms and Qualities* (which contains the clearest exposition of the corpuscular theory) only 'perfunctory because I was looking for experiments' (*Haller Papers*, Ms. 39, fo. 61r). For a brief discussion of the diverging interpretations of Boyle see P. Alexander, 'Robert Boyle', in A. Pyle (ed.), *The Dictionary of Seventeenth-Century British Philosophers* (2 vols., Bristol: Thoemmes, 2000), i, 109–16.

66. 'Ists möglich S. 240 daß der Hr. W[hytt] in Britannien es für eine unphilosophische Handlung ansieht, wann man einer blossen Materie würksame Kräfte zuschreibt.' *GGA*, 1752, 323. Haller alluded, of course, to Newton for whom, however, forces were mathematical powers with an ontological status of their own. They had no material existence and were more closely linked to the space in which they operate. See P. Heimann and J.E. McGuire, 'Newtonian Forces and Lockean Powers: Concepts of Matter in the Eigtheenth-Century Thought', *Historical Studies in the Physical Sciences*, 3 (1971), 233–306.
67. *Relationes de Libris Novis*, 3 (1752), 175.
68. *Bibliothèque Raisonnée*, 46/1 (1751), 87–8 (from the review of Vol. 2 of Buffon's *Histoire Naturelle*).
69. 'Physica causa est, latet in intima fabrica ...' *De Partibus*, 154.
70. See eg. his letter to Réaumur, 9 June 1752: 'Il m'a paru, que l'irritabilité se fait mecaniquement, sans la participation d'une ame ...' (*Haller Papers*).
71. Haller, *op. cit.* (note 14), § 1–5.
72. *De Partibus*, 153–4.
73. Duchesneau, *Physiologie* (note 4), 152.
74. *De Partibus*, 151–2.
75. 'Aliam causam majoris, qua cor gaudet, ad irritationem mobilitas, si quis proferet, auscultabo facilis.' *Elementa*, i (1757), 489.
76. *De Partibus*, 139, 156.
77. *Ibid.*, 126–9. On these experiments, see Chapter 4.
78. See *Praelectiones*, iii (1741), § 412.
79. *De Partibus*, 118–22.
80. *Irritable Parts*, 3, translation of *De Partibus*, 115.
81. *GGA*, 1765, 1036. For a similar statement see *Elementa*, viii (1766), v.
82. To Bonnet, 4 March 1760; *Corr. Bonnet*, 193.
83. *Mémoires*, i, 255–7; the Latin original was presented to the Royal Society of Göttingen in 1755.

84. '...mortuae irritabilitatis genus'; 'fortior... contractionis mortuae gradus', *Elementa*, iv (1762), 444, 514.
85. *Elementa*, iv (1762), 534–5. Haller stressed this notion, also shared by Caldani and Fontana, especially in his later writings.
86. *Primae Lineae Physiologiae in Usum Praelectionum Academicarum: Tertio Auctae et Emendatae* (Göttingen: Vandenhoeck, 1765), § 400; *Elementa*, vii (1765), preface, xi: 'Contractilis naturae in glutine sunt *rudimenta*: a contractili natura scala motus animalis ad vim insitam ascendit, ab ea ad nerveam. Harum virium motricium limites accuratissime definivi ....'
87. 'Peut-être même la force morte sert-elle de base à tout mouvement animal, & qu'elle devient *irritabilité* dans la fibre musculaire, uniquement parce que dans cette fibre l'aptitude à la contraction est plus forte que dans la fibre simplement cellulaire. La force nerveuse n'y ajoute peut-être encore qu'une liqueur stimulante, qui excite la fibre musculaire à se contracter. Cette fibre a dans les muscles volontaires besoin de ce secours pour agir, au lieu que dans les muscles vitaux, cette même force agit sans être aidée par le stimulus du suc nerveux.' 'Irritabilité', *Encyclopédie, Suppl.*, iii (1777), 665. See also the entry 'Fibre' (iii, 35): 'Toute *fibre* animale est donc élastique... sa contraction est le premier commencement de la force motrice des animaux.'
88. 'Il y a dans les fibres animales trois forces contractives, ou, si l'on veut, *trois dégrés, mais très-distincts, de la même force*; la force morte, l'*irritabilité* & le mouvement nerveux.' 'Irritabilité', *Encyclopédie, Suppl.*, iii (1777), 663 (my italics).
89. For the agreement with Whytt see Haller's *Opera Minora Emendata, Aucta et Renovata* (3 vols., Lausanne: Grasset, 1763–68), i, 488; A. von Haller, *Ad Nuperum Scriptum Roberti Whyttij Apologia: Leopoldi M. Antonij Caldani ad Albertum Allerum [sic] Epistola* (Yverdon: n.p., 1764), 21–2; *Elementa*, vii (1765), preface.
90. *GGA*, 1774, 307–8; for Unzer, see Chapter 5. See also *Auctarium ad Alberti Halleri Elementa Physiologiae Corporis Humani, Excerptum ex Nova Editione et Adaptatum Veteri* (Leipzig and Frankfurt am Main: n.p., 1780), 73 where Haller used the term 'integrity'. In a similar manner in the *GGA*, 1778, Zugabe, 649–50 he talked of the nerves' 'Antheil an dem guten Zustand der Fleischfaser'.
91. 'Nerf', *Encyclopédie, Suppl.*, iv (1777), 31. For similar statements see the articles 'Fibre' (iii, 1777, 36) and 'Muscle' (iii, 1777, 977). It has to be noted that Haller talked here about the nerves as structural units and not about their helping to nourish the tissue, an idea he supported in his earlier writings (*Praelectiones*, iii, 661) but not in the later works.
92. Haller, *op. cit.* (note 14), § 555.

93. *Elementa*, iv (1762), 321–2; he repeated this view in *GGA*, 1772, Zugabe, 218.
94. See 'Fluide Nerveux', *Encyclopédie, Suppl.*, iii (1777), 58 and especially 'nerf' (iv, 29).
95. 'Irritabilité', *Ibid.*, , iii (1777), 663 (my italics).
96. 'Sensibilité', *Ibid.*, iv (1777), 776.
97. 'Animal', *Ibid.*, i (1776), 435.
98. '...Robertus Whytt, qui sectam mechanicam, cui me non immerito adnumerat...' Haller, *Opera minora* (note 89), i, 469 (from the revised reply to Whytt).
99. 'Ein gereizter Muskel, sagen die neuen Mechaniker, erregt eine grössere Bewegung, als die reizende Ursache selber gehabt.' *GGA*, 1762, 611.
100. 'Faculté Vitale', *Encyclopedie d'Yverdon*, xviii (1772), 244. Haller might have had this idea – ie. to conceive faculties as mathematical variables instead of looking for their causes – from François Boissier de Sauvages's *Nosologia methodica* (2 vols., Amsterdam: de Tournes, 1768), § 209.
101. *Elementa*, iv (1762), 514.
102. 'Méchanisme', *Encyclopédie, Suppl.*, iii (1777), 876; Haller alludes to the optic laws which inform us about the mechanisms which take place in the eye. See 'Oeconomie Animale', iv (1777), 105.
103. Haller, *op. cit.* (note 14), § 415. See the introductory passage to this chapter.
104. 'Oeconomie Animale', *op. cit.* (note 102), 105.
105. 'L'irritabilité est en verité une force particuliere a la glu animale: Je ne crois pas qu'elle soit soumise aux lois communes du mouvement, aussi peu que les explosions de liqueurs chymiques, ou de terribles mouvemens naissent de la chute equivalente a une dragme, qui tombe d'une demi pouce.' Haller to Bonnet, 15 March 1755, *Corr. Bonnet*, 63.
106. *Elementa*, iv (1762), 557–8.
107. 'Faculté Vitale', *op. cit.* (note 100), 250. See also *GGA*, 1776, 517: 'Nun können wir doch nicht einsehn, daß die Schnellkraft, die Kraft des Brausens, des Gährens, des Feuers, der Electricität des Magnets, nicht ihre Quelle im Körper habe, und daß man den Ursprung dieser Kräfte in einem Geiste finden müsse.'
108. He used the term *vis insita* explicitly to denote irritability first in *Elementa*, iv (1762), 464; in Haller, *op. cit.* (note 14), § 409 he had used the term to designate elasticity.
109. Letter from 7 July 1771; *Corr. Bonnet*, 948.
110. See his *Briefe über einige Einwürfe Nochlebender Freygeister wider die Offenbarung* (3 vols, Bern: Typographische Gesellschaft, 1775–7), iii, 234.
111. 'Si verus finis vitae a me requiratur, dicerem tunc mortem adesse, quando cordis irritabilis natura periit ....' *Elementa*, viii/2 (1766), 123; the same

statement is to be found in the article 'Coeur', *Encyclopédie, Suppl.*, ii (1776), 496. For the use of the term 'vital faculty' or 'vital force' see *Elementa,* i (1757), 72; 'Foetus' and 'Sommeil', *Encyclopédie, Suppl.*, iii (1777), 66 and iv (1777), 809.

112. 'Vim vitalem maluerunt nuperi Cl. viri vocare, quae vox non perinde placet, cum vis nostra vitae aliquantum supervivat. Insitam adeo sive propriam musculi dicere malim.' *Elementa,* iv (1762), 464. In my view, the expression 'non perinde' has been misinterpreted by Monti and Toellner; it does not mean that Haller likes the term *vis vitalis* 'not at all' but rather 'not particularly' or 'not as much' as the term *vis innata.* See M.T. Monti, 'Les Dynamismes du Corps et Les Forces du Vivant dans La Physiologie de Haller', in G. Cimino and F. Duchesneau (eds), *Vitalisms: From Haller to the Cell Theory* (Florence: Olschki, 1997), 41–66: 54–5. R. Toellner, 'Principle and Forces of Life in Haller', in Cimino and Duchesneau, *ibid.*, 31–9: 37.

113. Roe, *op. cit.* (note 11). Some aspects of Haller's Newtonianism and the positions of some authors are discussed by Monti, *op. cit.* (note 4), 83–94, with whose general assessment I agree. Simone de Angelis stresses the importance of 'sGravesande for Haller as transmitter of Newtonian ideas; see his *Von Newton zu Haller: Studien zum Naturbegriff zwischen Empirismus und Deduktiver Methode in der Schweizer Frühaufklärung* (Tübingen: Niemeyer, 2003).

114. This was the view of Samuel-Auguste Tissot. See his 'Discours Préliminaire du Traducteur', in A. von Haller, *Dissertation sur les Parties Irritables et Sensibles des Animaux...*, trad. du Latin par M. Tissot (Lausanne: Bousquet, 1755), iii–xlix: v and Chapter. 4.

115. Duchesneau, *Physiologie* (note 4), 154.

116. See R. French, 'Ether and Physiology', in G.N. Cantor and M.S. Hodge (eds), *Conceptions of Ether: Studies in the History of Ether Theories 1740–1900* (Cambridge: Cambridge University Press, 1981), 110–34.

117. 'La gravitation, l'attraction, l'élasticité, l'effervescence, l'irritabilité, sont autant de sources de mouvement, auxquelles aucune ame n'a de part & qui produisent leurs effets, sans que l'on puisse soupçonner un être pensant, d'être l'auteur de ses mouvemens.' 'Faculté Vitale', *op, cit.* (note 100), 250.

118. See J.P. Wright, 'Substance versus Function Dualism in Eighteenth-Century Medicine', in J.P. Wright and P. Potter (eds), *Psyche and Soma: Physicians and Metaphysicians on the Mind-Body Problem from Antiquity to Enlightenment* (Oxford: Clarendon, 2000), 237–54.

119. See *Elementa,* iv (1762), 557–60; R. Toellner, 'Haller und Leibniz: Zwei Universalgelehrte der Aufklärung', *Studia Leibnitiana, Supplementa,* 1 (1973–5), 249–60; J.P. Wright, 'Boerhaave on Minds, Human Beings, and Mental Diseases', *Studies in Eighteenth-Century Culture,* 20 (1990), 289–302.

PART II
THE EUROPEAN CONTROVERSY

4

The Uses of Experiment

Haller's treatise on irritability and sensibility had three main features which made it likely to create a great stir in the medical community: it was written by a well-known authority, it contained important new findings that contradicted traditional beliefs, and it was based on experimental results that seemed to be unequivocal.[1] Although Haller is nowadays mainly remembered thanks to his later physiological and embryological works, he had already gained a high reputation in the 1740s with his edition of Boerhaave's commentaries, the anatomical plates, and the physiological textbook.[2] These, and not his later writings, were continuously quoted in the *Encyclopédie*, in which the Göttingen professor was the most frequently mentioned living medical author. His renown as one of the foremost botanists of the day and as main editor of the highly acclaimed Göttingen review journal (*Göttingische Gelehrte Anzeigen*), and his manifold other activities all enhanced his status as a scientific authority. Furthermore, his were the most widely read German poems of the 1730s and 1740s which, according to Friedrich Melchior Grimm (1723–1807), in their French translation (1750) had a 'very brilliant success' even in Paris.[3] The philosophical insights displayed in these didactic poems fostered the reception of Haller as a man of superior knowledge and understanding. Any major publication from such a figure had to be taken seriously, even if it contradicted established doctrines, which his treatise on irritability and sensibility certainly did. Antoine Portal (1742–1832) stated: 'What Mr Haller has put forward on sensibility is far removed from accepted opinions: but what he has written on irritabiliy is even more so.'[4] Many others stressed that Haller's assertions challenged the testimony of all former authorities. The Rouen surgeon Claude-Nicolas Le Cat (1700–68) asked shortly after Haller's results became public: 'Where is the surgeon, the physician who doubted before Mr Haller the intense pains and terrible accidents which follow the pricking of the tendons?'[5] Nowhere, of course, was his answer. For Giambattista Bianchi (1681–1761), head of the medical council of Sardinia and retired professor of anatomy in Turin, even more was at stake. If Haller's results were proved to be true, the entire system of theoretical and practical medicine would fall apart.[6] But the great number of experiments performed by Haller, and soon afterwards by many supporters, could not simply be swept aside by the weight of tradition. One author in the *Journal de Médecine* stated:

> The well-merited reputation of this savant, the infinite number of animals he has sacrified in his tests, and the illustrious figures who have renewed with some success the experiments of this great man, are since long time motives powerful enough to balance the votes of all the physicians of Europe.[7]

It is this balance that made it possible for an enduring debate to take place. But there were not, as the quotation might suggest, a few Hallerians opposed to the entire community of physicians, but rather two groups of similar size. The Hallerians were mainly united through Haller who was their undoubted leader and who directed and coordinated their actions and reactions to a considerable extent. This control he achieved largely through his extensive correspondence network. The map (Figure 4.1) shows how this network is linked to the debate on irritability and sensibility. Haller had direct contact only to a third (48) of all the persons (144) who took an active part in the debate either by experimenting or publishing. But he received further information on the controversy from additional fifty-seven persons who were often based at places of intense discussion. Being seen as the discoverer of irritability and sensibility, surpassing his adherents in fame and experience and having important scientific contacts throughout Europe, Haller was considered by many of his followers as their patron, who was allowed to define the strategy to be pursued.[8] Their adversaries were less united through personal relations but simply by their rejection of the new experimental results and their implications. They were, however, not simply old-fashioned opponents of new discoveries. They advanced, just as the Hallerians, a wide variety of arguments to support their position. Haller was not only confronted with differing theories and arguments of authority but also with contradicting experiments and with an evolving debate about the use, value and limits of experiment as well.

Note for Figure 4.1

Haller acted as the 'headquarter' of the Hallerians. From the majority of the important centers of research on irritability and sensibility he received direct information (⊙). In several places, however, research was conducted, without that Haller would have been updated directly by local sources (●). On the other hand, he was furnished with news from locations where nobody performed research on the topic (○). In total, 144 persons from 60 different places were actively engaged in the debate, i.e. either performed experiments or published on the subject. Haller corresponded with 93 persons from 50 different locations on the subject. The incoming information allowed him to react promptly and accurately to criticism and to coordinate the answer with his supporters.

Figure 4.1
The co-ordination of a European controversy

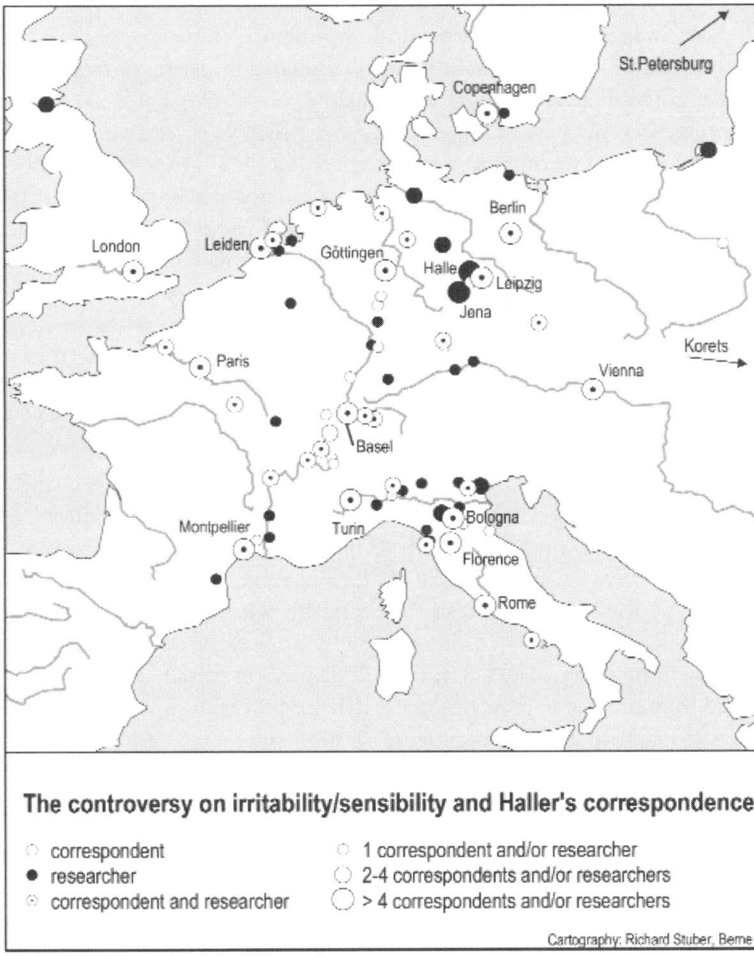

Approximately 120 dissertations and treatises devoted exclusively or to a large extent to the topic were published in the period to Haller's death in 1777. These publications represent what I call the 'core debate'. They not only treat questions of muscular or nervous activity in a general manner, but in some way or another are all concerned with Haller's specific notion of irritability and sensibility. Often they were direct reactions to other publications and were also seen by contemporaries as a specific corpus of controversial writings, and they all accepted the priority of experience. The large collections of experiments and observations published by Haller and the Bolognese physician Fabri are documents of this supremacy of empirical 'facts'.[9] Many authors, of course, had not made any trials and relied heavily in their judgements on theoretical arguments, but experiments and observations were still the main point of reference for their conjectures. On these methodological grounds a speculative book like Louis de Lacaze's *Specimen Novi Medicinae Conspectus* (1749), which was closely linked to the topic as it stated a membrano-nervous system as the source of all movement and sensation, was of no argumentative weight for the 'core debate'.[10] In a review, Haller displayed his astonishment that anybody could publish such an odd work. In his view, it was based on theories which had not the least correspondence with anatomy or animal experiments and was written in a style that was considered obsolete.[11] Bordeu and Barthez, whose concepts were both indebted to Lacaze, were not considered as directly engaged in the dispute because they did not discuss the observations and experiments on irritability and sensibility in any detail, but the surgeon and physician Pierre Fabre (1716–91) who, according to Haller, was a member of the 'sect of Lacaze and Bordeu', was regarded as a direct opponent of the new theory because he explicitly and extensively dealt with the trials in question.[12] Contemporary accounts of the debate listed the 'contestants' assembled in the two collective works, edited by Haller and Fabri, as well as other authors who had performed further tests.[13] This is also what Haller did in his own reviews.[14] Thus, the presentation of a new, controversial, and experimentally established finding created a realm of dispute in which experience was pivotal. But after some years of intense debate this realm slowly disappeared. The vast majority of the experiments on both sides were performed during the seven years following Haller's treatise and more than half of the publications of the 'core debate' appeared in this time. In 1760 the discussion was not yet settled, but many considered it useless to continue with the trials. The focus of attention shifted from methodological problems to the more general question of how the properties of irritability and sensibility – either understood in Haller's or in another sense – could be integrated into a broader concept of animal economy. Increasingly the discussion about the two properties were not necessarily conducted within the reference system of the 'core debate'. Bordeu and Barthez or Reil and Cullen had their own reference systems and did not constantly have to explain whether and in what manner their notions of irritability and sensibility differed from those of Haller, and which exactly were the experiments and observations that supported their view.

This chapter concentrates on methodology and the first period of the debate. It is mainly concerned with experiments on sensibility because these tests were more often debated and repeated than those on irritability. This discussion was nevertheless important for the whole controversy as it was crucial for the different concepts to establish whether certain organs were sensible or not. If no part of the body was devoid of sensibility, Haller's distinction between the two qualities became questionable. The next chapter will focus on physiological concepts and the later years of the controversy. Here, Haller's notion of irritability as an innate bodily power is of more interest, as it was at the core of many of the evolving theories. Sensibility, however, is of equal importance, because many physiologists considered it as connected or even identical with irritability.

The spread of experiment

Preliminaries

In 1744 a controversy was sparked off between Haller and Georg Erhard Hamberger (1697–1755), a leading iatromathematician and professor of medicine in Jena.[15] The dispute, which centred on the mechanics of respiration, has already briefly been mentioned in Chapter 2. Based on anatomical knowledge, physical models of attraction, and mathematical demonstration Hamberger, in an earlier dissertation, had argued that the external intercostal muscles elevated the ribs and thus were responsible for inspiration whereas the internal intercostal muscles lowered the ribs und thus caused exspiration. An observation made on a wounded dog furthermore assured him in his assumption that the pleural cavity (the space between lungs and chest lined with a membrane, the pleura) contains air. In 1744 Haller contested these views on anatomical and experimental grounds and, after Hamberger's defence, repeated the trials with more technical accuracy and performed new types of experiments in order to ascertain these facts. He used human corpses as models to simulate the motion of the chest in respiration. And, of more importance, he examined the movement of the uncovered chest in living animals. In the most 'famous' of his cruel trials he drowned dogs, opened their pleural cavity, and observed that no bubbles appeared. All these experiments left Haller with no doubt about the facts. In 1748 he summarised the debate for a wider public in the *Bibliothèque Germanique* and used this platform to demonstrate, as he saw it, the superiority of experimental physiology over mechanical hypotheses.[16] Further publications on both sides appeared, in which Hamberger repeated and varied his arguments and Haller reiterated his experiments. The controversy attracted considerable attention not least because it was regarded as a rivalry between the old Jena and the new Göttingen university. This reading was heightened by the fact that both professors recruited their students for their own purposes and continued the often heated dispute with blunt statements in the university-linked review journals that they both edited (*Göttingische* and *Jenaische Gelehrte Anzeigen*, respectively). Although in 1752 a pupil of Hamberger performed a number of experiments that contradicted the numerous Göttingen trials, Haller's portrait of the debate as a clash of methods in which Hamberger succumbed was generally adopted. Antoine Portal, for instance, in his already mentioned *Histoire de l'Anatomie et de la Chirurgie* in 1770 stated that 'Mr von Haller has talked only according to the experiments made on animals and... Mr Hamberger has consulted only his imagination and then wanted to correct his aberrations by performing experiments on artificial parts, which have not succeeded.'[17] After Hamberger's death in 1755 his faction increasingly lost ground and received its final blow in 1768 when Ernst Gottfried Baldinger (1738–1804), an ardent supporter of Haller, was called to Jena. He based his teaching of physiology upon the writings of Haller, to whom he triumphantly wrote that 'Hamberger's name is now wiped out'.[18]

Haller's controversy with Hamberger was well-known in the German-speaking countries and in the medical centres of France and Italy, too. But nobody outside of Göttingen and Jena actively joined the debate. No one felt compelled to enlarge upon the topic, as it was not a major issue of animal economy, nor were the experiments repeated in any great amount.[19] The few contradicting experiments performed in Jena were presented at a time when the dispute for many seemed already to be settled, and only induced Haller to repeat his trials but could not generate a broader discussion about the role and use of experiment. The main importance of the controversy probably lies in the fact that it encouraged Haller to perform many more experiments on other topics, notably on irritability and sensibility. These new experiments urged quite a few physicians in Europe to enter the debate because they dealt with main topics of animal economy and contradicted important traditional beliefs. For many, this was the first time that they had performed animal experiments or reflected upon methodological questions of experimentation.

Experimenting on irritability and sensibility

The first news about the experiments performed in Göttingen was disseminated by letters.[20] Haller often informed his correspondents briefly about his current research. Already in mid-1751, even before his pupil Zimmermann had finished his research, Werlhof in Hanover, Morgagni in Padua, and the physician Johann Stephan Bernard (1718–93) in Amsterdam were told that tendons and membranes were insensible. Many others presumably were given similar notices in letters that have not survived. In particular, those among Haller's pupils who remained in contact with each other were quickly notified about the teacher's activities. As the topic was controversial and many questions still uncertain even to Haller, the experimental curiosity instilled in them in Göttingen was aroused again. In 1751, Georg Christian Oeder (1728–91) in Copenhagen, Jakob Christoph Ramspeck (1722–97) in Basel, and Johann Gottfried Zinn (1727–59) with Johann Friedrich Meckel (1724–74) in Berlin all repeated the trials, with partly contradictory results. With the review of Zimmermann's dissertation in the *Göttingische Gelehrte Anzeigen* in 1751, the thesis became publicly known in the German countries where this journal was widely read. Although this caused initial reactions to Haller's work, it seems not to have encouraged anybody to repeat the experiments.

After the submission of his dissertation in August 1751, Zimmermann left for Leiden where he met, among others, Jerome Gaub and Frederick Winter. Both professors immediately discussed the dissertation in their lectures and Gaub pointed out its importance. This attracted the attention of his pupil Wouter van Doeveren (1730–83), who already in October 1751 repeated the experiments together with other students and partly in the presence of Gaub, Winter, and many others.[21] Doeveren would become the initiator of most of the experiments performed in the Netherlands, and even carried out some trials together with French physicians during his stay as a student in Paris in 1753. He confirmed that especially the muscular organs were irritable, but contradicted Zimmermann as he found the dura mater and the tendons to be sensible. The appearance of Haller's own treatise cast new doubts upon Doeveren's results and encouraged him to repeat his trials in October 1753, now in Utrecht together with Haller's former pupil and professor of experimental physics, Johann David Hahn (1729–84). In this case the tendons seemed to be insensible. During his tenure as professor of medicine, surgery and obstetrics in Groningen, Doeveren repeatedly performed experiments on the topic: in 1754, with equivocal results, and again in 1758 and 1764–5 together with several pupils. These later trials ultimately led him to the conclusion that the tendons and membranes were sensible, and that Haller was wrong. He published a circumstantial account of his experimental activities only in 1765.[22] In the meantime two other Leiden students, pupils of Albinus and Gaub, had already in 1757 published the results of their experiments, with roughly the same outcome.[23]

From his short stay in Leiden, Zimmermann moved on to Paris in late August 1751, where he easily gained access to scientific and literary circles thanks to Haller's letters of recommendation and books which he brought along. He distributed his own dissertation among various scholars such as Jean Baptiste Senac (1693–1770), private physician to the King; Antoine Ferrein (1693–1769), professor of anatomy at the *Collège Royal* and the

Jardin des Plantes; Jean-Baptiste-Louis Chomel (1709–65), professor at the medical faculty; and the famous Réaumur. Knowledge of the Göttingen experiments quickly spread in Paris and was presumably enhanced by the reading of a letter in a meeting of the *Académie des Sciences*, in which Haller discussed his research. The topic was considered as important, and already in January 1752 Zimmermann's notions on irritability were used as a basis for further speculations in a dissertation at the medical faculty.[24] Regarding sensibility, scepticism seems to have prevailed. In October 1751, Senac asked Haller to repeat the trials, as surgical experience seemed to prove the sensibility of the tendons. In Paris, the first experiments were performed on 25 January 1752 by the physician Claude-François Grandclas (b. 1725) together with Jean Sue (1699–1762), prevost of the *Collège Royale de Chirurgie*, and in the presence of several other persons. These experiments were repeated during the greater part of 1752, but presumably sporadically and not in great numbers. Although they were never published, their outcome, unfavourable to Haller, was well-known in Paris.[25] Experiments with similar results were performed by Le Cat in Rouen and already somewhat earlier, in late-1751 and early-1752, in Montpellier, by François Bourguignon de Lamure (1717–87) and François Boissier de Sauvages (1706–67), both professors at the medical faculty.

The lengthy review of Zimmermann's thesis in the *Journal des Sçavans* in September 1752 – an honour very rarely bestowed upon a foreign medical dissertation – definitely made irritability and sensibility a topic of public knowledge. But as the journal mentioned the opposing results of Grandclas, the Göttingen trials appeared doubtful. They gained new authority and also a different significance with Haller's treatise, which arrived as an early offprint in Paris in March 1753. Two months later Grandclas, together with the student Doeveren from Leiden, repeated the experiments. He was followed by the physicians Anne-Charles Lorry (1726–83) and Charles-Augustin Vandermonde (1727–62). All these experiments mainly contradicted Haller's denial of the sensibility of the tendons and membranes, but also contested his notions of irritability. In autumn 1754, with Haller's election as one of the eight *associés étrangers* of the *Académie des Sciences* and with the French translation of his treatise – made by a correspondent of Haller, the Lausanne physician Samuel-Auguste Tissot (1728–97) – and its subsequent review in the *Journal des Sçavans*, the topic received fresh attention. This led to a new series of experiments, now in Montpellier, performed in summer 1755 by Antoine Tandon (1717–1806) and Lamure. The sessions in the amphitheatre of the Medical Faculty were joined by many other people including the student Etienne-Jean-Pierre Housset (1733–1810), who furnished Haller with detailed accounts. Housset accused the professors of lack of experimental prudence and in three consecutive meetings of the Royal Society of Sciences of Montpellier, he delivered a critical account explaining why they had failed to yield the same results as Haller. In his own experiments he reached the same conclusions as the Göttingen professor.

Up to this moment none of the experiments performed in France had been published. In 1756, the edition of the first volume of Haller's *Mémoires*, which contained descriptions of no less than 567 experiments, seems to have generated a certain pressure to publicly report such trials. Vandermonde, the editor of the *Journal de Médecine*, published some Italian, as well as his own and Lorry's experiments but, as Haller's writings had raised his doubts about their validity, he also edited a selection of the Göttingen trials. Within one year the journal had devoted a third of its volume and more than 150 pages to the topic. Vandermonde received several other writings, but quite a few of his readers considered the issue as dealt with sufficiently.[26] After another report of experiments from a medical dissertation from Paris, only a few further accounts appeared in the journal.[27] Apart from the *Journal de Médecine*, experiments were published in the above mentioned dissertation which was also presented in a meeting of the *Académie des Sciences*.[28] Toussaint Bordenave (1728–82), a surgeon and professor of physiology at the *Collège Royal de Chirurgie*, recounted his examinations in the *Mercure de France*, and in a dissertation from Montpellier further experiments were presented.[29] The trials from the early 1750s conducted by Grandclas in Paris and Lamure in Montpellier were never published. Only those from 1755, performed by Tandon and Lamure, were edited, but as late as 1787 and in the critical version of their adversary Housset.[30] After 1760, experiments were carried out and published only occasionally: 1760–62 by the surgeon Jean-Jacques-Louis Hoin (1722–72) in Dijon, in 1771 by Portal in a course of experimental physiology in Paris, in the same year by the student Arthaud, and in 1775–6 by the professor Félix Vicq d'Azyr (1748–94).[31] All these later publications confirmed Haller's results but were not decisive contributions to the debate. The French medical community remained divided.

In Italy, the controversy developed in a manner quite different from France. It was less linked to medical faculties and arose later, but reached a much greater intensity.[32] Although Morgagni received a brief account of Haller's research as early as 1751, nothing of this knowledge spread. The Italians seem first to have heard about the Göttingen experiments through the reviews of Zimmermann's dissertation in the *Journal des Sçavans* (September 1752) and through the *Commentarii de Rebus in Scientia Naturali et Medicina Gestis* from Leipzig, where both Zimmermann's and Haller's treatise were discussed in 1753. Very few copies of these writings arrived in Italy. In early 1755, the French translation of Haller's work was available and immediately translated into Italian, both in Rome and in Naples. The person who presumably was the first to pick up the topic was the Roman physician Giambattista Bassani (d. 1768). He arranged for the translation of Haller's treatise and encouraged his friends to repeat the experiments. His most important ally was the Jesuit Urbano Tosetti (1714–68) who taught philosophy and experimental physics at the *Collegio Nazareno* in Rome. Tosetti performed his first experiments in April 1755, and published these and later trials in 1755–6. He also included reports of the animal experiments of some physicians and surgeons, Giuseppe Bianchi from Florence, Marcuzzi from Lucca, Morandi from Modena, Palliani and Piazza from Rome, and Vari from Ferrara. All these experiments basically supported Haller's views but they were contradicted by other trials, especially by those of Giambattista Bianchi, but also by those of the physicians Barbiellini in Rome and Sanseverini in Naples.

Thus, within half a year of Tosetti's first trials, an experimentation fever had gripped the country. The Roman professor of mathematics, Cesareo Pozzi (1718–82) was confronted with the following situation when he arrived in Florence in September 1755:

> ...I saw in all corners limping dogs, on which experiments on the insensibility of the tendons had been made but, as I have been told, without any or at least with doubtful success. The scholars were divided. It seemed incredible that the learned and highly esteemed Mr Haller could have been mistaken.[33]

And Pozzi did what in Italy appeared to be a perfectly normal reaction: he performed experiments himself and published them in a letter addressed to the Bolognese physician Tommaso Laghi (1709–64). Laghi, as expected, repeated and published the trials in his turn but rejected the views of the Hallerian, Pozzi. He edited his results also in the journal of the Academy of Sciences of Bologna (*Istituto di Bologna*) and was promptly countered by Leopoldo Marcantonio Caldani (1725–1813), another member of the academy. The physician Caldani was assisted by Felice Fontana (1730–1805) in his experiments, which were witnessed by many prestigious members of the *Istituto* like Francesco Algarotti (1712–64), Laura Bassi (1711–78) and her husband, the physician Giuseppe Veratti (1707–93). The secretary of the academy, Francesco Maria Zanotti (1692–1777), did not want to edit any further writings on this debate which, in his view, was conducted too impetuously. He delivered an account of the actual state of discussion and concluded his essay with the words:

> Thus the dissension spread and from day to day increased so that there seems to be a kind of general irritation of the whole of Italy [*communis quaedam totius Italiae irritatio*]. For us it is enough to have indicated these many studies; the subject itself we leave in the balance.[34]

Although Zanotti had quite obviously favoured Caldani, with this statement the academy itself was able to remain in a neutral position. As in the case of the *Journal de Médecine*, it did not want to let its journal be dominated by the controversy. Zanotti, presumably, was relieved to hear that the Bolognese physician and surgeon Giacinto Bartolomeo Fabri planned to edit a collection of all the writings on the topic.[35] It was Fabri who was now addressed by other zealous authors. His work, published in 1757–59, presented the writings of the Hallerians in the first volume, those of his opponents in the second, and further texts in a two-part supplement.[36] With fifty-five contributions, the greater part from Italy, it contained all the main Italian writings of these early years. Less than half of these publications included reports of experiments. Still, twenty-seven contributors – or rather groups and often quite numerous assemblies – in twelve towns performed experiments, far more than in any other country. After 1760, as in France, only a few authors in Italy carried on with trials. But, in contrast to all the transalpine countries, two major figures continued their experimental inquiries: Caldani and Fontana. They went beyond the simple question of which parts were sensible or irritable and proceeded to disclose hitherto unkown aspects of irritability. Thus, they opened new areas of experimental physiology where they were joined by others, notably Lazzaro Spallanzani.

Outside the Netherlands, France, and Italy, and apart from Haller and his pupils, only a few people felt compelled to treat the topic experimentally, and fewer still proceeded systematically. The surgeon Georg Heuermann (1723–68) in Copenhagen tested the sensibility of the dura mater in a few experiments, as did Achilles Mieg (1731–99), a physician in Basel, who performed his three trials upon Haller's request. August Friedrich Pallas (1731–1812), professor at the *Collegium Medicum Chirurgicum* in Berlin, noticed the movement of isolated muscles, and similarly Jean-Emanuel Gilibert (1741–1814), professor of medicine in Grodno, observed these movements during his extended studies of tortoises.[37] With more perseverance, the Basel student Johann Rudolf Müller (1740–88) conducted studies on the irritability of the iris and the motion of the pupil. In like manner, Jakob Eberhard Andreae, a former student of Gaubius and now in Tübingen, proceeded to test all parts of the body for irritability. But none of these studies ever occupied any significant position in the debate. Of more importance were the experiments of the student Ignatius Radniczky, performed in Prague in 1756. They confirmed the sensibility of the dura mater, the tendons and the membranes, and served several authors in Prague and Vienna – notably Anton de Haen (1704–76), the famous Viennese clinician – as proofs of Haller's error. Joseph Thaddäus Klinkosch (1734–78), who, as a student, had witnessed these trials, later as a professor of anatomy in Prague repeated them, with similar outcome.

By far the most important person to perform any experiments outside Italy and France was Robert Whytt (1714–66), professor of the Institute of Medicine in Edinburgh. Already in the 1740s he had initiated investigations on living animals that were similar to those of the pupils of Boerhaave in the Netherlands. Based on these and further published observations he formulated his theory of the 'sentient principle' (1751), which brought him into conflict with Haller. During the subsequent controversy Whytt conducted further experiments, mostly regarding the effects of opium, but never in such quantities as his opponent. Although he was joined in his research by some of his students, he did not establish an experimental tradition in Edinburgh. The only ones who reportedly performed trials on living animals were the student Thomas Smith in 1765–7 and Alexander Monro (1733–1817) in the late 1760s or early 1770s. The trials of Smith served William Cullen (1710–90) as a basis for some of his own theses but did not encourage him to carry out any similar studies.[38] In England there seems to have been only one single episode of experimentation: in 1755 the London physician Richard Brocklesby (1722–97) published some trials in the *Philosophical Transactions*, which endorsed Haller's results.[39]

The experimental tradition

The picture given of the spread of experiment thus far is, of course, incomplete. Some experimenters have undoubtedly escaped my attention, others presumably have left no trace, and many more witnessed the trials. However, I think that we now have a fairly complete list (see Appendix) of those who were of at least some importance within the debate. Seventy persons, or rather groups, carried out animal experiments, 27 in Italy, 16 in France, 10 in the German-speaking countries, 6 in the Netherlands, 5 in Great Britain, and 6 elsewhere. They did not constitute one large 'experimental community' in the sense that Haller and his pupils in Göttingen had done. Although for the most part surgeons and physicians, they came from varied educational, social, and professional backgrounds, and in their research were driven by various motives. Their only common ground was the performance of experiments. Some of them carried out only a few, others hundreds. The overall amount of animal experiments and their repetition all over Europe was a new phenomenon. No previous debate – not even Harvey's discovery – had provoked such widespread experimental investigation.[40] When he published the fourth volume of his collection of experiments in 1760, Haller declared with undisguised pride: 'I am presenting the most extensive collection of experiments which probably has ever been published in order to prove a physical truth.'[41] Besides his own trials, well over 500, he edited an equal number of experiments performed by his pupils and supporters. Haller had sacrificed almost 200 animals to his research, a similar amount had been used in the trials of Fontana, and many more in the investigations of the others. His opponents could 'boast' far smaller numbers but had still conducted a considerable amount of tests.

Already since the mid-seventeenth century, a great number of animal experiments had been performed in all countries of Europe. The best way to get an idea of how many persons were part of this tradition is to peruse Max Neuburger's classic description of 1897, *The Historical Development of Experimental Brain and Spinal Cord Physiology before Flourens.*[42] In this astonishing book, based on extensive reading of all the major and minor sources, Neuburger argued against the prevalent idea that progress in experimental physiology was almost entirely an achievement of the nineteenth century. But, despite the vast amount of evidence he found for early experimental activity, Neuburger was much too sensible a historian to simply present the opposite picture of an uninterrupted and widespread experimental tradition. Although he created a kind of historical alignment of all the single 'spots' of experimentation, he was well aware that the majority of writings did not promote this method. In the mid-eighteenth century, it was still the exception to approach a physiological problem by experimental investigation. In 1753, when the Academy of Sciences of Berlin formulated its prize question on the cause of muscular movement, it received seventeen contributions, but only one was partly based on experiments.[43] However, there was an awareness of the importance of experimental evidence: it was precisely this treatise, written by Claude-Nicolas Le Cat, which received the award.

Haller himself was part of the experimental tradition. As Richard Toellner has stressed, his intellectual home was the early Enlightenment of the Netherlands with Boerhaave as its master.[44] Iatromechanism without rigidity and the encouragement of the empirical approach were the two main elements Haller adopted as his own fundaments of physiology. The detailed study of Newton's and Boyle's work after his medical training reinforced this position and shaped the outline of Haller's world of science before he had performed any considerable research himself. The experimental tradition in Leiden, Boerhaave's call for empirical demonstration, and Albinus's actual performance of experiments added to the prestige and stressed the necessity and practicability of this approach. Haller, together with the Dutch pupils of Boerhaave, continued this tradition. Robert Whytt's experimental activities probably have their roots in Leiden too. His interests in the action of the heart may have been roused during his stay in Leiden in 1736, when Boerhaave gave his lectures on the heart and Frederick Winter performed his experiments on muscular motion.

Haller started with single trials as early as the 1730s, but developed his experimental research mainly in the 1740s, increasing its intensity dramatically during his investigations into irritability and sensibility. The transformation from rather incidental or isolated trials to a more determined and repetitive performance took place somewhat before that in the realm of lower animals. Bonnet's work on the aphids (1741) and Trembley's on the hydra (1744) may be described, following Marino Buscaglia, as 'systematic extensions of previous protocols by Redi, Vallisnieri, and Réaumur.'[45] A similar 'systematic extension' took place with Haller in the field of human or mammal physiology. Certainly, several earlier researchers such as Stephen Hales (1677–1761) had pursued their experiments in a similarly determined manner to Haller. But their work had not provoked a widespread debate and had thus not sharpened the public awareness of the experimental procedure, unlike the work on irritability and sensibility. Haller's absorption in his own research seems to have somewhat blurred his perception of the actual status of animal experimentation in the scientific community. On Tissot's preface, which enlarged upon the importance of this approach, he commented that in his view it had become superfluous to stress the advantages of experimental investigation.[46] But the experimental approach in physiology was not yet as firmly established as in the physical sciences.[47] Experiments had mainly been performed to explain mechanical processes, and the living body could not be reduced entirely to the mechanico–mathematical model. Even authors like Harvey, Borelli, Baglivi and Boerhaave, who were known as pioneers of a mechanical description of circulation and motion, believed in powers specific to life. Despite the prosperity of mechanical philosophy, a certain reservation as to the use of the experimental approach in the biological sciences seems to have prevailed. One could only overcome it if experimentation on living animals was considered a method in its own right. Haller adopted such a view. Animal experimentation, for him, was the method specific to physiology. Anatomical investigations were important, but not sufficient; as the living body moved, it had to be studied in motion. The experiments did not only, however, reveal simple mechanical laws of motion. Many processes in the living body could not be described by these laws; the flow of the humours was often less diminished than calculations predicted, and violent motions were produced where no adequate cause could be discovered.[48] The experiments gave evidence of processes peculiar to the living body. Animal experimentation was not a method that neglected the specificity of life, quite the contrary, it was the only method that took it into consideration. This was the view Haller promoted in his physiological works in general and in his writings on irritability and sensibility in particular. The following sections furnish some explanations as to why the majority of his contemporaries could not adopt it.

Methods, techniques, and rhetorics

General reflections about the methods of the 'biological' sciences are dispersed in many writings of the eighteenth century but there was no comprehensive treatment of the topic before Jean Senebier's *Art d'Observer* from 1775.[49] Senebier stated that 'it is astonishing that in a century which pretends to be that of philosophy one has not thought of establishing the rules of the *art of observation.*'[50] He had, as he said, to gain his insights less from methodological treatises than from scientific publications themselves, notably from the works of Réaumur, Haller, Duhamel, Trembley, and Bonnet. Bonnet had himself contemplated writing a similar book some twenty years earlier. In a letter to Haller he asserted that Haller's writings could serve as a model for the right methods and that Haller could actually compose a much better essay.[51] Haller, however, never intended to write such a book; his only methodological treatise dealt with the use of hypothesis and served as a preface to the German translation of Buffon's *Histoire Naturelle.*[52] Bonnet was quite right, Haller's methodology is plainly visible in his writings, where it is occasionally also discussed explicitly. It is tied to the actual performance of experiments and less concerned with general epistemological questions. The main problem for Haller and his colleagues was how to proceed in their research, which method to choose, whether, when and how to perform an experiment, and how to interpret the results. *Which* experiment to carry out had been defined by Haller's treatise. The only question asked by nearly everyone was, whether a part was sensible, irritable, or not. Caldani, Fontana, and Whytt were the only people to invent and perform other trials. As many persons made the same experiments but reached different results, the question of how these had been carried out became essential and encouraged Haller to publish his laboratory notes. In his four-volume edition, the *Mémoires*, he established some rules of animal experimentation and made further, more general methodological remarks. The works of the Hallerians, united in these volumes, were presented as guided by the same rules, and the performance of experiments appeared as their common ground, although not all the authors had contributed any trials. Their opponents were united through their rejection of Haller's doctrines but not through a common procedure.

They formulated criticisms of Haller's methods from disparate points of view and in return were blamed by the Hallerians for various methodological violations. The main aspects of the dispute are discussed in the next four sections on experiment and observation, validity of animal experiment, techniques and procedures, and anatomical demonstration.

Experiment and observation

There were two possible ways to study sensibility empirically: by experiment on animals, and by observation on patients. Irritability could rarely be studied in man, but wounds to the head, extremities, and the bowel allowed the physicians and surgeons to test the sensibility of certain parts. These trials were usually called 'observations' as the tendon or membrane did not need to be prepared and could be touched or 'observed' directly. Animal experiment, on the other hand, required deliberate denudation and the infliction of an injury. The difference between these two methods consists of two elements, the test object – which will be discussed in the next section – and the procedure itself.

To stress the distinction between experiment and observation was close to a commonplace, but where to put the emphasis was a matter of dispute. Haller stated: 'In fact, not everything we see is experimentation [*expérience*]. It deserves this name only when we have wanted to see and have helped nature to show herself.'[53] This was the version that favoured experiment as the best way to investigate nature. Haller played down the invasive nature of experimentation in the dedication of his *Mémoires* to Réaumur: 'The experiments, which I have the honour to present to you, have almost only required the effort of looking. Nature has offered herself to the physician, she has not made him to buy her favours.'[54] Thus seen, experiment was very close to observation, but in contrast to the latter it was guided by a well-defined arrangement and procedure. Observations on patients might be useful, but they were 'mostly necessarily vague and indecisive.'[55]

D'Alembert's version of 'experiment and observation' was quite different:

> Experimental physics is based on two points which should not be confused, genuine experimentation and observation. The latter is less artificial [*recherchée*] and less subtle, restricts itself to the facts before its eyes, to look properly and to specify the phenomena of all kind which the spectacle of nature presents. The former, on the contrary, seeks to penetrate it [nature] more profoundly, to steal what it conceals and to create, in a certain way, new phenomena by various combinations of bodies, in order to study them; in a word, it does not confine itself to listen to nature, but to interrogate and to press it.[56]

Whereas for Haller the controlled procedure was the main characteristic distinguishing experimentation from observation, for d'Alembert it was its invasive, creative power. The experimenter pressed nature, produced new phenomena, and disclosed hidden facts. This picture of the powerful scientist did not correspond to the physico–theology of Haller. The scientist could not press God's nature nor rise to be the creator of new phenomena. It is true, on one occasion Haller adopted the Baconian description of torturing nature and said that Harvey had forced nature to answer, but generally he described experimentation rather as helping nature to reveal herself, taking her by surprise, or asking her with a simple voice.[57]

D'Alembert's distinction served many as a model by which to formulate their own versions. Of particular interest are the French vitalists, as they expressed a more fundamental critique of experiments than anybody else.[58] This was done most explicitly by Jean-Jacques Ménuret de Chambaud (1733/9–1815) in the *Encyclopédie.* In the article entitled *Observateur* he stressed the denaturalising effect of experimentation:

> The name of observator has been given to the physician who confines himself to examine the phenomena in the manner nature presents them to him; he differs from the experimental physician who combines himself and who sees only the results of his own combinations; the latter never sees nature as it really is, he pretends to render her more sensible by his work, to take off the mask which conceals it from our eyes, [but] he often disfigures it and renders it unrecognizable.[59]

The problem of experiment was that it neglected the unity of the body. As Ménuret remarked in the article *Observation*, the physiologists have 'only seeked to draw up a list of the functions of man and then have explained them seperately, just as if they would not each act and influence on the other reciprocally.'[60] Both these aspects – of denaturalization and of violation of the unity of the body – were put forward by another vitalist, the Montpellerian physician Henri Fouquet (1727–1806), in his attack on Haller in the article *Sensibilité.* He particularly stressed the importance of the *consensus* of the whole body. The irritation of a certain part somehow accumulated sensibility in this department and the second part, when irritated afterwards, appeared as less sensible. The Hallerians in their attempt to isolate organs and functions neglected this and other mechanisms by which the body in its entirety was ruled. Ultimately, even the experiments performed with utmost care were insufficient to advance knowledge in such delicate matters.[61] Given these different assessments of experimentation it comes as no surprise to hear that Haller used Newton and Harvey to illustrate the fruitfulness of experimentation but called Descartes a storyteller, whereas the vitalists considered Newton as a perspicacious observer and Descartes as a misleading experimenter.[62]

Ménuret's and Fouquet's view may have been shared by several physicians who watched the debate on irritability and sensibility from a distance, but a fundamental critique of experiment was very rarely put forward by those actively engaged in the controversy.[63] Heinrich Friedrich Delius (1720–91) and Anton de Haen, professors of medicine in Erlangen and Vienna respectively, argued that the violence, unnatural circumstance, and coercion exerted in the trials would lead to results from which the natural state could not be securely concluded.[64] Another physician, Charles Lorry from Paris, reasoned similarly, but still considered it worthwhile publishing his own experiments.[65] Both the student Coutavoz from Montpellier and Johann Friedrich Gmelin (1748–1804), professor of medicine in Tübingen, stressed that the results gained in experiments had to be compatible with less misleading observations.[66] These were the most fundamental criticisms and even they acknowledged a certain heuristic value in experimenting. The vast majority of authors did not consider experimentation as a danger to the unity of the body and the harmony of nature. Whilst they objected that the animals were in a state of terror and reacted differently after the infliction of several wounds, this was not an issue that annihilated the validity of animal experiment but rather a technical problem, which had to be tackled by specific modes of procedure. Nevertheless, there was still one fundamental question to be solved: was *animal* experimentation a valid means of investigating *human* physiology?

Man and animal: the validity of animal experimentation

Haller's Latin treatise bears the title *De Partibus Corporis Humani Sensilibus et Irritabilibus.* It thus claims to give evidence for properties of the *human* body. Accordingly, in this essay, Haller continuously talked about the human body, although he referred exclusively to animal experiments. He did not even attempt to justify his method. The title of the French edition of his treatise, which used the word 'animaux' instead of 'hommes', had been chosen by the translator Tissot, and was subsequently adopted by Haller who, in his collection of experiments, employed the term 'corps animal'.[67] But Haller never doubted the validity of his experiments, and in all further Latin and German works kept to the adjective 'human'. He only used observations on man as additional support; the experiments carried the main argumentative weight. Not everybody shared his reliance on comparative anatomy and physiology, but leading anatomists such as Albinus in Leiden, James Douglas (1675–1742) in London, Alexander Monro (1697–1767) in Edinburgh, Antoine Petit (1718–94) in Paris, and Morgagni in Padua trusted in analogy at least in parts of their work.[68] Haller's greater confidence was further nourished by his Leiden background and especially by his own research. He had internalised the need to perform animal experiments, so to speak, and even expected that a comparison of the brains of animals with their behaviour might be of use for the investigation of the human mental faculties.[69] From this point of view, Haller considered the argument of the non-validity of animal experimentation as an excuse put forward after the experiments themselves had favoured his position.[70] Indeed, even for most of the anti-Hallerians, the validity of animal studies was not in dispute, but half a dozen used the argument to refuse Haller's claims.

Ernst Anton Nicolai (1722–1802), professor of medicine in Jena, argued that observations made on animals could not be transferred to man because these were two different species.[71] But such a simple statement unaccompanied by any explanation as to why the analogy could not be drawn was out of touch with the detailed debate at hand, and Nicolai's position was not taken into account. With more precision, Giambattista Bianchi based his objection on the fact that man had a much bigger brain, many more ramifications of nerves, and therefore a much more acute feeling than animals.[72] This was an argument which could at least be discussed and which, as expected, was refuted. Haller did not deny the larger anatomical differences, but he stressed the similarities on the functional and the more subtle anatomical level. The skin of quadrupeds, he argued, was furnished with as many nerves as our own, and the horse which reacts to the touch of flies demonstrates its acute sensibility. In the animal experiment, brutes felt the lesion of nerves as well as humans. Furthermore, inebriated animals staggered just as drunken men. Since they reacted similarly, it was likely that their nervous system was of a similar structure.[73] As Haller said himself, he:

> more and more realized that the elements of the body, and whatever is of a more subtle kind, is wholly of the same structure [*fabrica*] in various quadrupeds. But the larger and rougher parts vary according to the duties [*munera*] that the Creator has assigned to each species of animals.[74]

Although Haller was a physico–theologian and talked here of God's plans, it was not on such a basis that he postulated the likeness of the structures. Microscopic investigations showed the similarity of nervous and muscular fibres in man and animal, despite their variety in thickness.[75] Although their bodies showed visible differences, they had the same elementary configuration.

Anton de Haen, the famous clinical professor from Vienna, doubted this principle of correspondence. Man and animal differed in their whole nature (*tota natura*). Haller himself had conceded that experiments on birds had to be excluded because their reaction contrasted too widely with those of other brutes. If animals themselves were so essentially (*essentialiter*) unlike, the deduction to man was even less permissible.[76] De Haen was countered, of course, and most explicitly by Tissot, who repeated Haller's position.[77] The debate, he said, was about parts of the body whose texture, function, and use were clearly demonstrated to be the same in many animals. In such a case, analogy retained all its potency. If comparative anatomy was not admissible in this instance, it was of no use at all. Haller furnished a further argument, widely used by the Hallerians. Only through animal experiment had the circulation of the blood been discovered, and only by the same methods could insights into the mechanics of respiration, the motion of the muscles, and the digestion of food be gained.[78]

But de Haen argued from another point of view, too. As both sides produced a similar amount of opposing experiments, their value was annihilated. This was not, in his opinion, the case with the observations on man, where many more trials confirmed the sensibility of tendons and the dura mater.[79] In this line of argument he was especially supported by Le Cat, who collected a great many surgical observations. These alone, in his view, could decide the question.[80] In a similar manner to Haller and his experiments, he tried to overwhelm his opponents with a vast amount of observations. But de Haen's and Le Cat's strategy did not work out. The Hallerians presented a similar amount of contradictory observations. Neither in the case of the experiments nor of the observations was an absolute superiority achieved.

Technique and experience: the procedure of experimentation

Several authors argued that the disparities were due to the variability of nature; every man and also every animal reacted differently, and even the same individual varied in its responses.[81] But, Fontana countered, this could certainly not explain why both sides produced contrary results with notable regularity.[82] Quite soon it became obvious that *how* exactly these experiments were performed varied considerably and thus yielded different outcomes. Haller, in his treatise, had not yet emphasised the importance of the technical aspects of experimentation, but the issue was put forward in 1755, notably in the publications of Pozzi, Tosetti, and in Tissot's preface.[83] From then on, the details of the experimental procedure were essential elements and often at the core of the controversy. An elaborate description of all the many details at stake is probably less suited to conveying an idea of the discussion than an account of an experimental session in which the manner of operation was disputed. This is one of the reports of Jean-Pierre Housset, who informed Haller about the experiments performed in Montpellier in

1755–6. Housset, although a student at Montpellier, considered Haller as his real master, and thought that his own professor and *praeses* of his dissertation, Lamure, was prejudiced in his longing to contradict the great physiologist. The trials were carried out by the surgeon Tandon at the request of Lamure in the amphitheatre of the university and were witnessed by many medical doctors, anatomists, and surgeons:

> Today 28 April [1756] Mr Tandon has performed experiments on a dog of medium size which seemed to be old. He cut the general integuments of the head, removed the margins of the wound with his fingers in order to test the sensibility of the pericranium. He wanted to use vitriolic oil, a caustic which Mr Sauvages had considered as causing inconveniences, undoubtedly because it penetrates, extends too far and may cause ravages in the adjacent parts. Before performing the test on the membrane, I explained to Mr T[andon] that the manner to operate on a bloody animal – a victim of a recent useful cruelty – seemed to me not to be regular. I proposed to him that [manner] of Mr Pozzi which I consider excellent, ie. to uncover the part and to take off the neighbouring parts in the morning, for instance at eleven, and then to cover it with a bandage and to seek truth in the evening by the suitable mechanical or chemical means. He promises to employ this method and seems, some moments after, to refrain, advancing the argument that he had not changed the performance for five years. That is why this anatomist, after having let the dog rest for three or four minutes, slits the pericranium repeatedly with the scalpel, in the space in between the temporalis muscles. Thereupon the animal at one time shows signs of pain, the next time it makes us to see none, especially at the beginning of each test. It is true that the sheath which covers the temporalis muscle aroused the suspicion whether there was any sensibility, which one could have thought to be occasioned by the muscles involved, but reiterated experimentation on the membranous part adhering to the occipital bone switches the votes of all the spectators in favour of his [Tandon's] opinion [of the sensibility of the pericranium]. They have no doubts anymore and therefore sign. I do not, however, dare to sign these results hoping to receive in the continuation the clarifications necessary to raise the veil which [makes me] suspend my judgement.
>
> The experiment made the same day on the dura mater does not convince me of its sensibility. Mr Tandon uses the same dog, trepans it at the place of the temporalis muscle, reveals the dura mater, presses and pricks it, as some of his assistants, among others also Mr Sauvages, do. The dog testifies by his cries and movements that he suffers. I seize the instrument [a hooklet] at once and move its point across the membrane without penetrating it. Everybody is surprised to see no sensation [*sentiment*] of the animal. Mr S[auvages] wants to clarify the matter and hooks on to the dura mater, the

> pain is unequivocal. I take up again therefore the instrument [*machine*] and proceed in the manner I had done. I get again the same result as before but somebody of the group objects that I did not push the hooklet into the substance of the dura mater. I admit that and say that it would be sufficient to irritate the surface in order to ascertain truth. But, in order to satisfy what has been requested from me, I lower [*fonce*] the instrument and have to agree that there is pain, the cry uttered by the animal is its proof. The assistants testify in protocol the sensibility of this sheath of the brain. As to myself, I have testified that what I have seen without to determine anything for sure. Furthermore, I reserve the right [to make] remarks on these experiments and propose to repeat them. This will have an influence on the signing I can make.[84]

Housset did not agree with Tandon over the proper way to proceed and therefore signed the protocols only with reservations. His report makes clear that the exact way in which an experiment was performed could determine its outcome. Was the use of oil of turpentine permissible, how long had the animal to calm down, what had to be counted as a proper irritation? There was no accepted regulation that decided these questions. Whoever could set the standards of experimentation would control the debate and its outcome. Tosetti referred to the earlier debate over the nature of light and pointed out that as soon as the French had adopted Newton's method they yielded the same results as the English, and the controversy was over.[85]

But who was entitled to judge these matters? Those who had proved their experience and experimental and practical skill. Of course, traditional hierarchies were not simply swept away. Spallanzani, for instance, considered the Paduan physician Domenico Vandelli's (1732–1815) critique of Haller as the attack of a 'pygmy' on a 'giant'.[86] Such a judgement certainly reflected not only the greater amount of experiments Haller had performed but also his general reputation in the Republic of Letters. However, the display of practical experience became a central issue in the rhetoric of the contestants. Haller stressed that from youth he had used the scalpel, and repeatedly pointed to the many dissections and experiments he had performed.[87] Conversely, he argued, none of his opponents had made any anatomical discoveries. Bianchi even became famous for his anatomical errors, and Radniczky and Vandelli had revealed their lack of skill, as they were not even able to observe how the brain moved in accordance with respiration.[88] Vandelli replied that Haller was mainly supported by clerics and other unknown persons who were misled by his fame, whereas his opponents were mostly experienced surgeons and physicians.[89] Delius took a similar course. He stated that Haller's views were adopted in the monasteries in Italy and added that 'it is easy to imagine how a crowd of monks with totally different duties than that of dissection may be persuaded of the irritability and sensibility of animals.'[90] For those who denied the validity of animal experiments, a different kind of experience was required. Anton de Haen admitted that Haller was a skilful anatomist and an indefatigable inventor of experiments but he lacked *clinical* practice,[91] an allegation which, in its turn, was refuted by Tissot.[92]

The most important way of demonstrating experience, skill, and proper procedure was, of course, the performance of experiments and the publication of a report. Very rarely a trial was carried out by a single person; usually at least one other person assisted, and very often many others witnessed the operation. These witnesses, mostly physicians, surgeons, and students, but often also other interested people, were frequently mentioned in the publications. For the author, their function was not simply to testify the result of the trial. Nobody doubted, as we have seen, that the author actually had seen what he described. The purpose of mentioning expert witnesses was rather to confirm that the experiment had been carried out in a proper manner. Housset's account shows clearly that he was aware that signing the report also meant approving the procedure. It was thus of importance to note *how* the experiments had been performed. Most authors recounted some trials in detail and summarised the results of those with a similar outcome. The majority did not confine themselves to the relation of results but described the general circumstances and specific details of their procedure. The controversy, with its contradicting results, seems to have produced the insight that Abraham Trembley had formulated somewhat earlier: 'It is therefore not sufficient to say that we have seen a certain thing. This means saying nothing if we do not at the same time point out how we have seen it.'[93] The authors not only related details of the performance but also discussed them and explained why their own proceedings should be preferred to the one of their opponents. Haller took a similar view; as the trials were all of a similar nature, it was sufficient to denote the general procedure he had followed and the precautions he had taken. In his opinion, a brief account of each experiment, however, had to be furnished. Initially he had thought that he could do without it. After a vast number of experiments and still engaged in his continuous research, he was assured of the correctness of his results to such a degree that he could not conceive how anybody could doubt them. Nevertheless, contrary results urged him to publish his notes so that the reader could see 'the march of nature has used to convince' him.[94] Haller had already found out the usefulness of this procedure in his debate with Hamberger. In the edition of his experiments on respiration from 1751 he commented as follows:

> It is a useful custom, adopted by Boyle and nowadays by many, and also by Bonnet, to publish the whole series of experiments in order to confirm a certain ambiguous and not yet accepted opinion about facts. The reader informed about the controversy thereby himself acts as a judge and listens to the reports which nature delivers.[95]

The extensive report puts the reader in the position of a witness or – to use Steven Shapin's and Simon Schaffer's term – of a 'virtual witness'.[96] In contrast to Boyle, however, it is not the detailed narration but the repeated account of similar procedures, which produces the image of the experimental scene in the reader's head. Haller's stress on the repetition of the trials reflects a general current. In the late-seventeenth century, reports of experiments gained their credibility through the details of the account and the reputation of the witnesses present. As Christian Licoppe has shown, this type of report was increasingly replaced by descriptions in the manner of Newton and Réaumur, who emphasised the stability of the phenomena and thus the repeatability of the experiment.[97] This does not mean that the details of the experimental procedure became negligible. But they were less recounted to gain credibility than to allow the reader to repeat the trials. This was especially the case for Haller. Together with the edition of his 567 experiments he issued – as we have seen in the second chapter – exact instructions, which had to be followed for the test of sensibility. Thus, he explicitly tried to establish a standard of experimentation. Had he got all the participants in the debate to agree upon his standard, he would have succeeded with his demarcations of the realms of sensibility and irritability. But the experimental procedure was a matter of dispute. It centred on several details that we must now turn to.

The anti-Hallerians raised three arguments for why the lack of reaction was not proof of lack of sensibility. First, the animal may feel pain without showing it.[98] Considering the intensity of irritation this seems not to have been an objection Haller thought worth answering. Secondly, the animal was in a state of intense fear and shock, which diminished its sensibility.[99] To this, Haller ironically replied that therefore the best way to prepare patients for operations was to frighten them. But, of course, this was not his proper answer. The animal had to lie in a position as little painful as possible and it must calm down. Haller added, in these circumstances, the animals showed clear signs of pain when the skin was irritated.[100] His opponents also tried to cope with these difficulties. One student, Jausserand, personally fed the dogs several days before the trials, let them relax after each irritation, and blindfolded them so that they could not see when the torture proceeded.[101] Thirdly, and this was the most often-heard critique, a greater pain inflicted by an earlier wound could overshadow the pain suffered from a new irritation.[102] Furthermore, the animal might become somehow accustomed to the suffering.[103] Haller retorted that it was not he but rather his opponents who performed too many experiments on the same animal. He stressed that the animals must regain calm before a new irritation.[104] But, as Housset's report has shown, it was a matter of dispute how long the animal had to rest. Tandon waited three to four minutes whereas Pozzi and Housset proposed waiting a whole day.

Probably the most important issue of the controversy was the question of how far a part of the body had to be prepared in order to be properly irritated. Comparison of Haller's and Zimmermann's technique of irritation has revealed how the Göttingen professor tried to examine the organs in their 'pure' form and to isolate their specific structures in order to determine their specific functions. The problem was discussed particularly in the case of the tendon which, according to the Hallerians, had to be totally uncovered from its sheath.[105] Laghi admitted that upon total denudation the tendon often appeared to be insensible.[106] But, he maintained, this was too severe a procedure because the tendons would thereby be deprived of their nerves and rendered insensible, just as other parts were after a ligature of their nerves. Moreover, the Hallerians could always argue that their opponents had not laid bare the tendons sufficiently.[107] In fact, this was a difficult issue: were the nervous filaments that the Hallerians destroyed in their 'purification' really only part of the surrounding structures and not of the tendon itself? Caldani thought that he had proved insensibility by another technique, and without denudation. He cut the tendon and inserted a needle into its proximal ending and still the animal showed no signs of pain.[108] One Michel Angiolo Grima, on the other hand, argued that the section of the tendon would equally remove nervous filaments.[109] In the end, no agreement upon the topic was reached and everybody continued to lacerate the tendons in the manner he considered appropriate.

Many other, minor problems kept the debate going. They touched questions such as the alterations due to inflammation – which will be discussed in Chapter 6 – the inadvertent irritation of adjacent parts or the use of caustics to test irritability. Although Haller was quite aware of the different procedures of experimentation he still argued that a sufficient repetition of the experiments would furnish unequivocal results. In his view, his opponents did not produce constant but often vague and contradictory outcomes, which sometimes even supported Haller's position.[110] Many accidental causes must have intervened that 'disguised the habit of nature.'[111] Haller himself had performed ambiguous experiments and published them as a sign of his candour. He had always been surprised to see, he said, that certain researchers saw only that what they wanted to see. Only in fictionwere the heroes always victorious.[112] But through repetition the essence could be distilled from its distorting contamination:

> The experiments which I have made and which my friends have added to mine are extremely numerous, and this number is not irrelevant. Foreign causes may introduce error into experiments which are not verified, but these causes are separated increasingly as the same fact is repeated even if these [foreign causes] are absent. After a certain number of verifications only those results remain which emerge from the nature of the things.[113]

Whereas for the encyclopedist Fouquet, the variability of nature and of technical circumstances was a reason to declare the futility of repetition, for Haller this was precisely why the experiments had to be reiterated.[114] This was the only way to avoid errors, not only in Haller's but also in many of his followers' views, notably Caldani and Fontana.[115] Having performed so many experiments, they could not err. 'No other physical truth is confirmed by that many testimonies' Haller declared.[116] Their superior experience argued for their veracity. Nevertheless, the trials of their opponents were not negligible, and not of such a small number as the Hallerians would have liked to claim. The numerical argument was less of an epistemological than of a rhetorical nature. Haller had, as he remarked, performed his many experiments 'in order to overwhelm, so to speak, the incredulous by the number of concordant testimonies'.[117] The Paris physician François Thierry (1719–93) correspondingly asked: 'What to answer to 1,200 experiments?'[118] Besides, repetition was essential for Haller's own research, which followed detailed standards of experimentation. It may also have served him, like others, as a means of self-conviction. But within the debate, where these standards were not fixed, repetition simply yielded two large piles of contrary results. One of these piles might be higher but this was not necessarily read as a sign of trustworthiness, it could also be seen as a mark of missionary zeal.

Thus, the repetition of experiments could not solve the problems originating from different standards of procedure. For this, a transfer of technique and skills would have been necessary. The detailed description of manoeuvre delivered by some authors, could only impart this practical knowledge in small degree. However, as the case of Tandon and Housset has shown, presumably even direct confrontation between the experimenters would not have resulted in an assimilation of method. The defects of the opponents were known in advance and would only have been confirmed. Laghi, one of the most unbiased and self-critical of all those involved in the debate, originally intended to perform experiments together with his opponent Caldani, but in the end, despite repeated invitations of the latter, never appeared.[119] The authors continued to collect their experiments and arguments. One of them, the physician Domenico Sanseverini from Naples, realised that the dispute could not be solved in this way. Another method had to be followed in order to find truth: anatomy.[120]

The ambiguity of anatomy

Haller declared in his treatise that only those parts of the body furnished with nerves are sensible. Hardly anyone doubted this statement. It was obvious, therefore, that anatomy, like animal experimentation, should be an adequate means of determining whether a part of the body was sensible or not. Haller had exclusively used the experimental approach but he backed up his findings with the assertion that parts of a weak sensibility were supplied by only few nerves. In particular, it could not be shown that the small nervous filaments that crept upon the surface of the tendons and the dura mater entered these membranes. This notion was contested by Laghi and Vandelli, who both published anatomical tables clearly showing that small nervous branches penetrated the Achilles tendon and dura mater (Figure 4.2 and 4.3, overleaf).[121] Laghi added, several anatomists who had witnessed the dissection confirmed the presence of these nerves. The Hallerians quickly countered this. Tosetti searched for these nerves but could not find them, even with a microscope. His plate, therefore, displays no such structures (Figure 4.4).[122] Besides, he argued, the authorities which Laghi quoted had not explicitly confirmed that nerves would enter the Achilles tendon. Picking up Haller's point he said that we should distinguish between *intrinsically* sensible parts, into which the nerves entered, and *extrinsically* sensible parts, the substance of which were not penetrated by nerves. If the latter were to be considered as truly sensible even bones and fat had to be called sensible, an absurd idea in Tosetti's view. Haller himself did not exclude the possibility that some very tiny nerves, not even visible with the aid of a microscope, might enter the membranes. The tendons and dura mater might therefore be sensible but to an almost imperceptible degree.[123] Fontana added that the argument of invisible nerves should not be allowed. 'Anatomy and, indeed, the whole of medicine would be lost if it was allowed to oppose contrary conjectures to facts.'[124] He and Caldani put forward additional arguments. The descriptions of Laghi and Vandelli were utterly contradictory. Whereas Laghi could, with great care, distinguish only three small nervous filaments penetrating the Achilles tendon, Vandelli described not less than twenty-three nerves which, allegedly, were plainly visible (Figure 4.2 and 4.3). Moreover, Laghi's illustration of the dura mater was inaccurate, as several nerves were missing, and the transparency of the filaments he described was a clear sign that these were not nerves but cellular fibres.[125]

Figures 4.2 – 4.4
The visualisation of (in-)sensibility

Figure 4.2
Domenico Vandelli (1756).

Vandelli shows the Achilles tendon (3) with its sheath removed to the left. One nerve (6) perforates the muscle and with several branches and filaments (8-15) leads to and lastly inserts in the tendon. Two other nerves (16, 17) penetrate the sheath and enter the tendon with many filaments (23, 24, 26, 31, 33).

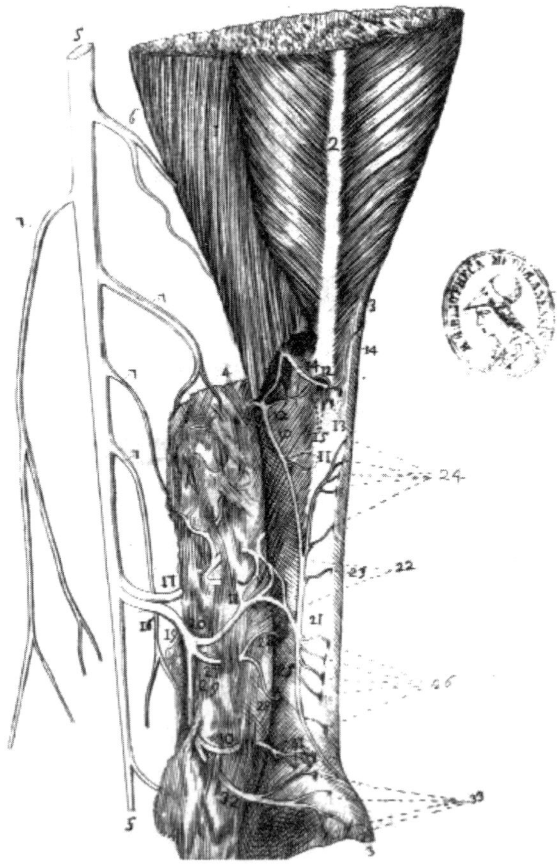

Figure 4.3
Tommaso Laghi (1757).

Laghi describes three small nervous filaments (g, g, g) entering the Achilles tendon (B).

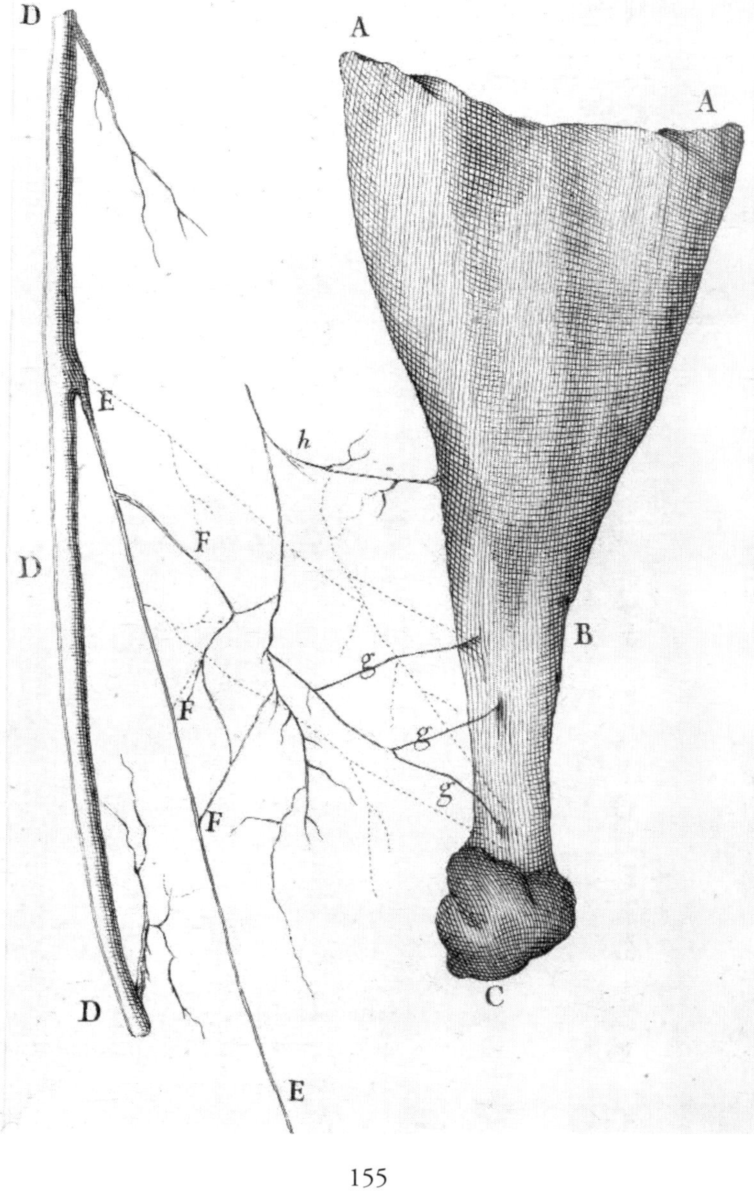

Figure 4.4
Urbano Tosetti (1757).

On fig. I. we see small nervous branches (G) pervading the membrane covering the Achilles tendon (H). Fig. II. demonstrates that these branches (H) do not enter the Achilles tendon (A).

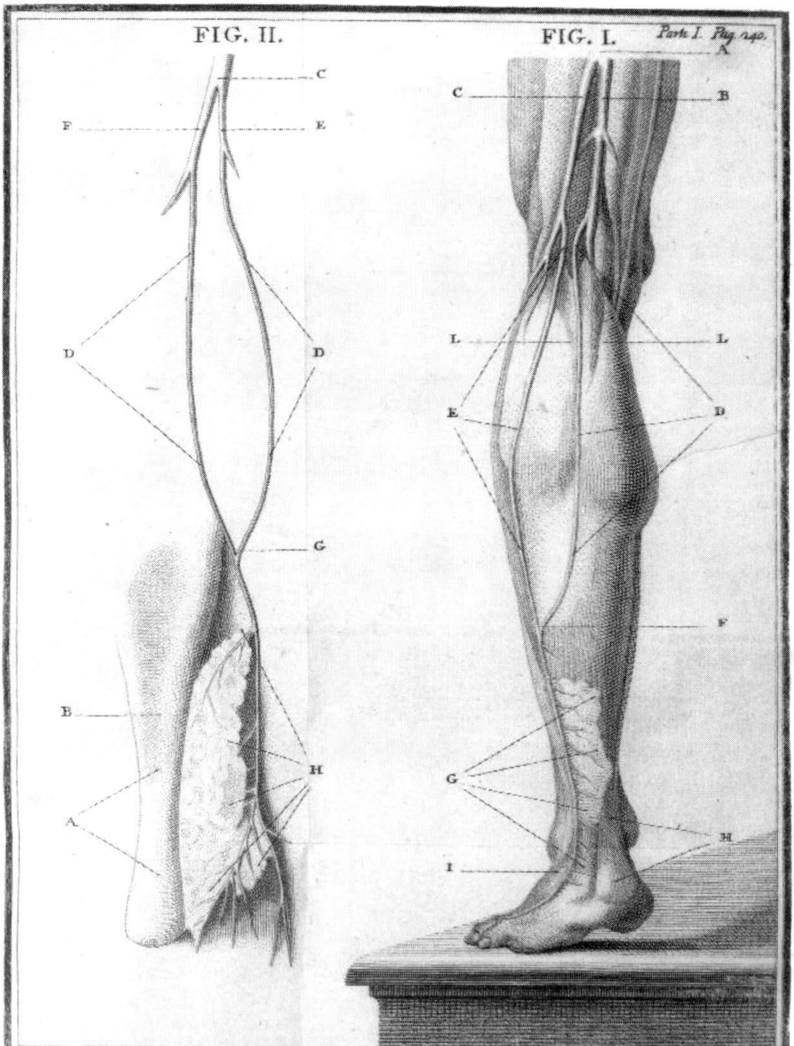

Nevertheless, even if illustrations seemed insufficient to decide the matter, at least the anatomical examples should be able to. Fabri invited Fontana to witness his own specimen of the dura mater with its nerves. Twice, however, he was absent when Fontana and his friends appeared, and the third time, he informed his guests that he no longer possessed the specimen.[126] But, Fabri assured them, the nerves had been plainly visible.[127] It took almost ten years for a similar event to happen. In 1765 Domenico Cotugno (1736–1822), anatomist and physician from Naples, arrived in Padua and revealed to several physicians that he had found a nerve entering the dura mater.[128] Haller asked his correspondent Johann Friedrich Lobstein (1736–84) from Strasbourg to examine the matter, and was duly informed that Cotugno's nerve was, in fact, a small artery. But the famous Morgagni was not of this opinion. On 24 February 1768, in his public lecture, he presented a specimen of the dura mater, prepared by his prosector Giambattista del Covolo (d.1768), which revealed this nerve. After the lecture, a student stole this specimen and carried it to the house of Caldani, where he asked the latter in the presence of sixty students whether the filaments in question where nerves or not. Caldani, after some reflection, declared that these were not nerves but branches of a small artery. Covolo, in reply, prepared a second specimen, but neither he nor Morgagni wanted Caldani to see it. The whole debate took various further turns, including some publications on both sides in the early 1770s, then interest for these detailed anatomico–physiological questions seems to have diminished. Caldani and Haller, with the support of two German anatomical studies, held a somewhat stronger position but the case was never properly closed. Not even anatomy could produce definitive arguments for one or the other side.

The aporia of physiology

If the performance of experiments had been standardised and the outcome unequivocal, there would still have remained a discussion about their interpretation. Many considered Haller's experiments on irritability as correct but denied that they demonstrated a quality independent of the nerves. Haller, on the other hand, argued that he could accept all the experiments of Whytt without having to change his own concept.[129] In his view, his opponents were misled in their interpretations by unascertained theories. The experiments largely spoke for themselves, and in a debate on recent discoveries he called for restraint. In a review of 1753, he declared: 'Mr Haller rather is desirous to ensure that, in this new field, one does not pass beyond experience and seize new but weak hypotheses.'[130] In his essay on hypotheses (1750), he did not reject the use of assumptions but he demanded a clear distinction between ascertained facts and plausible argumentation.[131] Hypotheses were useful as they connected the single elements of knowledge that would otherwise remain fragments. They led to novelties and truth, and posed questions which would not have occurred to us and which called for experimental testing. Haller advocated the *esprit systèmatique*, the building of theories based on experience, but he refused the *esprit de système* which was solely nourished by conjectures.[132]

Thus, although Haller stressed that we should stick to experimental results, he was well aware that these 'fragments' did not compensate for a comprehensive physiological and medical concept which, however uncertain, could at least satisfy the 'explicit interests' of the physicians, ie. to explain the functions of the body and propose rules for treatment.[133] He saw how the results – those of the Hallerians and those of their opponents – were indeed quickly used for the establishment or for the confirmation of medical systems. 'As soon as my experiments became known', he said 'irritability gained such an expansion that almost all vital and involuntary motions of the body were traced to it.'[134] If medicine could not do without systems, should he not at least support a version which maintained the core of his concept, the clear separation of irritability and sensibility? Even for the empiricist Haller the establishment of a doctrine had a certain appeal:

> A doctrine [*Lehrgebäude*] that should bear our own name, an idea which has arisen from our life-forces, does for a learned man what ambition did for Alexander. Effort, expense, time, experience, art and instruments, all the powers of the will and understanding, are employed freely and without protest, if we have a goal, and if our doctrine thereby becomes more probable, more certain and more acceptable.[135]

As we saw in the previous chapter, Haller did not himself present a comprehensive medical system with irritability and sensibility as its cornerstones. But he was quite flattered by Tissot's preface to the French edition, considering it as 'too obliging to me', and he published it without alterations.[136] For Tissot, irritability was one of the fundamental qualities of animated bodies and 'probably the basis of their life.'[137] In his view, Haller had found one of the 'keys of nature':

> NATURE, this very famous word in medicine, the word about which we talk that often and which we understand so little, will finally be determined; it is the sum total of the powers of the vital principle, a principle which has not been known until now.[138]

This was of importance for clinical medicine, too:

> The theory of fevers, that of inflammations and, in one word, of all the diseases which depend on an increase of circulation, shall from now on be determined because the knowledge of the cause of circulation [ie. irritability] leads to the knowledge of those causes which can increase or reduce it.[139]

He concluded that 'irritability has come to open a new area of research, a new source of solutions.... We owe physics to England, we will owe physiology to Switzerland, and the treatise on irritability will be its immovable foundation.'[140] Haller as the Newton of physiology – this was certainly a picture by which the 'discoverer' of irritability was flattered. Although Tissot indulged in far-reaching and unascertained conjectures, he still maintained the separation of irritability and sensibility and restricted these qualities to those parts of the body designated by Haller.

Along similar lines were the essay of Gian Francesco Cigna (1734–90), physician in Turin, and the preface to the Italian edition of Haller's treatise, written by Giovanni Vincenzo Petrini (1725–1814), reader in philosophy and mathematics at the *Collegio Nazareno* in Rome.[141] All three texts – Tissot's, Cigna's and Petrini's – demonstrated how efficient an encompassing theory of irritability could be in the explanation of vital phenomena. They were all re-edited in Haller's *Mémoires,* once more with Tissot's preface as the introduction heading the whole collection. As editor, Haller thereby supported the systems based on his own, less ambitious concept. He used the speculative theories along with the experiments as arguments against his opponents:

> There are other critics who have not wavered to pronounce that – whatever the outcome of my experiments would be – the result seemed to them to be fairly useless, and that we equally would not gain any benefit for the art of healing.... It is to the criticisms of this kind that father Petrini has answered....[142]

The utility of his concept served as an argument for its veracity, the theory supported the experiment. Nevertheless, if necessary, Haller could easily dissociate himself from such speculations, and especially in the case of his dispute with de Haen, he stressed his purely empirical position.[143]

The four-volume *Mémoires* are an impressive demonstration of the experimental foundation of Haller's concept. The twenty-four authors assembled in this collection are grouped around Haller, who defines the rules of experimentation. Thanks to these common rules the authors appear as an 'experimental community' and their individual statements as collective.[144] The countless experiments confirm that the members respected the rules. Even the purely theoretical writings of Tissot, Petrini, and Cigna appear as a part of this experimental project. Interestingly, Haller did not edit any treatises that contained both descriptions of experiments and speculative conclusions drawn therefrom. From Zimmermann's dissertation he excerpted only the experiments and not the pupil's conjectures on irritability as a vital force, which had served Tissot and Petrini as a starting point for their own theories. Haller carefully separated experiment from theory and eliminated the necessarily speculative path leading from one to the other. The theories seemed to be conclusions emerging directly from the experiments. He also used this technique for the presentation of his own results (see Figure 4.5, opposite).

Thus Haller used two strategies to propagate his notion of irritability and sensibility. On the one hand he created an 'experimental community' and postulated the primacy and independence of experiment. On the other hand he supported – at least to some extent – medical systems that accepted his separation and anatomical demarcation of these qualities. The experiments conveyed credibility and the theories an explanation of the body that could serve as a point of reference for physiology and therapeutics. The fact that Haller sought support from speculative, clinically relevant theories which went far beyond any conjecture he himself ever made, demonstrates the weakness of experiment in a period when physiology was not yet an independent science.

The debate on irritability and sensibility itself also revealed an ambiguous picture of experiment. On the one hand, Haller's research had drawn attention to animal experiment as a promising method to investigate fundamental physiological problems, and had encouraged many to follow his path. However, it also had produced a large amount of contradictory results, which demonstrated the problematic nature of this approach. Those who already had their doubts about the validity of the experimental method saw their reservations confirmed. Those who performed experiments realised the necessity of a standard of experimentation, but could not agree upon a standard. They recognised that experimental results did not necessarily display a pure image of nature. The experimenters lost their naive trust in the innocence of experiment and surrounded the presentation of their trials with a whole series of arguments. Every censure of their performance was disproved, for each argument a counter-argument was raised. In a similar manner to Lakatos' research programme, the whole debate about experiment served as a protective belt that blocked all possible critiques and secured the core statement, the experimental proof of presence or absence of sensibility and irritability.[145] In 1763, Haller needed fifty-five quarto-pages of his famously concise Latin style to present all his arguments.[146] The experiment lost its immediacy, and did not appear as sharply opposed to theory and hypothesis as expected. It was not a rigid test in the sense we perceive it

Figure 4.5
The separation of experiment from theory.

At the end of the account of the experiments, Haller presents conclusions, which seem to follow directly therefrom. Conclusions like 'il y a trois forces contractives dans les muscles' are, however, based on several theoretical assumptions, which Haller does not deal with. Mémoires, *i, 254–5.*

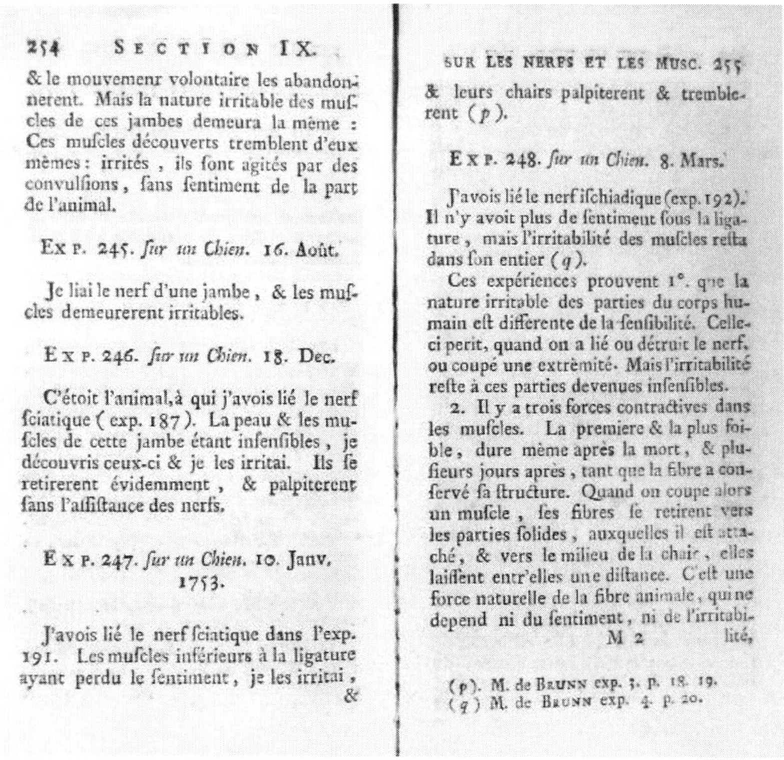

254 SECTION IX.

& le mouvemenr volontaire les abandonnerent. Mais la nature irritable des muſcles de ces jambes demeura la même : Ces muſcles découverts tremblent d'eux mêmes : irrités , ils ſont agités par des convulſions , ſans ſentiment de la part de l'animal.

EXP. 245. *ſur un Chien.* 16. Août.

Je liai le nerf d'une jambe , & les muſcles demeurerent irritables.

EXP. 246. *ſur un Chien.* 18. Dec.

C'étoit l'animal,à qui j'avois lié le nerf ſciatique (exp. 187). La peau & les muſcles de cette jambe étant inſenſibles , je découvris ceux-ci & je les irritai. Ils ſe retirerent évidemment , & palpiterent ſans l'aſſiſtance des nerfs.

EXP. 247. *ſur un Chien.* 10. Janv. 1753.

J'avois lié le nerf ſciatique dans l'exp. 191. Les muſcles inférieurs à la ligature ayant perdu le ſentiment , je les irritai , &

SUR LES NERFS ET LES MUSC. 255

& leurs chairs palpiterent & tremblerent (*p*).

EXP. 248. *ſur un Chien.* 8. Mars.

J'avois lié le nerf iſchiadique (exp. 192). Il n'y avoit plus de ſentiment ſous la ligature , mais l'irritabilité des muſcles reſta dans ſon entier (*q*).

Ces expériences prouvent 1°. que la nature irritable des parties du corps humain eſt differente de la ſenſibilité. Celle-ci perit, quand on a lié ou détruit le nerf, ou coupé une extrêmité. Mais l'irritabilité reſte à ces parties devenues inſenſibles.

2. Il y a trois forces contractives dans les muſcles. La premiere & la plus foible , dure même après la mort , & pluſieurs jours après , tant que la fibre a conſervé ſa ſtructure. Quand on coupe alors un muſcle , ſes fibres ſe retirent vers les parties ſolides , auxquelles il eſt attaché, & vers le milieu de la chair , elles laiſſent entr'elles une diſtance. C'eſt une force naturelle de la fibre animale , qui ne depend ni du ſentiment , ni de l'irritabilité,

M 2

(*p*). M. de BRUNN exp. 3. p. 18. 19.
(*q*) M. de BRUNN exp. 4. p. 20.

nowadays but was used jointly with other forms of argument. Eighteenth-century physiologists easily wandered between experimental, clinical, anatomical, and theoretical reasoning and the different methods were considered as different only by grades. A plurality of methods persisted and if, following Roselyne Rey, we describe experiment as the method proper to physiology, and observation as the characteristic approach of animal economy, we must assert that physiology was not yet firmly established as a science on its own.[147] Eulogists like Condorcet may have praised Haller's methodological approach and hailed him as 'the creator of physiology', but this does not change the fact that the first time these methods were tested on a large scale all over Europe, they produced contradictory results.[148] The debate on irritability and sensibility had directed the focus toward specific, investigable properties of the animated body and thus had helped to shape the notion of physiology as a proper branch of knowledge. But at the same time, it demonstrated the aporia of this new science, as it could not secure an agreement upon the epistemological status and procedural aspects of its own method.

The status of experimentation did not change considerably over the entire century. Experiments were performed all over Europe, but not in exceptional quantities and only rarely in a systematic manner. In Edinburgh they were used as a means of confirming existing hypotheses, and only in a few instances as a method of research.[149] The same is true of England and is illustrated by the few experimental publications in the *Philosophical Transactions*.[150] Not even in Germany, where Haller had worked, did a continuous experimental tradition emerge. Most of his pupils did not pursue an academic career nor did they perform any substantial research after their graduation. The two exceptions, Johann Gottfried Zinn and Johann Friedrich Meckel, were excellent anatomists but did not continue their experimental investigations. Besides, both died fairly early, Zinn in 1759, Meckel in 1774. There was no Hallerian school of experimental physiology after Haller's return to Switzerland. Many supported his views but rarely did anybody perform experiments. One author remarked in 1778 that the physicians now denied or, more precisely, neglected experiments and argued with unconnected, single and unverified observations.[151] The most notable physiologists of the next generation, Johann Friedrich Blumenbach (1752–1840), Johann Christian Reil (1759–1813), and Carl Friedrich Kielmeyer (1765–1844) did not carry out experiments of any importance. Nor did *Naturphilosophie*, which constituted the theoretical background for many physiologists of this period, particularly promote experimentation. It did not refute the experimental approach but it stressed that a mistaken idea about nature that contradicted reason could lead to incorrect experiments and erroneous results. The experiment was carried out within a well defined framework and had to reveal the order of nature that the researcher had to some extent anticipated.[152] In Italy – thanks to the still-living Galilean tradition and its renovation in the Newtonian movement – experimentation was more highly esteemed and more experiments were performed than in any other country.[153] Caldani and Fontana had not only repeated Haller's trials but also asked new questions and thus extended the area of experimental investigation. In their efforts they were even surpassed by Spallanzani's works on generation, circulation, and digestion. Finally, the

Galvani–Volta controversy in the 1780s and 1790s focused the attention of physicians and naturalists upon animal experiment in a similar degree to the debate on irritability and sensibility in the 1750s and 1760s.[154] But it too had a controversial outcome, was only properly settled in the 1830s, and was not able to provoke a definitive shift to the experimental approach either. Francesco Vaccà Berlinghieri (1732–1812), for instance, who wrote an important physiological textbook (1783, 2nd edn 1795), stressed the importance of clinical experience and considered animal experimentation as an artificial method, which would yield reliable results only in exceptional cases.[155] Positions in France were similarly divided. Several physicians and surgeons in Paris, who had also worked on sensibility, continued with experimental investigations. Notably, Antoine Portal in the 1770s and Félix Vicq d'Azyr in the 1780s gave courses that included, or at least propagated animal experimentation.[156] In Montpellier, however, after the repetition of Haller's trials in the 1750s, interest in experiments lessened. The vitalist concept that dominated the Faculty in the latter half of the century and its main representative Paul-Joseph Barthez did not encourage the experimental approach but, taking into consideration the complexity of the links between the parts of the organism, preferred observation.

The eighteenth century did not establish a continuous tradition of animal experimentation.[157] Several eminent figures pursued investigations of a systematic nature unlike any before, but a physiologist did not necessarily have to follow their path. The researchers acted as individuals, and not as members of a community or institution that would have expected the performance of experiments. Universities and other institutions of medical education did not demand original research from their teachers. Only in 1794, with the foundation of the *Ecoles de Santé* of Paris, Montpellier and Strasbourg, did this change. The decree of the *Ecoles* regulated that, besides teaching, the professors had 'as their further purpose the most extensive research into all branches of the art of healing [and] as their aim the advancement of all the sciences which can shed light on the physics of life.'[158] This innovative approach was reflected in the teaching itself, which included theoretical and experimental physiology, notably in the theses of the pupils. More than in former times, the dissertations were concerned with pathological anatomy, often presented results of chemical analysis and were, as we noted earlier, in twelve per cent of the cases based on animal experiments.[159] Conditions similarly favourable to experimentation were created by Xavier Bichat (1771–1802) at the *Hôtel-Dieu.* He served as the model to follow and to emulate for an emerging generation of experimental physiologists that included Julien Legallois (1770–1814), Pierre Nysten (1771–1818), Guillaume Dupuytren (1777–1835), and lastly, François

Magendie (1783–1855). With Magendie in the 1820s – now at the Medical Faculty – physiology was definitively established as a science in its own right, with experiment as its primary method. Magendie was joined by many researchers in other European countries but remained the dominating figure, whose model was quickly adopted in Germany, and later in Italy, England, and the United States.[160]

John Lesch has argued that the French Revolution created the appropriate context for the emergence of experimental physiology as a scientific field.[161] Although Paris medicine focused on clinical observation and pathological anatomy, the reformation of the medical schools with the unification of medicine and surgery, and thus the clinical experience and operative facility of the physicians, and an institutionalised commitment to research combined to produce a new generation of physicians and scientists who were stimulated and supported in their experimental investigations.[162] Such a development was not possible in the older, tradition-bound medical faculties. Lesch's thesis is essentially right, but we should not thereby conceive the idea of the sudden appearance of an independent science. Lesch quite correctly noted that Bichat's physiology was still founded on two different modes of research: firstly, on the primarily classificatory approach based on vitalism, and secondly, on the experimental method aiming at the operative knowledge of functions. The rise of experimental physiology was a continuous event, it looked back to a tradition maintained by many individuals and, thanks to its institutionalisation, in France it gradually superseded other modes of research. But it could only become a true science in our modern sense when the researchers established standards of experimentation in order to yield identical results. This is what Magendie and Johannes Müller (1801–58), amongst others, did in the 1820s and 1830s, and this is what had been lacking in the debate on irritability and sensibility.[163]

Notes

1. Some aspects of this chapter I have discussed in my paper, '"Die Ehre des Rechthabens": Experiment und Theorie im Streit um die Lehre von der Irritabilität', *Sudhoffs Archiv*, 82 (1998), 141–69.
2. *Praelectiones*; *Primae Lineae Physiologiae in Usum Praelectionum Academicarum* (Göttingen: Vandenhoeck, 1747); *Icones Anatomicae* (7 fascicles, Göttingen: Vandenhoeck, 1743–56).
3. Grimm to Haller, 7 September 1752 (*Haller Papers*).
4. A. Portal, *Histoire de l'Anatomie et de la Chirurgie* (7 vols, Paris: Didot, 1770), iv, 709.
5. C.-N. Le Cat, *Dissertation qui a Remporté le Prix Proposé par l'Académie*

Royale des Sciences et Belles-Lettres de Prusse, sur le Principe de l'Action des Muscles avec les Pieces qui ont Concouru (Berlin: Haude und Spener, 1753), 127.

6. G.B. Bianchi, 'Lettera Seconda... alli... Signori Lettori Pubblici dell' Università di Bologna', *Opuscoli*, ii, 25–52: 36.
7. 'Si l'Irritabilité Dépend des Nerfs Comme la Sensibilité? Extrait d'une Thése Soutenue aux Ecoles de Médecine de Paris... par M. Girard de Villars', *Journal de Médecine*, 6 (1758), 310–15: 310.
8. A more thorough exposition of the role of the correspondence in the debate on irritability and sensibility I have given in the paper 'Der Patron im Netz. Die Rolle des Briefwechsels in Wissenschaftlichen Kontroversen', in M. Stuber, S. Hächler and L. Lienhard (eds), *Hallers Netz. Ein Gelehrtenbriefwechsel im Zeitalter der Aufklärung* (Basel: Schwabe, 2005) (in press).
9. *Mémoires, Opuscoli.*
10. On Lacaze see R. Rey, 'Vitalism, Disease and Society', in R. Porter (ed.), *Medicine in the Enlightenment* (Amsterdam: Rodopi, 1995), 274–88.
11. *GGA*, 1754, 361–4.
12. P. Fabre, *Essais sur Différens Points de Physiologie, de Pathologie et de Thérapeutique* (Paris: Didot, 1770). See Haller's review in the *GGA*, 1771, Zugabe, ccxlvi–ccxlviii.
13. See for instance D. Vandelli, *Apologia Contra... Hallerum* (Padua: Conzatti, 1760), 11–12; Portal, *Histoire* (note 4), iv, 712–13; N.-F.-J. Eloy, *Dictionnaire Historique de la Médecine Ancienne et Moderne* (4 vols, Mons: Hoyois, 1778), ii, 439.
14. See eg. A. von Haller, 'Sensibilité', *Encyclopédie Suppl.*, iv, 776–9.
15. The most detailed, although biased account of the controversy is given in J.C. Blasche, *Das Leben des Herrn Hofraths und Professors Georg Erhard Hambergers, nebst einer Nachricht von seinen Schriften, und Gelehrten Streitigkeiten* (Jena: Güth, 1758). For a very brief summary see H. Buess, 'Experientia Majorum: Ein Physiologenstreit vor 200 Jahren', *Experientia*, 2/4 (1946), pp. 150–1.
16. A. Haller, 'Mémoire sur une Controverse au Sujet de la Respiration', *Nouvelle Bibliothèque Germanique,* (1748), 412–28.
17. Portal, *op. cit.* (note 4), iv, 703.
18. Baldinger to Haller, 29 July 1768 (*Haller Papers*).
19. Some experiments were repeated by Haller's pupil Johann Georg Runge (1726–1781) in Bremen and some in Montpellier by Sauvages and Lamure alongside with experiments on irritability in 1752/3. See the letters from Runge (31 December 1753), Sauvages (1 March 1752) and Daniel Rudolf Ith (11 April 1753) to Haller (*Haller Papers*).

20. The following description is based on Haller's correspondence and on contemporary writings on irritability and sensibility, to which I will refer only occasionally. A full list of the sources is given in the bibliography.
21. Haller received a description of these experiments (Biblioteca Nazionale Braidense, Milano, A. XIV. 2432/29). The author of this manuscript and how it reached Haller is unknown.
22. W. van Doeveren, *Specimen Observationum Academicarum ad Monstrorum Historiam, Anatomen, Pathologiam, et Artem Obstetriciam, praecipue Spectantium* (Groningen and Leiden: Bolt, Luchtmans, 1765), 208–54.
23. I.J. van den Bos (1731–1788) and L. Bikker (1732–1801). The results were published in the former's dissertation *De Vivis Humani Corporis Solidis* (Leiden: Luchtmans, Moens, 1757). See also G. Van der Waa, '*De Irritabilitate*': *Een Onderzoek naar de Betekenis van het Irritabiliteitsbegrip in de Geschiedenis van de Achttiende-Eeuwse Nederlandse Fysiologie* (Rotterdam: Erasmus, 1992), Chapter 7.
24. J. Gourlez De La Motte, *Quaestio Medica... Discutienda in Scholis Medicorum Die Decimo-tertio Januarii MDCCLII. Guillelmo De Magny Praeside: 'An a Vasorum Aucta aut Imminuta Irritabilitate Omnis Morbus?'* (M.D. thesis, Paris: Quillau, 1752).
25. The experiments are mentioned in the letters of Nicolaus Himsel (7 August 1752), August Friedrich Pallas (23 December 1752), Louis-Marie Girard de Villars (31 March 1757) to Haller (*Haller Papers*) and in the book of Doeveren, *op. cit.* (note 22), 225.
26. See the editorial note in the *Journal de Médecine*, 6 (1757), 315.
27. Girard de Villars, *op. cit.* (note 7).
28. L.-M. Girard de Villars, 'Mémoire sur la Sensibilité des Parties des Animaux [lue le 20 janvier 1759]', *Mémoires de Mathématique et de Physique, Présentés à l'Académie Royale des Sciences, par divers Savans, & Lus dans les Assemblés*, 4 (1763), 580–96.
29. T. Bordenave, 'Remarques sur l'Insensibilité de quelques Parties, Établie par la Pratique', *Mercure de France*, Juin 1757, 137–54; J.P. de Jausserand, *De Sensibilitate et Irritabilitate Partium Corporis Humani Experimenta* (M.D. thesis, Montpellier: Rochard, 1758).
30. E.-J.-P. Housset, *Mémoires Physiologiques et d'Histoire Naturelle* (2 vols, Auxerre: Fournier, 1787), i, 91–134.
31. J.J.L. Hoin, 'Observations et Expériences sur les Plaies du Tendon d'Achille', *Journal de Médecine*, 30 (1769), 56–78; C. Arthaud, *Dissertations sur la Dilatation des Artères, et sur la Sensibilité; Apuyées de Plusieurs Expériences, Faites sur les Animaux Vivans, Auxquelles on a Joint Deux Observations sur l'Hydropisie du Péritoine* (M.D. thesis, Paris: Cavelier, 1771); *Lettre de M. Collomb... sur un Cours de Physiologie Expérimentale, Fait cette Année 1771,*

au Collége de France, par M. Portal (Paris: Didot, 1771); 'Experiences sur la Sensibilité, la Respiration et l'Anatomie de la Matrice', in J.-L. Moreau de la Sarthe (ed.), *Œuvres de Vicq d'Azyr* (6 *vols* , Paris, 1805), v, 1–10.

32. For the Italian debate see R.G. Mazzolini and G. Ongaro (eds), *Epistolario di Felice Fontana, I. Carteggio con Leopoldo Marc'Antonio Caldani 1758–94* (Trent: Società di Studi Trentini di Scienze Storiche, 1980); G. Ongaro, 'Giambattista Morgagni e la Questione dell'Irritabilità', in V. Cappelletti and F. Di Trocchio (eds), *De Sedibus, et Causis: Morgagni nel Centenario* (Rome: Istituto della Enciclopedia Italiana, 1986), 27–45; G. Ongaro, 'L'irritabilità Halleriana in Italia', *Atti del Convegno: Lo Sviluppo Storico della Neurologia Italiana, Padova 1987* (Padua: La Garangola, 1990), 113–25; M.T. Monti, 'Inediti di Pietro Moscati. Un Episodio del Dibattito Italiano sull'Insensibilità della Dura Madre', *Nuncius*, 2 (1987), 157–70; A. Dini, *Vita e Oganismo: Le Origini della Fisiologia Sperimentale in Italia* (Florence: Olschki, 1991), 25–45; A. Dini, 'Spallanzani e la Teoria dell'Irritabilità di Haller', in W. Bernardi and M. Stefani (eds), *La Sfida della Modernità* (Florence: Olschki, 2000), 197–208; M. Cavazza, 'La Recezione della Teoria Halleriana dell'Irritabilità nell'Accademia delle Scienze di Bologna', *Nuncius*, 12 (1997), 359–77; M. Bresadola, *Medicina e Filosofia Naturale in Luigi Galvani (1737–1798)* (Florence Univ. D.Phil. thesis, 1998), Chapter 2.
33. C. Pozzi, 'Cl. Viro Thomae Laghi... [1755]', *Opuscoli*, i, 230–6: 231.
34. 'De quibusdam Animalium Partibus, an Sensu Sint Compotes, et unde Irritabiles', *De Bononiensi Scientiarum et Artium Instituto atque Academia Commentarii*, 4 (1757), 48–57: 57.
35. Walter Tega has suggested that Zanotti might have asked Fabri to do so. See his article 'Mens Agitat Molem: L'Accademia delle Scienze di Bologna (1711–1804)', in R. Cremante and W. Tega (eds), *Scienza e Letteratura nella Cultura Italiana del Settecento* (Bologna: Il Mulino, 1984), 65–108: 99.
36. *Opuscoli*.
37. For Gilibert see his letters to Haller (*Haller Papers*), and for Pallas see the dissertation of his pupil A.J. Güldenstedt, *Theoria Virium Corporis Humani Primitivarum* (M.D. thesis, Frankfurt am Main: Winter, 1767).
38. W.F. Bynum, 'Cullen and the Nervous System', in A. Doig *et al.* (eds), *William Cullen and the Eighteenth Century Medical World* (Edinburgh: Edinburgh University Press, 1993), 152–62: 158.
39. See W.S. Curran, 'Dr. Brocklesby of London (1722–1797): An 18th-Century Physician and Reformer', *Journal of the History of Medicine*, 17 (1962), 509–21.
40. 'Jamais, au reste, vérité expérimentale, pas même la circulation du sang, n'a été soumise à tant d'épreuves.' Haller, *op. cit.* (note 14), 777.
41. *Mémoires*, iv, 21.

42. M. Neuburger, *The Historical Development of Experimental Brain and Spinal Cord Physiology before Flourens*, transl. and ed. with additional material by E. Clarke (Baltimore: Johns Hopkins University Press, 1981; transl. of the original German edition from Stuttgart, 1897).
43. See J.F. Meckel's letter to Haller from 28 July 1753, edited in *Epistolae*, iii, 402–05.
44. R. Toellner, *Albrecht von Haller: Ueber die Einheit im Denken des Letzten Universalgelehrten* (Wiesbaden: Steiner, 1971), 195.
45. M. Buscaglia, 'The History of the Experimental Method in the Life Sciences as an Illustration of the Versatility in Interpretation', in J. Montangero *et al.* (eds), *Conceptions of Time over Change* (Cahiers de la Fondation Archives Jean Piaget, 13, 1994), 45–64: 58.
46. Haller to Bonnet, 15 March 1755. *Corr. Bonnet*, 62.
47. For a general description of the role of the experimental method in the seventeenth and eighteenth-century physics see T.S. Kuhn, 'Mathematical versus Experimental Traditions in the Development of Physical Science', in T.S. Kuhn, *The Essential Tension: Selected Studies in Scientific Tradition and Change* (Chicago: University of Chicago Press, 1977), 31–65.
48. *Elementa*, i: iii, v–vi.
49. Two other works that should be mentioned are P. van Musschenbroek's *Oratio de Methodo Instituendi Experimenta Physica* (Leiden: Verbeek, 1731) and J.G. Zimmermann's *Von der Erfahrung in der Arzneykunst* (2 vols, Zurich: Heidegger, 1763–4). They both provide examples from physiology too but, as their titles suggest, are more concerned with physics and practical medicine, respectively. There is a further work, W. van Doeveren's *De Imprudenti Ratiocinio ex Observationibus et Experimentis Medicis: Sermo Academicus* (Groningen: Crans, 1754). It seems not to have been widely known, not even Haller knew it, and I have not seen it either.
50. J. Senebier, *L'Art d'Observer* (Geneva: Chirol, 1775), xiv–xv.
51. Bonnet to Haller, 22 July 1757, *Corr. Bonnet*, 107.
52. On this essay see Chapter 2.
53. *Mémoires*, i, 115.
54. *Ibid.*, i, dedication to Réaumur, dated 24 January 1756.
55. A. von Haller, 'Oeconomie animale', *Encyclopédie Suppl.*, iv, 105.
56. J. le Rond d'Alembert, 'Expérimental (Philosophie nat.)', *Encyclopédie*, vi (1756), 298.
57. *GGA*, 1775, 419–23 (Harvey has 'die Natur zum antworten gezwungen'), *Bibliothèque Germanique*, 4 (1748), 425 ('surprendre la Nature sur le fait'), *Relationes de Libris Novis*, 4 (1752), 477–8 ('simplici voce naturam interrogare').
58. For a concise account of the methodological position of the vitalists see R.

Rey, *Naissance et Développement du Vitalisme en France de la Deuxième Moitié du 18e Siècle à la Fin du Premier Empire* (Oxford: Voltaire Foundation, 2000), esp. 144–7.

59. J.-J. Ménuret de Chambaud, 'Observateur', *Encyclopédie*, xi, 310–13: 310.
60. J.-J. Ménuret de Chambaud, 'Observation', *Ibid.*, xi, 313–23: 318.
61. H. Fouquet, 'Sensibilité, Sentiment', *Ibid.*, , vi, 38–52: 52.
62. Ménuret de Chambaud, *Observateur* (note 59), 310; *idem*, *Observation* (note 60), 314. *GGA*, 1775, 419–23 (Haller on Harvey and Newton); *Nouvelle Bibliothèque Germanique*, 4 (1748), 416 (Haller on Descartes).
63. Claire Salomon-Bayet's assertion that the French 18th century generally preferred observation to experiment is somewhat exaggerated, as it presumes the dominance of vitalist thought. See her *L'Institution de la Science et l'Expérience du Vivant: Méthode et Expérience à l'Académie Royale des Sciences 1666–1793* (Paris: Flammarion, 1978), 337.
64. H.F. Delius, *Animadversiones in Doctrinam de Irritabilitate, Tono, Sensatione et Motu Corporis Humani* (Erlangen: Camerarius, 1752), xxiii; A. de Haen, *Vindiciae Difficultatum circa Modernorum Systema de Sensibilitate et Irritabilitate Humani Corporis, contra... Alberti v. Haller ad easdem Difficultates Apologiam* (Vienna: Krüchten, 1762), 102–4.
65. C. Lorry, 'Nouvelles Expériences sur l'Irritabilité', *Recueil Périodique d'Observations de Médecine, Chirurgie, Pharmacie &c.*, vol. 5 (Paris, 1756), 322–40, 403–421 and vol. 6 (1757), 7–19.
66. J.-A. Coutavoz, *Quaestio Medica Quodlibetariis Disputationibus Mane Discutienda in Scholis Medicorum Die Vigesima-sexta Februarii MDCCLXVII. Henrico Raymond Praeside 'Num Arteriae ut Sensibiles sic Irritabiles?'* (M.D. thesis, Paris: Quillau, 1767), 3; Gmelin to Haller, 16 September 1776 (*Haller Papers*).
67. See Haller's letter to Tissot, 7 September 1754, *Corr. Tissot*, 30.
68. On the general discussion of the validity of animal experimentation in the seventeenth and eighteenth centuries see A.-H. Maehle, *Kritik und Verteidigung des Tierversuchs: Die Anfänge der Diskussion im 17. und 18. Jahrhundert* (Stuttgart: Steiner, 1992).
69. *Elementa*, vi, 529. M.T. Monti has drawn the attention to this passage; see her *Congettura ed Esperienza nella Fisiologia di Haller: La Riforma dell'Anatomia Animata e il Sistema della Generazione* (Florence: Olschki, 1990), 101.
70. A. von Haller, 'Ad Objectiones contra Experimenta sua Propositas Responsio', *Opera Minora Emendata, Aucta et Renovata* (3 vols, Lausanne: Grasset, 1763–68), i, 441–502, here 475; *Mémoires*, iv, 75.
71. E.A. Nicolai, *Programma de Sensatione ac Sensibilitate* (Jena: Fickelscher, 1758), 22.

72. Bianchi, *op. cit.* (note 6), 19. Some other authors argued similarly.
73. Haller, *Ad Objectiones* (note 70), i, 477.
74. *Elementa*, i, praefatio, xi.
75. *Ibid.*, iv, 187, 409–13.
76. A. de Haen, *Difficultates circa Modernorum Systema de Sensibilitate et Irritabilitate Humani Corporis* (Vienna: Krüchten, 1761), 19–20.
77. S.-A. Tissot, *Lettre a Monsieur Hirzel... sur quelques Critiques de M. de Haen* (Lausanne: Grasset, 1762), 100–6.
78. A. von Haller, *Ad Viri Illustris Antonii de Haen Difficultates Apologia* (n.p.: n.p., 1761), 14–15; similarly *GGA*, 1763, 809 and 1775, 419.
79. Haen, *op. cit.* (note 64), 67–70.
80. Le Cat, *op. cit.* (note 5), 122. The same point is made in several of Le Cat's works which are listed in the bibliography.
81. Eg. Fouquet, *op. cit.* (note 61), 51, Bianchi, *op. cit.* (note 6), 4–5, and others.
82. F. Fontana, 'Dissertation Épistolaire de Mr. L'Abbé Felice Fontana... Adressée au R.P. Urbain Tosetti [23 May 1757]', *Mémoires*, iii, 157–243: 171–2.
83. Pozzi, *op. cit.* (note 33); S.-A. Tissot, 'Discours Préliminaire du Traducteur', in Albrecht von Haller, *Dissertation sur les Parties Irritables et Sensibles des Animaux...* trad. du Latin par M. Tissot (Lausanne: Bousquet, 1755), iii–xlix: xxxvi–xxxviii; U. Tosetti, 'Sull'Insensibilità di alcune Parti degli Animali: Lettera Prima [-terza] al Signor Dottore Giuseppe Valdambrini', in G.V. Petrini (ed.), *Sull'Insensibilità e Irritabilità di alcune Parti degli Animali* (Rome: Zempel, 1755), 181–220.
84. Housset to Haller, 7 May 1756 (*Haller Papers*). These experiments were never published.
85. U. Tosetti, 'Sull'Insensibilità di alcune parti degli Animali: Lettera Quarta... al Signor Dottore Giuseppe Valdembrini', *Opuscoli*, i, 237–67: 246.
86. In his letter to Antontio Vallisneri junior, 15 November 1762, in P. di Pietro (ed.), *Edizione Nazionale delle Opere di Lazzaro Spallanzani: Parte Prima. Carteggi* (12 vols, Modena: Mucchi, 1984–90), x, 245.
87. See esp. Haller, *Ad Objectiones* (note 70), 459.
88. *Mémoires*, iv, 51–2; *GGA*, 1758, 782 and 1760, 950.
89. Vandelli, *op. cit.* (note 13), 5, 10–17.
90. 'Physiologische Anmerkungen', *Fränkische Sammlungen*, 3 (1758), 406–26, here 410.
91. Haen, *op. cit.* (note 64), 27, 112, 118, 184.
92. Tissot, *op. cit.* (note 77), 48–50.
93. A. Trembley, *Mémoires pour Servir à l'Histoire d'un Genre de Polypes d'Eau Douce, à Bras en Forme de Cornes* (2 vols, Paris: Durand, 1744), 2.
94. *Mémoires*, i, 105.

95. A. von Haller, *Opuscula sua Anatomica* (Göttingen: Vandenhoeck, 1751), 104, preface to the experiments, dated 12 December 1750.
96. S. Shapin and S. Schaffer, *Leviathan and the Air-Pump: Hobbes, Boyle, and the Experimental Life* (Princeton: Princeton University Press, 1985), 61–5.
97. C. Licoppe, *La Formation de la Pratique Scientifique: Le Discours de l'Expérience en France et en Angleterre (1630–1820)* (Paris: Editions la découverte, 1996).
98. Eg. D. Vandelli, *Epistola de Sensibilitate Pericranii, Periostii, Medullae, Durae Meningis, Corneae, et Tendinum* (Padua: Conzatti, 1756), iv–v.
99. Eg. [F. Boissier de Sauvages], 'Von der Reitzbarkeit der Theile des Menschlichen Körpers: Auszug eines Schreibens. Aus dem Lateinischen', *Fränkische Sammlungen*, 1 (1756), 146–50.
100. *GGA*, 1757, 180–4; *Mémoires*, i, 108–9 and iv, 76–7.
101. Jausserand, *op. cit.* (note 29), 1–4.
102. Eg. *Abhandlung des Herrn von Haller von den Empfindlichen und Reizbaren Theilen des Menschlichen Leibes.* Verdeutscht und geprüft von C.C. Krausen (Leipzig: Jacobi, 1756), 53. R. Whytt, *Physiological Essays* (2nd edn, Edinburgh: Hamilton, Balfour and Neill, 1761), 254.
103. T. Laghi, 'De Insensibilitate, atque Irritabilitate Halleriana', *De Bononiensi Scientiarum et Artium Instituto atque Academia Commentarii*, 4 (1757), 208–17: 215–6.
104. *Mémoires*, iv, 75–6; *GGA*, 1757, 1002–05 and 1766, 73–5; Haller, *op. cit.* (note 14), 777.
105. See eg. *Mémoires*, i, 108 and Fontana, *op. cit.* (note 82), 167.
106. T. Laghi, *De Sensivitate, atque Irritabilitate Halleriana: Sermo Alter* (Bologna: Vulpe, 1757), 1–2.
107. Laghi, *op. cit.* (note 103), 215.
108. L.M. Caldani, 'Sull'Insensitività, ed Irritabilità di alcune Parti degli Animali: Lettera Scritta al... Signore Alberto Haller [25 November 1756]', *Opuscoli*, i, 269–336: 279.
109. *Mémoire sur la Sensibilité des Tendons, Prononcé en Italien à l'Académie des Apathistes* (Paris: n.p., 1760), 21–2. Grima's essay appeared also in the *Journal Etranger*, août 1760, 169–97.
110. *Mémoires*, iv, 41–7.
111. See eg. *GGA*, 1769, 1359.
112. *Mémoires*, i, 114–5.
113. *Ibid.*, iv, S. 25.
114. Fouquet, *op. cit.* (note 61), 51.
115. See Haller to Bonnet, 1 September 1757. *Corr. Bonnet*, 109. Caldani and Fontana both performed a vast amount of experiments.
116. Letter to Bonnet, 10 August 1758. *Corr. Bonnet*, 145.

117. *De Partibus*, 115.
118. Letter to Haller, 28 January 1758 (*Haller Papers*).
119. See Caldani to Haller, 16 May, 19 July and 17 October 1757. *Epistolae*, iv, letters 34, 39, 45.
120. 'De Fibrarum Sensibilitate atque Irritabilitate', *Opuscoli*, ii, 67–78: 72.
121. Laghi, *op. cit.* (note 103), 211, tab. I, II; Vandelli, *op. cit.* (note 98).
122. Tosetti, *op. cit.* (note 85), 238–44, 254–7, tab. 240.
123. *Mémoires*, iv, 32–3.
124. Fontana, *op. cit.* (note 82), 198.
125. *Ibid.*, 182–94; Caldani, *op. cit.* (note 108), 279–82; id., 'Sur l'Insensibilité et l'Irritabilité de Mr. Haller: Seconde Lettre... à Mr de Haller', *Mémoires*, iii, 343–485: 394–437.
126. Caldani, *Sur l'Insensibilité, op. cit.* (note 125), 421.
127. *Opuscoli*, iv, 143.
128. For this story see Ongaro, *op. cit.* (note 32), 38–40, with references to the sources.
129. *GGA*, 1764, 193–200.
130. *GGA*, 1753, 69.
131. On Haller's concept of hypothesis, see Chapter 2.
132. These two methods were distinguished by J. le Rond d'Alembert, 'Discours Préliminiaire', *Encyclopédie*, i, i–xlv, here vi, xxxi.
133. The term is Bruno Latour's; see his *Science in Action: How to Follow Scientists and Engineers through Society* (Cambridge, Mass.: Harvard University Press, 1987), 108.
134. *Elementa*, iv, 462.
135. 'The Haller Preface to the German Translation of the Histoire Naturelle (1750)', J. Lyon and P.R. Sloan (eds), *From Natural History to the History of Nature: Readings from Buffon and his Critics* (Notre Dame, Ind.: Notre Dame University Press, 1981), 295–310: 303. I have slightly corrected the translation.
136. Haller to Tissot, 29 September 1754, *Corr. Tissot*, 31. For Tissot's position see also my article 'Tissot Traducteur de Haller: De l'Expérience à la Théorie', in V. Barras and M. Louis-Courvoisier (eds), *La Médecine des Lumières: Tout Autour de Tissot* (Chêne-Bourg: Georg, 2001), 103–11.
137. Tissot, *op. cit.* (note 83), v.
138. *Ibid.*, xx.
139. *Ibid.*, xxxiii.
140. *Ibid.*, xlviii, xiv.
141. G.F. Cigna, *Johannes Franciscus Cigna... Publice Disputabat... MDCCLVII. Die XIV. Aprilis* (Turin: Typographia Regia, 1757); G.V. Petrini, 'Prefazione del Traduttore Italiano', Tosetti, *op. cit.* (note 83), ix–xxix.

142. *Mémoires*, iv, 21–2.
143. For further details of this debate see U. Boschung, ' "Multa pro Nostra Innocentia": L'Implication de Tissot dans la Querelle Haller-de Haen', in Barras and Louis-Courvoisier, *op. cit.* (note 136), 113–47.
144. 'Experimental community' and 'collective' statement are terms used by Shapin and Schaffer. They are part of the 'social technology' used in scientific disputes. Shapin and Schaffer, *op. cit.* (note 96), 69–79.
145. I. Lakatos, 'Falsification and the Methodology of Scientific Research Programmes', in I. Lakatos and A. Musgrave (eds), *Criticism and the Growth of Knowledge* (Cambridge: Cambridge University Press, 1970), 91–196.
146. Haller, *Ad Objectiones* (note 70).
147. Rey, *op. cit.* (note 58), 146
148. M.-J.-A.-N. Caritat de Condorcet, 'Eloge de M. de Haller', *Histoire de l'Academie Royale des Sciences*, année 1777 (Paris: Impr. Royale, 1780), 127–154: 135–6.
149. L. Rosner, 'Eighteenth-Century Medical Education and the Didactic Model of Experiment', in P. Dear (ed.), *The Literary Structure of Scientific Argument: Historical Studies* (Philadelphia: University of Pennsylvania Press, 1991), 182–94.
150. C. Bazerman, *Shaping Written Kinowledge: The Genre and Activity of the Experimental Article in Science* (Madison: University of Wisconsin Press, 1988), 67–8.
151. *GGA*, 1778, 433–4 (review of the second edition of Haller's *Elementa*). The author is probably Haller himself.
152. B. Lohff, *Die Suche nach der Wissenschaftlichkeit der Physiologie in der Zeit der Romantik* (Stuttgart and New York: Fischer, 1990), Chapter 7.
153. V. Ferrone, *The Intellectual Roots of the Italian Enlightenment: Newtonian Science, Religion, and Politics in the Early Eighteenth Century*, Eng. trans. (Atlantic Highlands, N.J.: Humanities Press, 1995).
154. The debate is described by Pera and, with more attention to the general physiological background, by Bernardi. M. Pera, *The Ambiguous Frog: The Galvani-Volta Controversy on Animal Electricity*, transl. by J. Mandelbaum (Princeton: Princeton University Press, 1992). W. Bernardi, *I Fluidi della Vita: Alle Origni della Controversia sull'Elettricità Animale* (Florence: Olschki, 1992).
155. Dini, *Vita* (note 32), 63–8.
156. See note 30 and L. Brockliss and C. Jones, *The Medical World of Early Modern France* (Oxford: Oxford University Press, 1997), 426.
157. As Andreas-Holger Maehle has shown, experimentation with drugs was not unusual; but the tests were mainly performed *in vitro* or on human subjects and less on animals. See his *Drugs on Trial: Experimental Pharmacology and*

Therapeutic Innovation in the Eighteenth Century (Amsterdam: Rodopi, 1999).

158. *Rapport et Décret de la Convention Nationale sur les Ecoles de Santé de Paris, Montpellier et Strasbourg,* 14 Frimaire, Year III (Paris, 1794), 17–8. Quoted after L. Brockliss, 'Before the Clinic: French Medical Teaching in the Eighteenth Century', in C. Hannaway and A. La Berge (eds), *Constructing Paris Medicine* (Amsterdam: Rodopi, 1998), 71–115: 93.
159. See R. Rey, 'L'École de Santé de Paris sous la Révolution', *Histoire de l'éducation*, 57 (1993), 23–57: 38–41.
160. A survey of animal experimentation in the first decades of the nineteenth century is given by P.W. Lund, *Physiologische Resultate der Vivisectionen Neuerer Zeit: Eine von der Kopenhagener Universität Gekrönte Preisschrift: Aus dem Dänischen Übersetzt* (Copenhagen: Brummer, 1825). For Magendie, see W.R. Albury, 'Experiment and Explanation in the Physiology of Bichat and Magendie', *Studies in History of Biology*, 1 (1977), 47-131.
161. J.E. Lesch, *Science and Medicine in France: The Emergence of Experimental Physiology, 1790–1855* (Cambridge: Harvard University Press, 1984), Chapters 1 and 2.
162. This aspect has been underestimated by E.H. Ackerknecht, *Medicine at the Paris Hospital 1794–1848* (Baltimore: Johns Hopkins University Press, 1967).
163. See Albury, *op. cit.* (note 160) and B. Lohff, 'Johannes Müller und das Physiologische Experiment', in M. Hagner and B. Wahrig-Schmidt (eds), *Johannes Müller und die Philosophie* (Berlin: Akademie-Verlag, 1993), 105–23. For the spread of experimental physiology in the nineteenth century see N.A. Rupke (ed.), *Vivisection in Historical Perspective* (London and New York: Routledge, 1987); W. Coleman and F.L. Holmes (eds), *The Investigative Enterprise: Experimental Physiology in Nineteenth-Century Medicine* (Berkeley: University of California Press, 1988); A. Cunningham and P. Willliams (eds), *The Laboratory Revolution in Medicine* (Cambridge: Cambridge University Press, 1992).

5

Irritability, Sensibility, and Medical Philosophy

In the second half of the eighteenth century it was common to distinguish three general medical systems: mechanism, animism, and a third alternative which considered the basic physiological properties and operations as specific actions of the organic body or of vital forces. This third notion in particular was continuously developed and presented in such manifold ways that its common ground was not a precise concept anymore, but lay rather in its opposition to the other two systems. Haller's contemporaries clearly conceived his position as a refutation of animism but besides that located it in various ways. Johannes Weise, for instance, doctoral candidate in Jena in 1772, defined the three medical systems as mechanism, animism and a third group consisting of the Hallerians and the Dutch school of Gorter, Gaub and their pupils.[1] Apart from some small differences, the members of the third faction would agree perfectly in their descriptions of irritability and vital forces.

Friedrich Casimir Medicus (1736–1808), court physician in Mannheim, drew a somewhat different picture in his oration *On the Vital Force*, held in 1774.[2] In his view, there were two systems that divided the scholars. The first was animism, formerly supported mainly by Stahl and now by Sauvages in Montpellier and Whytt in Edinburgh. The second was mechanism, which had won almost universal recognition thanks to the works of Hoffmann and especially Boerhaave. Their theories were still taught in the Netherlands and Austria. With his concept of irritability, Haller had introduced some changes in this system, and most of the Germans, Italians and also some French and English had adopted his views. Haller's position now was the predominant view of the mechanical philosophers and physicians. Both these systems, animism and mechanism – in either its Boerhaavian or Hallerian form – Medicus rejected, and he proposed a third, allegedly completely different explanation, based on the notion of a vital force that was neither part of the soul nor of the organic body.

Still another version was offered by Ernst Platner (1744–1818), professor of physiology in Leipzig. In his reflections *On Some Difficulties of the Hallerian System* (1781) he maintained that Haller had considered Boerhaave's explanation as too mechanical and Stahl's as too metaphysical.[3] Haller therefore created a new system that stressed the specific character of animal nature and of corporeal properties. According to Platner, Haller's view reigned in the German schools with almost unrestricted power and the whole science of pathology was restructured according to the notion of irritability.

Weise, Medicus, and Platner represent the three common ways of situating Haller in the field of medical theories. They either aligned him with vitalist concepts or described him as a member of the mechanical school or as the father of a new system. Modern scholarship in its careful evaluation has merged these views and admits the validity of all three versions without playing off one against the other.[4] Haller's contemporaries, engaged in an actual debate about medical theory and practice, were less cautious and often interested in providing a particular account of his theories. It is the aim of this chapter to describe their different reactions and the general reception and transformation of the concept of irritability and sensibility in the eighteenth century. An exact survey, however, of the geographical and chronological dimensions of the adoption and appropriation of Haller's ideas cannot be provided. The authors of the eighteenth century disagreed about the spread of 'Hallerianism' themselves. Medicus and Platner considered it as the dominating system in Germany and Italy but Ernst Gottfried Baldinger, a staunch Hallerian and professor of medicine in Göttingen, felt compelled to defend Haller against the criticisms of several German physicians.[5] Only a few dogmas, he remarked in a plaintive tone, were shared by the physicians of this critical century. In Italy, scepticism about Haller's novelties continued to be articulated after the more intense years of debate in the 1750s. Germano Azzoguidi (1740–1814), professor of medicine in Bologna, wrote to Haller in 1773 that several people, especially in Bologna, would refute the theory of irritability and follow Stahl or Boerhaave or argue in the manner of the Dutch vitalist Jan de Gorter.[6] For the establishment of a detailed notion of the spread of Hallerianism, medical dissertations, lecture notes and handbooks should be studied in great numbers.[7] Inevitably, the scope of this chapter has to be more modest. I have tried to identify representative modes of reception, appropriation and rejection and to determine some general shifts in physiology effected or affected by Haller's theories. For such an approach, the traditional arrangement according to the three concepts of mechanism, animism and vitalism is still useful, although the lines between the systems will often turn out to be rather blurred.

Mechanism: innovation and tradition

Even before his treatise on irritability and sensibility, Haller had given rise to a discussion about his position on the mechanical system. In his edition of

Boerhaave's lectures (1739–44) he had dared to add numerous critical remarks that questioned the anatomical and physiological knowledge of his teacher and had presented explanations that contradicted the master's doctrines. Unlike Haller, another famous pupil, Gerard van Swieten (1700–72), physician in Leiden and later professor of medicine in Vienna, followed Boerhaave closely in his edition of the lectures on practical medicine and denied himself any criticism. In 1744, a review of Haller's edition appeared in the *Bibliothèque Raisonnée*, a renowned review journal published in Amsterdam. The anonymous critic – actually Haller himself – stated that Boerhaave had not been well versed in anatomy and in his last twenty years had not followed new developments in medicine closely enough. Haller therefore had to correct and amend many of his master's observations; and he had been able to do so because he followed only the path of nature and did not have – as Boerhaave – 'to adorn hypotheses and to defend a system.' He even disclosed a *bon mot* of a friend of his:

> Mr van Swieten, inseperably attached to his master, has adopted all his systems and hypotheses. Mr Haller, full of veneration for the same master, admits however only those which he considers right, and he opposes – although with respect – to the smallest brilliant error which could blind him. ...Mr van Swieten comments in a Catholic and Mr Haller in a Protestant manner.[8]

Haller heightened the contrast between the two different approaches, just as four years later in his dispute with Hamberger he stressed the difference between his own experimental and the latter's mathematical method. Quite deliberately he set himself in opposition to the Boerhaavians of strict adherence. His review created quite a stir. According to van Swieten, the description of Boerhaave as an inexperienced anatomist and inventor of hypotheses was received with unanimous indignation. A refutation appeared in the *Journal Britannique*, which in its turn was attacked by Haller in the *Bibliothèque Raisonnée*. The Göttingen professor had thus already gained the reputation of a critical, if not harmful, heir of the Boerhaavian legacy.[9]

Haller's orations on irritability and sensibility were considered by several physicians to be a further blow to Boerhaave's model of mechanism. Only a few, however, openly described their refutation of Haller as a defence of Boerhaavianism. The Milanese physician Giambattista Fè, for instance, entitled his essay *Saggio Critico... in Difesa di Ermanno Boerhaave*.[10] Anton de Haen, the famous Viennese clinician, complained that Hippocrates' and Boerhaave's untouchable principles of therapy were replaced with some simple notions of increased or decreased irritability.[11] De Haen's argument shows that it was not necessarily Haller's own description of the properties that prompted criticism but often rather the consequences which might result from this. Tissot's preface – which preceded all the major editions of Haller's text – indeed drew conclusions that called for a re-alignment of theoretical and practical medicine. For Tissot, irritability as the basis of muscular contraction and thus of circulation was the key to the explanation of the theory of fever, inflammation and many other diseases. Although he extended the realms of irritability beyond the limits set by Haller, his ideas were still based on the notion of a specific corporeal property responsible for movement and independent of the soul. It was this notion that was repudiated by strict mechanists and was conceived as the weak foundation of a misled system; they seem not to have realised that such a concept was not necessarily a break with Boerhaavian tradition and could also be seen as a development of the non-reductionist elements of the late ideas of the Leiden master. In the exact manner of their refutation they differed. De Haen, for one, accepted Haller's distinction of irritability and sensibility and he even seems to have accepted the notion of irritability as independent of the nerves.[12] But he considered it as a kind of mechanical force that was not able to produce the motions of the body alone. There was something else responsible for life and motion, something much stronger than irritability. It acted in all parts of the body, but its nature was unintelligible to man.[13] One could gain some general notion of the vital functions as they were performed according to mechanical and hydraulic laws, but no more could be known.[14] Haller's project to explain the resuscitation of drowned persons through irritation, and thus to conceive irritability as a prerequisite for life, was dangerous, as these were the ideas upon which La Mettrie had established his materialistic theory.[15] Another quite frequent and simple mechanistic argument against Haller's concept was put forward by the Bolognese medical professor, Tommaso Laghi. As all the irritable parts of the body were furnished with nerves, one had to infer that even in extracted muscles a certain amount of nervous fluid remained active and caused the contraction.[16] Fè, de Haen and Laghi represent the faction of rigid mechanists who upheld a strict dualist viewpoint.[17] The phenomena of irritability, post-mortem movements and movements displayed in excised parts were considered as purely mechanical processes, but the action and motion of the living body in its entirety depended on the presence of the soul. The category of specific physiological innate properties had no place in this scheme. The important role of the soul was acknowledged and was always at the back of their minds, but it was not stressed since its nature and actions were beyond the reach of research.

Other mechanists, by contrast, softened Cartesian dualism and emphasised the union of body and soul. This was the position of Claude-

Nicolas Le Cat, the well-known surgeon from Rouen.[18] Like most mechanists, he considered the nervous fluid to be the physical cause of muscular contraction. But in his view this fluid was not simply a material liquor but 'a mediatory substance between the soul and the body'. It was matter but 'affected by its creator with a superior nuance which connects it with the immaterial Being'. The nourishing juice was physically linked to the animal spirit, which in its turn in all its parts was united by God with the original seat of sensation and movement, the soul.[19] Le Cat thus favoured a kind of occasionalism postulating that God guaranteed the continuous unification of body and soul. Muscular contraction was caused by an 'expansive movement' of the nervous fluid which was triggered by the soul.[20] This mechanism also worked in a muscle detached from the body, as long as it contained some nervous juice. Thus, the soul was active in all different separated parts. As an immaterial substance, though, it had no extension and was not *present* in these parts although it *acted* through the animated nervous fluid, which Le Cat also called 'âme sensitive'.[21] His conception of physiological processes as either entirely mechanical or as co-directed by the soul did not allow for the notion of active corporeal properties. He could conceive Haller's irritability only as the description of 'an entirely mechanical effect of the prick'.[22] For Le Cat, instead, the reaction of the muscular fibre to the irritation included sensation and therefore the action of the soul. In his view, Haller had created a metaphysical, incomprehensible distinction between the irritable and sensible, he had postulated irritability without sensation and thus without irritation.[23]

Le Cat had to stress the influence of the soul in order to repudiate a purely mechanical explanation of irritability, which in his opinion tended towards materialism. As a result, he came close to the position of the animists. To what extent his physiological views were shared by others is difficult to say. Le Cat was a celebrated figure and his writings were well-known. His essay on muscular movement was awarded the first prize by the Academy of Berlin in 1753. But in the German countries it was rather critically received and many of his theories were considered as hypothetical and somewhat odd.[24] The French press generally extolled his works but his particular conception of the nervous fluid seems not to have gained many followers. Of greater importance was his refutation of Haller on the experimental level (see Chapter 4), which made him one of Haller's most important opponents in France.

Apart from those who criticised Haller from a Boerhaavian or occasionalist position, there was a wide variety of mechanists who welcomed his research. Their common ground was the description of irritability and sensibility as separate specific corporeal properties. We may distinguish three groups of these non-reductionist mechanists. First, there were those faithful adherents who fully adopted Haller's views. For the most part these were authors who joined the debate actively and defended his position openly, either by experiments or by arguments. These were chiefly 'minor' authors like the physicians Housset from Auxerre, Brocklesby from London, Hirzel from Zürich and Sarcone from Naples. Few of these rigid Hallerians wrote systematic physiological works or even textbooks, exceptions being Johann Peter Eberhard (1757) and Germano Azzoguidi (1775).[25] Only one of them gained a European reputation as a physiologist: Marcantonio Caldani. He was Haller's most important and most faithful disciple and the head of the Hallerians in Italy. In several publications on irritability and sensibility and with his physiological textbook, he supported and spread the views of the famous physiologist.[26]

The second group we might call the 'semi-Hallerians'. They were closer to Haller than those who held more rigid mechanical concepts but still disagreed with him on various accounts. The majority of these authors were Germans. The adaptation of the new theory into the traditional scheme of mechanism is best illustrated by the writings of one German medical professor, Andreas Elias Büchner (1701–69).[27] Büchner was a student of Friedrich Hoffmann and, after his appointment as medical professor at the University of Halle, published a textbook of physiology, based on his master's works (1746–8). As president of the *Leopoldina* (1735), editor of various periodicals, and acclaimed physician, he had established a considerable reputation. He published few of his own works but presided over 300 dissertations, and four of these deal explicitly with the subject of irritability and sensibility. Karl Wilhelm Schultz's thesis of 1755 is still embedded in the vocabulary of traditional mechanism.[28] Although it is essentially a description of Haller's doctrine, Schultz tried to avoid the notion of active corporeal faculties and thus did not talk of irritability and sensibility but of irritation and sensation. The ability of the muscle to contract upon irritation was due to its special 'disposition', its 'flexibility', ie. 'a certain mechanical structure of the solids' that allowed the parts of the fibres to move together. Haller's description of irritability as a quality particular to animated bodies is, in this dissertation, reduced to a mere arrangement of physical parts. Gabriel Valentin Köhler in 1763 still described the body as a machine, but as a machine moved by certain powers.[29] He accepted Haller's distinction of elasticity, *vis nervosa,* and irritability and also the definition of the latter as an 'innate force' (*vis insita*). Irritability, in his view, was the most powerful of all bodily forces. Köhler agreed with Haller that the presence of gluten was a necessary prerequisite for irritability but he did not consider it as a sufficient cause. In his view, irritability depended on the cohesion of the glutinous and earthen parts of the muscular fibre. An excess of cohesion impeded the movement of the fibre and thus destroyed its irritability. Increase of cohesion, and thus decrease of irritability, was especially caused by solidification, lack of humidity and reduction of the interstices in the fibres, which could be effected by the application of astringents. Two years after Köhler, the student Johann August Pakendorf defended his thesis *On the General Laws of Irritability*.[30] Like Haller, he described irritability as a *vis insita*, as a 'merely corporeal faculty', but he could not agree with Haller's reservations about the term *vis vitalis*.[31] Haller's argument that irritability lasted longer than life was, in his view, not convincing. A force could be called 'vital' if it helped to produce and maintain life to an eminent degree. And this certainly applied to irritability. As to the material cause of this faculty, he added a further element to Köhler's concept of cohesion: the nervous juice. Irritability was thus caused by the gluten in the muscular fibre, its grade of cohesion and its connection with the nervous liquid. After these general remarks, Pakendorf went on to define twelve general laws of irritability. Among other things, he maintained that a strong irritation excites sensibility and a strong sensation stimulates irritability. The two faculties had to be distinguished and could each endure without the presence of the other, but a certain interdependence was caused by the connection between motor and sensory nerves. A violent contraction of the muscles led to a pull on the motor and, consecutively, also on the sensory nerves. An intense sensation, on the other hand, implied a vehement action of the sensory and thus also of the motor nerves. Another law stipulated that the irritability of a person increases and decreases according to his or her sensibility. This was due to the fact that sensibility as well as irritability depended on the abundance of nervous liquid. In 1769 another student of Büchner, Gottlieb Christian Busse, published a dissertation on the topic.[32] He modified the notion of cohesion as introduced by Köhler. It was a general force present in all parts of animated bodies and thus not a characteristic of irritability. But he agreed with Pakendorf on the importance of the nervous liquid. Irritability was due to the mixture of gluten and the nervous juice in the muscular fibre. Thus it depended on the presence of nerves. Although this flatly contradicted Haller's theory, Busse supported his views with quotes from Haller's works, which he cited incessantly. The Göttingen professor had himself stated that the power of the heart was to some extent determined by its nerves. Busse quoted Haller quite correctly but he drew conclusions that went beyond the latter's cautious interpretation. Haller had accepted a modulatory effect of the nerves on the irritability of the heart, but he had not – as Busse suggested – declared the nerves as a prerequisite of irritability.

The dissertations directed by Büchner represent an appropriation of Haller's theory that seems to have been quite common among mechanists, especially in Germany. The discovery of irritability as a distinct property and its separation from sensibility was accepted at once, but it required a certain time until the notion of an active quality could be embraced as a principle of mechanist thought. From the 1760s, though, the description of irritability as a force, and even as a vital force, was familiar among this group of authors. Nevertheless, the Cartesian model of the hydraulic machine and the concept of contraction as a result of nervous influx still affected all ideas on muscular movement. Irritability thus could not be conceived as entirely independent of the nervous liquid. This is the position characteristic of non-reductionist mechanists such as Büchner, and it may be found in many variations in other publications as well.[33]

The third group of pro-Hallerian 'mechanists' presented ideas that tended towards vitalism. They extended either the realm of irritability or of sensibility and considered it as a basic principle of life. They may be described as Hallerians insofar as they regarded themselves as supporters of Haller and – with the exception of Zimmermann – accepted his distinction between irritability and sensibility. Less justified is the label 'mechanism' and it is purely for practical reasons that they are treated here together with the other Hallerians. With equal right they might be called 'vitalists' or 'semi-vitalists'. Their conjectures did not emerge from Haller's late, partly vitalistic ideas – which seem not have been widely recognised – but are rooted in Zimmermann's dissertation of 1751 and were supported by the interpretative vagueness of Haller's orations of 1752. As shown in Chapter 2, Zimmermann had not properly separated sensibility from irritability and had considered the latter quality as the principle responsible for life. But even after Haller's publication he continued to maintain his position. In a manuscript of the late 1750s he talked of the irritability of the nervous system and illustrated how the irritability of the whole body – and especially of the sexual organs – acted upon the soul in conjunction with sensibility and thus dominated not only the corporeal but also the mental operations.[34] Zimmermann's dissertation was the first detailed description of irritability from the Göttingen school and was thus considered as Haller's work. It continued to affect the reception of the teacher's theory, even though it contradicted it on several essential points. It was also the text through which Auguste Tissot took notice of the topic. In May 1754, when he finally received a copy of Haller's treatise, Zimmermann's ideas seem already to have been firmly fixed in his mind.[35] A former Montpellier student, Tissot seems to have been particularly receptive to a vitalist reading of Haller's theory. He continued to describe Zimmermann's dissertation as an important description of the new concept and even as late as 1797 stated that it contained the whole doctrine of irritability.[36] As described in the last chapter, in his preface to Haller's treatise he considered irritability as 'the sum total of the powers of the vital principle'. He accepted the distinction of irritability and sensibility but, in contrast to Haller, detached the notion of sensibility from conscious perception. Only man was able, he argued, to transform sensations into ideas; in animals, sensations were restricted to the mechanical corporeal realm.[37] Haller had not been very clear on this subject. He had called that part sensible 'which upon being touched transmits the impression of it to the soul'. But 'in brutes, in whom the existence of a soul is not so clear', he called 'those parts sensible, the Irritation of which occasions evident signs of pain and disquiet in the animal.'[38] Despite the undecided nature of the animal experiments as to the consciousness of the process, Haller – considering his research as a contribution to human physiology – could never understand the notion of unconscious sensation. Sensibility was thus linked to the soul and had all the more to be distinguished from the local property of irritability. For Tissot, in contrast, one could allow for the possibility that irritability might direct the sensations, as neither of these two processes involved the action of the soul.[39] Although he did not develop this thought in detail, his preface conveyed the picture of a unified bodily realm detached from the soul and governed by a single vital principle. The notion of a universal property of animated bodies was stressed by the analogy between Newton's force of attraction and Haller's irritability, which Tissot drew repeatedly in his essay. Just as celestial mechanics were ruled by the laws of attraction, the whole of animal mechanics was directed by the laws of irritability.[40] This extension of the realm of irritability upon the Newtonian model was typical for vitalist interpretations of Haller's concept.

Tissot's preface was particularly important because it was included in Haller's own French edition of his treatise, as well as in most other editions, and thus enjoyed a kind of official status as an interpretation accepted by the master himself. Tissot had stressed the explanatory power of the concept of irritability and the consequences to be drawn for medical therapy. Several others adopted a similar course. The Italians Giovanni Vincenzo Petrini and Giovanni Francesco Cigna followed Tissot quite closely.[41] Later authors presented variations, still within the Hallerian distinction of the properties of irritability and sensibility. Johannes Weise, for instance, the doctoral candidate from Jena mentioned above, stated that these were the only two vital properties of the body.[42] They had to be distinguished although they affected one another reciprocally. As diseases were caused by an excess or lack of motion, they had to be explained by the force of irritability, the motor of life. Félix Vicq d'Azyr, on the other hand, the famous French physician, envisaged a 'tonic movement' caused by a reciprocal influence of irritability and sensibility on the organs and a 'nervous turgescence' proportional to the 'vital energy', which increases and decreases according to diseases. This turgescence was nothing but an extension of irritability, which Haller had kept within too rigid boundaries.[43] In Vicq d'Azyr's concept irritability and sensibility assumed a special position as the 'principal modifications' of the 'vital principle', but they were only two of the nine general properties of life, the others being digestion, nutrition, circulation, respiration, secretion, ossification and generation.[44] Although Vicq d'Azyr objected strongly to vitalism as a philosophy, he adopted a vocabulary that made irritability and sensibility look like properties unexplainable in mechanical terms.

The concepts of the 'vitalist' or 'semi-vitalist' Hallerians from Tissot to Vicq d'Azyr display a latent reductionism. Weise stressed the mutual effect that irritability and sensibility had on each other, and Vicq d'Azyr described both properties as modifications of a single vital principle. There were two positions to which reductionism might lead. Both of them partly emerged in Tissot's text, but both were also developed prior to and independently of Haller. The first was a generalisation of irritability, which definitely acknowledged the supremacy of this property and degraded sensibility to a subordinate or secondary quality. Both Tissot and Weise were very close to such a position, which was mainly supported by the Dutch vitalists. They were – like Haller himself – heirs to the Boerhaave–Albinian tradition and thus considered movement as the basis of life. The second position stressed the pre-eminence of sensibility or sensation, which was conceived – in contrast to Haller – as an active and vital quality. Irritability instead was considered as a manifestation of sensibility. Such a view could only be endorsed if the notion of unconscious sensibility was accepted. In particular, Théophile Bordeu (1722–76) and the French vitalists promoted this idea. Felice Fontana was the most prominent Hallerian to adopt a similar course, in his later years. Less faithful to Haller than Caldani, he developed his own views and presented several laws of irritability (1767, 1775) that defined the domain of this property with more precision and clarity than Haller had done.[45] These contributions did not yet touch the fundamental doctrines of Hallerian irritability. Only in 1781 did he posit a 'sentiment obscur' which lasted in the limbs even after their separation from the rest of the body and the brain, and which might be responsible for their irritability.[46] Finally, in 1792, in light of Galvani's ground-breaking publication, he argued that it was conscious sensibility that normally controlled the muscular contraction. In Galvani's experiments with decapitated frogs, however, an unconscious sensibility of the local organs themselves continued to regulate the process.[47]

Sensibility had replaced irritability as the guiding principle of animal motion.

Animism: a system apart

At the end of his first important anti-mechanical work, the *Recherches Anatomiques sur la Position des Glandes et sur Leur Action* (1751), Théophile Bordeu drew a short personal portrait of Stahl and an account of the adoption of his views in Montpellier.[48] The work of the Montpellerian animists – he meant Sauvages and his pupils – had the merit of having clearly shown that mechanism could not be upheld in its traditional form as represented by Boerhaave. They had made laborious calculations in order to demonstrate that the human body could not work in the manner of an artificial machine. Their system was not Stahlianism anymore but rather an adoption of British animist ideas. Nevertheless, they still exaggerated the power of the soul and Bordeu, who in his book introduced some new vitalist ideas, had chosen not to follow them.

Bordeu's account points to the main elements of animism in the mid-eighteenth century. It deviated from Stahl's position of the beginning of the century. New forms of animism were elaborated especially in England and France, the two dominating figures being Robert Whytt and François Boissier de Sauvages, who both departed from their originally mechanistic viewpoints in the late 1730s. Authors of a similar standing did not emerge in Italy and the German countries. We shall nevertheless discuss the position of Heinrich Friedrich Delius, one of the few German animists of the second half of the century, in order to describe different animist responses to Haller's doctrines. All three authors, Whytt, Sauvages and Delius, dissociated themselves from Stahl. Sauvages stated that Stahl had 'distinguished himself among the animists only by exaggerated, barely probable opinions' and Whytt equally talked of the 'extravagant flights of Stahl and his followers'.[49] Although they acknowledged his importance for the development of alternatives to the dominant mechanist concept, the younger animists had an ambivalent attitude towards the great master and explicitly refused to be called Stahlians.[50] They did not agree with his scepticism about anatomical and physiological research and considered his notion of a 'wise soul' controlling all corporeal actions as too severe an exclusion of any other than a metaphysical approach to the body. Delius remarked that the Stahlians have 'taken too much care of the effects of soul and too little of the effects of the body'.[51] Stahl had driven his system too far and, according to Whytt, this had been the reason why the question of the influence of the soul on the body had 'rather for many years... been looked upon as a subject of ridicule, than deserving a serious and rational answer.'[52]

The animism of the mid-eighteenth century had departed from Stahl's position as much as Haller had diverged from Boerhaavian mechanism. To some extent this development encouraged a rapprochement between the two opposed systems. But new interpretations on both sides still provoked immediate refutation. The two systems were not reconcilable and the animist-mechanist controversy was never resolved. As Roger French has remarked, it was gradually displaced by the debate about various forms of vitalism or non-reductionist mechanism.[53] Despite the innovations of Sauvages and Whytt, animism never gained a weight in medical discourse equal to that of the other two systems. It nevertheless stimulated theoretical discussion, especially through its categorical rejection of mechanism. The following sections on Sauvages, Whytt and Delius present various aspects upon which no agreement was possible. The three animists all shared the conviction that matter is inert and motion thus has to be generated by an immaterial principle. But in the selection of arguments and the emphasis put on various aspects they varied greatly and, indeed, the whole outlook of their theories differed substantially.

François Boissier de Sauvages (1706–67)

Sauvages's dissatisfaction with a purely mechanical account of the motion of the body originated from his own application of mechanical and physical laws to physiology.[54] In a dissertation of 1740 he calculated the power exerted in the heart and in the flow of the blood and argued that circulation caused a loss of energy by friction.[55] This approach seems to have been particularly encouraged through reading Hales's *Haemastaticks* (1733) in which the author had calculated that the blood loses ninety per cent of its motive energy during circulation. Sauvages argued that a constant generation of power was needed, which could not be explained by an 'automatic', mechanical force but only by a 'motive potency', ie. nature as the power or the faculty of the soul. Julian Martin has shown that Sauvages's exposition is indebted to Newton and follows quite closely the argumentation of the *Principia*.[56] His anti-mechanism was directed against Descartes but not against Newton. His description of the 'automatic force' corresponded to Sir Isaac's *vis inertiae* and the 'motive potency' paralleled the latter's 'impressed force', which acts upon a body and sets it in motion. Martin thus argues that Sauvages could with no more right be called a 'vitalist' – or, more accurately, we should say 'animist' – than Newton or Pitcairne. Such reasoning is, in my view, hardly convincing. Recurrence to Newton was ubiquitous in medical literature of this time and Newtonianism or parallels to Newtonian thoughts and forces did not exclude a vitalist or animist position. Sauvages's invocation of the soul was not just a heuristic device, which could be replaced by Newtonian powers. The soul exerted its power thanks to motive potencies received from God. But the soul itself was free and all its operations were active. Sauvages defended the notion that the soul actually wants (*velle*) and performs (*operari*) even those motions of which it seems not to be aware, such as the action of the heart and intestines. He argued that the soul out of habit prescribes itself certain laws according to which it acts and which even the use of reason cannot violate.[57] It is evident that Sauvages, although he rejected the idea that the soul consciously performs all its duties, was very much indebted to Stahl and may without reservation be called an animist. That he did not envisage the soul as an undefined principle that could stand for other non-mechanical powers, is manifested even more clearly in his later writings and especially in his correspondence with Haller.

In late 1751, upon the reception of Zimmermann's dissertation, Sauvages explained his position regarding the new phenomena of irritability in a letter to Haller. In the dissertation of 1740 he had asked whether the motion of a heart detached from the body might not be due to a 'vital principle which dwells in the whole body and its principle parts'.[58] But, he had answered, it seemed more probable that the motion was caused by the heart's imperfect elasticity, which led to an incessant decline of the muscle's activity. Such a motion would in no way demonstrate that the heart was a machine (*automatum*), ie. capable of continuous self-movement. In 1751 he still admitted this as a possible explanation, but Zimmermann's descriptions of long-lasting and intense movements of isolated muscles seem to have shaken his conviction. In a similar manner to Le Cat, he argued that the soul had to be responsible for these movements:

> Neither religion nor psychology prevent us from thinking that the soul, which is neither here nor there, acts on a limb separated from the body as long as it is able to receive impressions from it. Saint Augustin, at least, says clearly that the soul of the lizard acts on the tail separated from the body. And in such profound obscurities I am always glad to be on the side of such a great man. If animals have a soul – which I don't doubt they have – then the polyp which lives separated from its father, lives just as the child lives separated from the mother. For me, it is sufficient that there is no contradiction here. As to clarity, I don't expect any.[59]

This is the view that Sauvages maintained also in his later works. In his letter to Haller he disclosed the religious background of his position in the debate on the power of the soul:

> I think even that this last dispute can not be settled between people of different religions. After having well thought about it, I believe that religion

> is a principle we have to start from. I know, Sir, that you have a lot of it, and we can not respect such an eminent virtue too much. But I have to say that those who have none – which is the majority – are by that authorised not to believe in the power of the soul over the body. And they are mostly materialists, fatalists, or Spinozists who believe that the soul has nothing more than the faculty to perceive and not the liberty or faculty to move.[59]

In a letter from 1753, Sauvages stressed again that religious differences had a great influence on the discussion. According to his religion he believed that the soul was the 'principle of life', but those who were not constrained to follow the decisions of the Catholic Church might well attribute not only the power of motion but also the faculty of sensation to the human machine. As certain authors had shown that the functions of the soul could be reduced to sensation, nothing would prevent the machine from performing all the functions which Sauvages ascribed to the soul. He agreed with Delius that the *machinistes* provided grist to the mill of La Mettrie. He concluded:

> God forbid me from attributing these errors to all those who explain the vital functions of man mechanically. I say only that we have to consider the consequences. These errors flow from their principles or their principles support these errors. Now, if they win, goodbye religion.[60]

Haller's reply to Sauvages's letter is not preserved. But it is not difficult to guess what he answered. It was not his concept but animist theories that paved the way for materialist interpretations. We will encounter the details of his argumentation in his debate with Delius where he developed them. As to the more general role of religion, Haller responded in 1762, answering a letter of Sauvages's, which, in its turn, is not preserved. Did Sauvages really believe that religion entered into the debate on the motive powers of the soul, Haller rhetorically asked. Was Stahlianism not born among Lutherans? Did it not have adherents in England? Had Italy produced many Stahlians? Haller did not believe that the bible (*la revelation*) taught of the spirituality of the soul. It simply told us that the self, capable of feeling and memory, survives the dead body. The self could be a flame, as the ancients thought, without contradicting the Bible. Haller had observed differences between the operations of the soul and the body and thus considered them to belong to two different classes. He added:

> I would agree, if you want, that your opinion seems to serve religion better than ours. But we are so blind! We grasp the chain from universal principles to remote consequences so badly! Should we make the majestic revealed truth depend in some manner on a phenomenon when the phenomena resist to our hypothesis even in the most simple sense?[61]

Haller certainly agreed that neither a phenomenon nor a correct theory could contradict religion. Like Sauvages he excluded certain explanations – notably the materialism of La Mettrie – as theologically and thus also scientifically impossible. But he did not agree with Sauvages' assertion that religion should provide the first principles. This contradicted his physico–theological belief. It was not religious doctrines that revealed the wonders of nature, but the study of nature that revealed the wonders which God had created. Whereas for Haller the confrontation between religion and science may disclose imprecise religious preconceptions, for Sauvages it exposes misled scientific theories. Although in practice these two modes of approach presumably did not differ as widely as one might think, there was still a difference in emphasis.

We have to agree with Haller that we cannot simply align different physiological theories with Protestant or Catholic beliefs. At least, there is no direct correspondence between the spreading of one denomination and a specific physiological system. On the personal level, however, religious beliefs certainly played a part, and Sauvages is a good case for that. Also, in his publications he repeatedly stated that his views were in conformity with faith and the Holy Fathers of the Church but he never presented his religious motives as openly as in his letters to Haller.

Since Sauvages defined the soul as the vital principle he could not conceive of human life as independent of it. Equally he could not imagine that matter is irritable or sensible. Irritation implied a disturbance perceived by a sensible being which reacted accordingly. It was against the common use of language to talk of an irritation of inanimate bodies.[62] Sauvages mentioned the experiments of Whytt in order to argue his case. The Edinburgh professor had observed that a decapitated frog invariably retracts his legs whatever part of it is irritated, be it the thigh, the foot or even the muscles responsible for extension. Whytt had argued that this could not be explained by a simple contraction of an irritated muscle but only by a reaction to an uneasy feeling which was perceived by a sentient principle. Sauvages argued similarly that this could only be understood when irritability was sensibility collaborating with the muscular motive faculty.[63] Sensibility, in this context, did not imply consciousness, but had to be relegated to a lower level of confused perceptions which nevertheless included the action of the soul. But it would be wrong, Sauvages said, to take things to the extreme. Although the soul played its part in the process of muscular contraction, blood circulation, and respiration, there were four other basic operations which were directed purely by mechanical forces. Secretion, digestion, nutrition and generation were functions exercised also by plants and thus could not be subjected to the power of the soul.[64] They had to be explained by the forces of gravity, cohesion, attraction, and elasticity and the effects depending thereupon, viz electricity, putrefaction, fermentation, heat, rarefaction, dissolution, condensation, etc.[65] Despite his animist conviction Sauvages adhered to a strictly mechanist description and calculation of many corporeal processes. This is what distinguished him considerably from Stahl. Although indebted to the great systematist from Halle, Sauvages's champion was, in fact, Borelli.[66] We have encountered Borelli in the first chapter and I would like to recall that his animism seems to have its origins in the very same difficulties as Sauvages's. His calculations of the flow of the blood and his Cartesian belief in the inertia of matter drove him to the conviction that the power of the heart had to be generated by an immaterial cause, the soul. Sauvages had drawn the same conclusions already in 1740 and his position was not significantly altered after studying the works of Haller and Whytt.[67] Although in a letter to Whytt he declared himself 'one of your admirers and a zealous partisan of your opinion', his religious motivation and his notion of the soul as an active, deliberately choosing principle place him in marked contrast to the Edinburgh professor.[68]

Robert Whytt (1714–66)

Whytt's main physiological contribution was the *Essay on the Vital and Other Involuntary Motions of Animals* of 1751, a work of which Haller commented that 'nothing more accurate and more valuable has hitherto been said' in favour of animism.[69] Quite in accord with this judgement, Whytt was in his time and is still nowadays considered as the most important and most original of Haller's animist opponents. He has therefore also received considerable attention.[70] A detailed exposition of his thoughts in this place is neither possible nor necessary, but some points should be highlighted in order to explain the singularity of his position.

As Whytt himself reported, in the late 1730s he began to be dissatisfied with common, ie. mechanical theories of respiration and the motion of the heart.[71] The origins of this dissatisfaction must be searched for in his early student years of 1730–1, when he followed the lectures of George Young (1692–1757) in Edinburgh, who doubted the exclusively mechanical character of muscular contraction and especially of all operations in which nerves and immaterial spirits were involved. In early 1736, Whytt briefly visited Leiden, where he attended the lectures of Albinus and Boerhaave. Here, he might have heard some of their non-reductionist views, which we have encountered in the first chapter. A further source that fostered Whytt's animist realignment was – as in the case of Sauvages – Hales's *Haemastaticks* (1733), which touched a sore point of mechanism: the loss of energy due to friction.[72] Also like Sauvages, Whytt rejected the notion of active powers inherent in matter. But, in contrast to the French author, he argued less from a religious and more from a philosophical or physical point of view. To Haller's early notion (1743) that some latent power or property seemed to reside in the muscular fibres, he answered that this was no explanation but rather 'a refuge of ignorance' and that 'besides it must appear greatly unphilosophical to attribute active powers to what, however modified or arranged, is yet no more than a system of mere matter.'[73] Obviously Whytt defended the philosophy of Newton, which considered forces as principles superadded to matter.[74] The notion that in the case of the human body this principle had to be the soul, was clearly stated in 1737 by William Porterfield (c. 1696–1771), Whytt's colleague in the Medical Faculty in Edinburgh and, besides Whytt, probably the most important English animist of this period.[75]

Inspired by all these sources, Whytt still departed markedly from them, presenting a new, original theory which he corroborated by observations and experiments. He rejected the typical Stahlian argument, also used by Sauvages and Porterfield, that vital motions had initially been performed voluntarily but by habit had been rendered independent of the will. Whytt agreed that we may have become accustomed to the continuous irritation of the vital organs and thus do not feel the stimuli. But this was also due to the gentleness of stimulation. Strong irritations of these organs, as in the case of purgation, were exquisitely sensed. As to the motions which followed the irritations, these had always been performed unconsciously, the reaction was due to the 'original constitution' of body and mind. Nevertheless, they were governed by the soul, which was not divided but performed its duties in two different capacities, as the rational, and as the sentient soul or, as Whytt usually called it, the 'sentient principle'. This principle perceived, albeit unconsciously, the irritations to which it responded accordingly. As such, the soul acted not freely but as a 'necessary agent':[76]

> The mind, therefore, in carrying on the vital and other involuntary motions, does not act as a rational, but as a sentient principle; which, without reasoning upon the matter, is as necessarily determined by an ungrateful sensation or *stimulus* affecting the organs, to exert its power, in bringing about these motions, as is a balance, while, from mechanical laws, it preponderates to that side where the greatest weight prevails.[77]

Vital motions, in this theory, are as rigidly determined as in any mechanical system, and Roger French therefore has quite aptly described Whytt's explanation as a 'quasi-mechanism'.[78] It is not the soul, as in the case of Sauvages and other animists, which imposes its laws upon itself, but the laws are given, and ultimately prescribed by God. One might ask to what extent this is still an animist position, and, indeed, Porterfield objected that there is no point in saying that motions are 'caused by the Mind' when in fact they are 'caused by a Law'.[79] But in Whytt's view law-like action did not exclude their being governed by the soul. Simple mechanical models could not explain why an irritation could cause contractions in remote muscles to which it was not applied, why very different stimuli caused the same reactions, and why the permanence of a stimulus did not result in an enduring constriction but in alternate contractions.[80] Whytt's most convincing experiment was the one which we have encountered in Sauvages' argumentations. A decollated frog invariably retracts its legs whatever part of it is irritated. Neither a mechanical model nor the concept of 'sympathy' could account for the fact that only a certain group of muscles reacted with contraction.[81] In his *Observations* of 1755, which were explicitly directed against Haller, Whytt extended the argument, adding the observation that an irritation of the toe was followed by a more vehement retraction of the leg than an irritation of the thigh. This happened simply because the toe was more sensible than the muscles of the thigh.[82] Reaction upon irritation involved sensation and thus the 'sentient principle'. Another important experiment pointed in the same direction. After decollation a frog remains almost motionless and regains his irritability only ten to fifteen minutes later. This observation served Whytt as a confirmation of the Hippocratic dictum that a greater pain destroys, to a considerable degree, the perception of a lesser one.[83] The experiment initially troubled Haller, as it seemed to prove the dependence of irritability on sensibility. He asked Caldani to repeat it, but although the Italian physician got more or less the same results as Whytt, Haller concluded that the experiment was uncertain and did not touch his own theory.[84]

Although the main agent in life for Whytt was an immaterial principle, he was still at pains to demonstrate the material level of vital actions. He agreed with Haller that only the muscular fibre was irritable, ie. capable of movement. All other parts were by their structure not fitted for motion. But although the muscle was necessary it was not sufficient to generate motion. Irritability always implied sensibility and it was always proportional to the sensibility of the muscle.[85] It thus depended upon the presence of nervous spirits which, in the case of dissected muscles, were still available for a certain time. This nervous juice or power was indispensable as it was the material agent upon which the 'sentient principle' could act. The soul was thus present in the whole body. In the brain it exerted the power of conscious sensation ('reflex consciousness') and reason, and in the those parts furnished with nerves it was only capable of feeling or simple sensation.[86] The soul or, more precisely, the sentient principle, was united with the different parts of the body, and gave them motion.[87] Even a dissected heart was therefore animated and alive.[88]

Heinrich Friedrich Delius (1720–91)

Unlike Sauvages and Whytt, Delius stemmed directly from the Stahlian school.[89] As a student of Johann Juncker (1679–1759) in Halle, he had adopted the animist position from the outset of his medical career, although not in its strictest form. In 1749 he was appointed professor of medicine at the University of Erlangen, where his research was primarily concerned with practical medicine and the chemical analysis of medicinal plants and spring water. The University itself was rather poorly equipped and although a considerable number of medical students matriculated there – between ten and twenty each year – it did not have a standing comparable to that of Göttingen or Halle. Nevertheless, Delius gained a certain reputation as a medical author and as editor of a well-known journal on medicine and natural history.[90] His election as president of the *Leopoldina* in 1788 reflects the general esteem he enjoyed.

Already before Haller's orations, Delius had published a small work on the topic of motion and sensation.[91] In this, he argued that each sensation caused a change in the body and that no change happened without a motion. Sensation thus induced motion and, using the vocabulary of mechanics, Delius asserted that the motion had to be proportional to the sensation. As the movement often could not be controlled by our will, the principle of proportionality had to be considered as a divine law, the regulations of which we could not wholly understand. Although the soul did not consciously realise all the changes happening in the body, it still envisaged (*sich vorstellen*) them. The movement of the extracted heart was a purely mechanical phenomenon and was not governed by the soul as there could be no purpose in such an action. Zimmermann's dissertation – which he considered to be a work of Haller – he regarded as dangerous, and immediately upon its publication he published a rebuttal in which he stressed that motion was not only proportional but also congruous to sensation.[92] Sensation always included a 'representation' in the mind and although the process was based on the principle of proportionality it could not be explained purely mechanically. The attribution of all movements to the faculty of irritability neglected the power of the soul. With great surprise, Delius stated, he had discovered that Haller's claims were very close to a confirmation of the theory of the *Homme Machine*. At this point it makes sense to insert a digression on La Mettrie.

Digression: La Mettrie

A lot has been written about the controversy between Haller and Julien Offray de La Mettrie (1709–51).[93] This was much more than a personal quarrel between two members of the Republic of Letters. It represented the clash of two opposite types of personality, philosophy and science: La Mettrie, the witty and radical philosopher against Haller, the pious and industrious scientist. The single events and the behaviour and motives of the protagonists cannot be discussed in detail here. We have, instead, to focus on the physiological aspects of the debate. The quarrel had its origin in La Mettrie's French edition of Boerhaave's lectures on physiology, published in 1743–50.[94] This was, in fact, nothing more than a translation of Haller's own edition and commentaries, adorned with a few further remarks from La Mettrie.[95] Although the French physician admitted to having included 'the excellent notes, dressed in French manner [*habillées à la française*]', he gave the impression that much of the work was his own. For Haller, who was always keen on stressing his own contributions, this was plagiarism. The case was aggravated with La Mettrie's *Histoire Naturelle de l'Âme* from 1745 which, again, borrowed heavily from Haller's commentaries without mentioning the source. But more serious than the copying of the notions of others were the materialist conclusions which were drawn from them.[96] The main argument for the materiality of the soul was deduced from Haller's assertion that different qualities are felt in different regions in the brain and that the *seat* of the soul therefore has a certain extension. La Mettrie inferred that the soul *itself* had to be extended and thus was material. Haller replied that in such a concept, different parts of the brain would feel separately, and that it could not explain how the different sensations merged into the unity of one soul and into one 'self'.[97] In the *Histoire Naturelle de l'Âme,* La Mettrie also adopted some of Haller's early ideas on irritability. He talked of an 'innate force in all fibrous elements', to be distinguished from elasticity, which was independent of the soul and responsible for many movements of the body.[98] In the *Homme Machine*, written in Leiden in autumn 1747, La Mettrie developed this notion and mentioned several observations and experiments which served as the *pièces de résistance* of his concept of the machine man. These were phenomena like the movement of the isolated heart and of isolated muscular fibres, which could be reactivated through irritation. Although most of the observations described by La Mettrie could be found in older medical literature – as well as in Haller's Boerhaave-edition – it is no coincidence that it was in Leiden where the French physician was attracted to their importance. It was the time when Albinus performed similar experiments, Gaub in a public lecture stressed the corporeal origin of motion, and Winter, the 'specialist' on irritability, arrived from Franeker. In fact, we know that La Mettrie attended Gaub's oration in February 1747 and that he was informed about Albinus's research.[99] The conclusions which he drew therefrom were to a certain point similar to those of his Dutch colleagues. He postulated that each fibre of organised bodies moves by a principle on its own, by an innate force the action of which is independent of the nerves. This force manifests itself as a continuous 'natural oscillation' of the fibre.[100] So far, La Mettrie quite faithfully repeated the thoughts of Albinus, which we encountered in the first chapter. The crucial point is that he argued that a principle of movement was all that was needed to generate motive *and* mental activities:

> Given the most simple principle of movement, the animated bodies will have all they need to move, feel, think, regret, and, in a word, to find their way in the physical and the moral world which depends therefrom.[101]

The processes taking place in the body acted in the same manner in the soul, which 'has its muscles to think as it has its legs to walk'. The soul was nothing but a principle of movement, a material sensible part of the brain and the mainspring (*ressort principal*) of the whole machine. La Mettrie called this active principle – like Albinus, Gaub and Kaau – *enormôn*. It animated all parts of the body, was the source of sensation, passion and reason and exerted its power on the whole body through the nerves. The relation between the *enormôn* and the innate forces of each single fibre was not made clear. It seems that the latter derived from or were activated by the former because all the springs (*ressorts*) of the body were only emanations of the *enormôn*. Although he did not clearly state it, La Mettrie thought that reaction upon irritation required a kind of (unconscious) feeling (*sentiment*). The experiments showed that 'movement and feeling excite each another' and that there was a 'combination (*réunion*) of feeling and movement'. For La Mettrie in the whole universe there was only one diversely modified substance. Organised matter was endowed with a motive principle which alone distinguishes it from unorganised bodies. Matter thus may become active and capable of movement, feeling and reason. How this happened, La Mettrie said, is unknown.[102]

Apart from the monist materialisation of the soul La Mettrie's physiological ideas matched quite closely those of Haller and his Dutch contemporaries of the 1740s.[103] These authors were embarrassed when they saw how their own notions were used for the construction of a materialist concept. The fact that La Mettrie considered his theory not as an established truth but rather as the most plausible hypothesis in view of our ignorance, did not alter the situation. The hypothetical character was not made very plain and readers failed to notice it.[104] Gaub was especially annoyed because several of La Mettrie's arguments were plainly borrowed from his oration of 1747. This did not induce him to change his position but to argue more carefully, to refute La Mettrie openly and – as late as 1772 – to defend his concept of a vital force against materialist interpretation.[105] Haller was in a special position because La Mettrie had ironically dedicated the *Homme Machine* to him and called himself a pupil and friend of the Göttingen professor.[106] In later writings, Haller was again the target of his sarcastic remarks. The serious and pious professor, incapable of dealing with his witty and meandering opponent, published official rebuttals in which he denied any personal connection with La Mettrie and dissociated himself from any materialist thought. He did not present scientific arguments against La Mettrie's views before the elaboration of his concept of irritability and sensibility. In his critique he assumed that according to La Mettrie, the soul or *enormôn* located in the brain was the mainspring and ultimate cause of all movement. But the experiments showed, he said, that irritability was totally independent of the nerves and the soul. To Delius' reproach that his theory was close to La Mettrie's, Haller replied that it was not his but in fact the animist's concept that paved the way for materialist interpretations, because they considered irritability as an act of the soul. As irritability was a visible phenomenon and the soul inaccessible to our senses it was only a small step to reduce the latter to the former.[107] Haller's argument as such is sound, but his rendering of La Mettrie's theory is questionable. As we have seen, La Mettrie's exposition was ambiguous. He postulated on the one hand an independent, innate force of movement in the corporeal fibres, and on the other hand the *enormôn* in the brain as the mainspring of movement. It is not clear whether all motion ultimately derives from the central principle. The *Homme Machine* was not a concise physiological treatise but a philosophical pamphlet which collected various observations and thoughts in order to argue for a specific idea. Haller never considered it a serious scientific publication requiring a detailed study, such as the works of Whytt. Nevertheless, he realised the power and attraction it might have for a wider audience and felt compelled to refute it. The arguments he raised against La Mettrie in his orations of 1752 were determined by the aim to disclaim any connection between himself and such an impious and impudent author. Haller did not discuss La Mettrie in his later physiological works but be did so in his theological writings.[108] Probably Haller realised that his argumentation was rather weak and that it made no sense to refute materialism with scientific arguments. Of course, he was happy to hear that John Tuberville Needham (1713–81) in 1769 considered Haller's discovery of the distinct qualities of irritability and sensibility as the best refutation of the materialists, who confounded the corporeal level of motion with the spiritual level of feeling.[109] Ultimately, the question of the nature of the soul was beyond physiological research. When Haller treated the internal senses, the actions of the conscious soul in the *Elementa*, he said that he had to enter into 'a reign of hypotheses and conjectures'.[110] The sensualist origin of our ideas and their material basis as movements of the fibres in the brain were unquestioned.[111] But the soul was only accessible through its activities, and metaphysical speculations about the relations between body and soul were useless. Nevertheless, there was no doubt that the soul is not reducible to matter.[112] The immortality of the soul was only a revealed truth but its immateriality was proven, as Haller said, 'mathematically'.[113] The soul was a unity but contained thousands of the most different impressions which, on a corporeal level, could never be united into one point. Based on this conviction Haller had no difficulty in arguing that experiments could not be recounted differently just because somebody feared atheistic interpretations.[114] The basic dualism of body and soul was not vulnerable. Unlike many others, such as the animists Delius and Sauvages, and the mechanists de Haen and Le Cat, Haller did not think that the theory of irritability fostered materialism.

The *Homme Machine* was unanimously condemned by the whole medical community. Radical materialism of La Mettrie's kind remained outside the mainstream of medical theory. This is not to say that it had no impact on the physiological discourse. The *succès de scandale* made the theory of the machine man widely known and seems to have forced several authors to stress their anti-materialist position and to keep their distance from any concept of irritability.

Delius, continued

Haller's expulsion of La Mettrie's theory from the domain of physiological research, and his restriction of the soul to the realm of the internal senses, conflicted with Delius's views. For him, by contrast, there was no physiology without the soul. We should not, he said, talk of the nature of bodies but only of the nature of man consisting of body and soul. Likewise the 'nature of man' should be considered as the effect of the sum of all the forces, of the body *and* the soul. Delius complained that, as in the times of Glisson, a second being (*Wesen*) besides the soul was now introduced, which governed the actions of man and animals, an *archaeus* which caused material life and which was now called, 'nature', 'irritability', 'oscillation', etc.[115] Ernst Platner, medical professor in Leipzig, who in the 1780s aimed at a renewal of animism, formulated a credo of this system:

> Animal movement, says the Stahlian, is something totally different from any other movement of unanimated bodies. The animal is a body unified with the soul. Whoever violates this basic law corrupts the whole natural history

> of animated bodies at once and distorts the angle of all its investigations and doctrines. Therefore, in the living organic body of an animal, there is no operation of any instrument, no movement of the smallest fibre, no action of an invisible element, in which the soul is not involved.[116]

Although Platner went somewhat further than Sauvages and Delius in his attribution of all movement to the soul, he was an animist of the newer generation who rejected Stahl's notion of the wise and conscious soul. His pointed statement illustrates the categorical animist refusal of any position which allowed manifestations of life independent of the soul.

Conclusion

Haller had always taken a strong stand against the 'Stahlian sect', as he called it, and considered his treatise on irritability and sensibility as a major if not definitive blow to animism. Although he was well aware of the diverging positions of Whytt, Sauvages and Delius, he still aligned them with the professor from Halle and called them Stahlians or, more frequently, 'members of the Stahlian sect' or 'semi-Stahlians'. In contrast to Stahl they did not believe that the soul directed the vital actions consciously. But they still agreed that matter was inert and that vital motions had to be derived from an immaterial principle, the soul. For Haller, who could not conceive the soul as an unconscious agent, this was an equally erroneous position as it still exaggerated the power of the soul and ignored the potential of the body. The younger generation of animists was thus a target of Haller's criticisms, just as Stahl had always been. Whytt, Sauvages and Delius had disclosed some of their animist thoughts already before Haller's treatise. They were aware at least of some of the phenomena – especially the motions of dissected muscles and the heart – which Haller would use to argue his position. Thus Haller's orations of 1752 did not confront the animists with a completely new problem, nor did they conceive Haller's explanations as a completely new solution. Just as Haller considered all animists as heirs of Stahl, they regarded him as heir to the mechanist tradition of Boerhaave. Because he rejected the influence of the soul on vital motions, in their view he proposed a mechanical model. Besides some smaller modifications the animists did not change their positions considerably in the second half of the eighteenth century. Neither Haller's work nor other non-reductionist or vitalist explanations of motion provoked a major shift within animist theory. But they endangered the animist position as a whole. Looking back to the development of the last three decades, Platner in 1781 realised that these concepts had posed a new threat to animism. He remarked that mechanism had never been able to destroy the Stahlian position but that with Haller's new theory, animism – at least in Germany – had lost all reputation.[117] The new models, which considered vital actions to be distinguished from mere mechanical processes and from the metaphysical, immaterial realm as well, seemed to be more successful in explaining the phenomena that riveted the attention of the public. After 1740, nature revealed an entirely new face.[118] In 1741, Charles Bonnet detected the parthenogenesis of the aphids, ie. the fact that a single animal was able to generate offspring without any intervention from outside. In the same year, Abraham Trembley observed the incredible capacity of regeneration of the polyp (*Hydra viridis*), which transformed into as many complete animals as the parts into which its body had been divided. Nature had reproduced itself. The phenomena which Haller described were not really new, but his work called for new attention and his description of irritability as an independent property of organic matter was the third important discovery that demanded a new interpretation of vital processes. Neither reductionist mechanism nor animism in its various forms seemed to be able to furnish convincing explanations and both these currents gradually lost importance. Religious arguments, which, for at least some of the animists, were an important reason to insist on the power of the soul, lost their significance. Both Sauvages and Whytt enjoyed considerable reputations but neither of them could establish an animist tradition. As Bordeu testified, Sauvages stimulated the anti-mechanical current in Montpellier. But eventually he was superseded by vitalists, and Barthez would become the champion of the University. In a similar manner, Whytt in Edinburgh was replaced by William Cullen. Cullen agreed that a sentient principle operated throughout the whole body but he rejected the notion of unconscious sensation and considered the vital motions as basically carried out independently of the soul. Delius in his turn was the leading medical figure at the University of Erlangen.[119] But Erlangen itself was not a leading university and Delius could not boast a large number of followers. Although Stahl's generation had produced a number of animists, after the mid-eighteenth century only a few chose to adopt this position. In Italy and the Netherlands, animism was even less supported. Very few, Haller reported, regarded the soul as the cause of vital motions.[120]

Vitalism: a new field of theories

In an inspiring paper written in 1977, Richard Toellner stated that Haller's theory of irritability and sensibility had been the 'clearly determinable reason' for a dramatic shift from mechanism to vitalism taking place in physiology in the mid-eighteenth century.[121] This development could be read as a Kuhnian scientific revolution with Haller's theory as the paradigmatic model but, Toellner argued, Kuhn's theory had its flaws because Haller as the creator of the revolution had retained a mechanistic position.[122] In view of recent research on Haller, mechanism, and vitalism, it must be noted that Toellner's portrayal overstates the swiftness and intensity of this shift, as well as Haller's contribution to it and the mechanistic character of his physiology. Most importantly, we have learnt to realise that there were many different forms of mechanism and, even more so, of vitalism. Although certain aspects of vitalist theories were indebted to the Swiss physiologist, others stood in marked contrast to his ideas. The relation between Hallerian physiology and vitalism therefore varies according to the different models of vitalist theories. Despite considerable divergences, we may describe these concepts in a very broad sense, following Hans Driesch, as theories according to which vital processes have their own character and are not reducible to a 'special constellation of factors'.[123] There are three aspects that distinguish them from Haller's concept. Vitalist theories postulated vital powers that may or may not be based on organic structures but that ultimately could not be reduced to these structures. In this respect, there is often only a very small difference from Haller's notion of irritability. For Haller, too, function could not simply be deduced in a geometrical or mechanical manner from structure. The correlation between the fibrillar structure and the organic function was known only to God who instilled these forces in the bodies. Nevertheless, in contrast to the vitalists, Haller stressed that this correlation existed and that irritability was a corporeal force emerging from the complex structure of the muscular fibre. The second point of distinction concerns the holistic approach of vitalism. This was expressed in the understanding of the organism as a unity determined by mutual contacts between the different organs and the interactions of body and soul, and often guided by one main vital principle. This leads to the third aspect of vitalism, ie. teleology, as the vital powers are directed towards one goal, the constitution and conservation of life. Such a description of the organism contrasts with Haller's separation of the bodily functions, which he maintained throughout his life, although in his later works he tended to connect the properties of irritability and sensibility. It contrasts with his physico–theology which viewed forces as acting according to laws given by God. Haller's teleology is a methodological tool which allows deductions from structure to function and *vice versa*. But it does not include the notion that vital powers can envisage a goal. They are not subordinated to a general aim of the organism. Quite typically, Haller's last book of the *Elementa*, entitled *Vita Humana et Mors*, contains no definition of life but a detailed description of the changes of the physiological processes during the whole lifespan. Life is seen as an expression rather than as an aim of organic functions. The neat separation between creator and creation allowed Haller to envisage nature as a machine that runs according to the prescribed laws of mechanics. For Haller, an aim can only be envisaged by a rational being. The realm of purpose is restricted to God and the conscious soul.

In order to describe the relations between the many vitalist – and semi-vitalist – authors and Haller, a selection as well as a grouping is necessary. François Duchesneau has identified three main schools with different models, each culminating in the theory of one outstanding representative: the French school with Barthez, the German with Blumenbach and the Scottish with John Hunter.[124] I would like to adopt this scheme, but prefer to talk of currents prevailing in different countries rather than of proper schools. Furthermore, we should add a fourth major centre of vitalism, the Netherlands, which is often neglected but particularly important in our case, as it is the one most closely linked to Haller.

The Netherlands

The early phase of Dutch vitalist thought was outlined in the first chapter with the discussion of the concepts of Bernard Siegfried Albinus, Jan de Gorter and Frederick Winter. Their views were described partly as a reaction against Boerhaavian mechanism and partly as a development of its non-reductionist elements. They postulated an innate vital principle – also called irritability or *enormôn* – responsible for movement, which acted in all parts of the body and especially in the muscles. The ontological character of this force was not very clear. It was either described as a subtle substance attached to inert matter (Albinus), as belonging to a category distinct from body and soul (Gorter), or as inherent in an undefined manner in the fibre (Winter). It was certainly independent of the soul but depended on the presence of nervous liquid which seems to have been conceived as a kind of carrier of the vital principle (Albinus, Gorter), or at least as a necessary further source of power for movement (Winter). These authors recognised the potentials of Zimmermann's dissertation, and their student Wouter van Doeveren immediately repeated the experiments of Haller's pupil (see Chapter 4). The consecutive and repeated experimental refutation of Haller's findings was crucial for the further development of Dutch vitalist thought. It confirmed the sensibility of membranes and tendons and the ubiquity of the vital force or irritability and thus allowed the upholding of the unitary Boerhaavian fibre theory that emphasised the uniformity of the elementary corporeal structures. Whereas van Doeveren mainly furnished the experimental evidence, Hieronymus David Gaub emerged as the main authority to formulate the theoretical consequences.

Gaub, who was only briefly mentioned in the first chapter, had already presented some of his ideas before Haller's treatise of 1753. In an oration *On the Regimen of the Mind*, delivered in 1747, he stressed the extent to which bodily processes affected the status of the mind.[125] He distinguished a bodily from a mental *enormôn* that acted in concordance. The corporeal *enormôn* or vital force as the principle of all movement derived from the origin of the nerves in the brain, was distributed through the nerves in all parts of the body and therefore might be called the 'neural man' (*homo nervosus*).[126] It was also responsible for the movements observed in parts separated from the body. In reaction to Haller's work, Gaub later placed less emphasis on the neural and more on the local origin of motion. His notions are displayed in the dissertations of his pupils,[127] but more importantly in his *Institutiones Pathologiae* of 1758, one of the most widely used medical textbooks of the eighteenth century, which ran through three original editions, was repeatedly reprinted, and translated into the major European languages.[128] Although devised as a textbook of pathology, it contains important physiological passages, since Gaub considered physiology to be the foundation of pathology.[129] In accordance with Haller's notion of irritability, Gaub now described the *enormôn* or vital force as that which contracted upon irritation. It had to be distinguished from any other force and was present in the whole body but especially in the heart and the muscles. It was located in the body itself but could not be derived from the mixture of its elements, and it might be due to a certain inherent subtle fluid. It was a principle of its own kind and acted according to laws specific to living bodies, which could only be described on an observational level. Besides the ubiquity of this faculty, Gaub differed from Haller on another main point, as he allowed for a kind of sentient faculty (*facultas quasi sentiendi*) in the fibre, which perceived the stimulus and had to be distinguished from the motive faculty.[130] Although this is not clearly stated, together they presumably constituted the *vis vitalis*. Gaub also detected a certain reciprocal interaction between the vital power of the body and the animal power of the soul and assumed lower, unconscious activities of the soul.[131] Thus, like all Dutch vitalist authors, his separation of movement and sensation, of body and soul, was less rigorous than Haller's. There was no doubt, he asserted, that there are corporeal powers independent of the soul. But, considering the endless disagreement between animists and mechanists about the source of vital actions, pathologists and physicians were best left to leave the question undecided until it could be resolved by the physiologists.[132]

Gaub's book fostered the dissemination of the concept of irritation and reaction as a basic model not only of physiology but, more importantly, also of pathology (see Chapter 6). Quite in line with his main interest, he did not accept Haller's terminology and considered irritability not as a physiological property but as an over-abundance of the vital force, which caused an excess of motion and a perturbation of the regular processes. The effects were vibrations, tensions, and convulsions which resulted in pain, anxiety, obstructions, inflammations, and disorders of circulation and secretion.[133] What is of interest for our present discussion of physiological concepts is Gaub's predominant concern with power and motion. This is what distinguished the Dutch authors especially from the French vitalists. Emerging from a mechanist tradition, their emphasis was on the responsive rather than on the perceptive aspect of the vital faculty, their key word was irritability, not sensibility. Haller's work did not considerably change the outlook of the Dutch theories, they remained essentially the same. The ubiquity of a vital principle responsible for movement, principally independent of the soul and still somehow connected with the nervous system, was postulated in the early works of Albinus, Gorter and Winter, and later by Gaub and his pupils van Doeveren and Verschuir as well.[134] Together, these six medical professors dominated the physiological curriculum in the Netherlands, notably in Leiden and Groningen, for half a century, from the late 1730s to the 1780s. There was an *unité de doctrine* in the Dutch school, and Haller was well aware of it.[135] His own work presumably fostered rather than prevented the spread of their thoughts, as it directed the general attention to independent corporeal forces and thereby paved the way for a broad reception of the work of Gaub and his colleagues.

France

French vitalism in its origin, development, and different styles is by far the best studied of the four 'schools' here presented.[136] My scandalously short survey cannot reflect the breadth of our actual state of knowledge.[137] I will focus on some aspects where the connections with and distinctions from Haller's notions are best visible. The sources of the French vitalists are to be located mainly in the seventeenth century, in the notions of a living fibre as revealed in the works of Glisson, Baglivi and others, which have been partly discussed in the first chapter. Van Helmont's *archaeus* and Stahl's animism were further stimulators for an anti-mechanist turn. The first comprehensive formulation of French vitalism was Louis de Lacaze's *Specimen Novi Medicinae Conspectus* from 1749, which served as an inspiring reservoir of thought for many of his followers. But vitalism was mainly spread through the writings of Bordeu, the *Encyclopédie* and the works of Barthez, which will now be discussed successively.

Bordeu's first major work, the *Recherches Anatomiques sur la Position des Glandes et Leur Action* of 1751, derives its strength from the failure of mechanism to explain excretion and secretion. Anatomy, in Bordeu's view, furnished no evidence for a simple compression of the glands (excretion), and the notion of microscopic sieves could not account for the production of specific humours in these organs (secretion). These processes had to be explained by the action of the gland or rather its nerves, which acquired power upon irritation.[138] The nerves of the glands had their specific tone (*ton*) and were only sensible to irritations that were in proportion to this tension. Secretion was therefore due to a kind of 'sensation', although this term was only a metaphor, as there were no available words to talk about the power that directed the movements in such processes.[139] There were two levels of life. The living animal as a whole was constituted by a general vital tonic movement in all parts of the body, which depended on a continuous irritation of the nerves through the movements of the brain. Besides that, each organ of the body had its specific kind of sensation (*sentiment*), 'tact' or disposition.[140] In order to stress the particularity of each part and its dependence on the others, Bordeu drew his famous comparison of the animal body with a beehive.[141]

Bordeu's model, with its conception of sensibility as due to a specific tension of the nerves and of life as due to their continuous central irritation, was still heavily indebted to mechanical images. But in its emphasis on the particular life of the organs and their interactions, and on the nerves and their local, independent reactions, it envisaged a new kind of living, organic body. Presented just one year before Haller's new concept, it did not touch the question of the relation between sensation and movement. In his later *Recherches sur le Pouls* (1756), Bordeu was more explicit on the topic:

> Each organic part of the living body has nerves which have a sensibility, a particular kind or degree of sensation [*sentiment*]. This sensibility makes the life of the nerves. It is the necessary result of their consitution... It mixes more or less with mobility or contractility. Functions in which motion or mobility is evidently visible, have less sensibility or sensation. In contrast, there is only little motion or mobility in functions executed only by sensation or sensibility.[142]

As with the main quality, sensibility, Bordeu seems also to have conceived movement as a sign of the activity of the nerve. In his *Recherches sur l'Histoire de la Médecine* (1764) he described life as consisting of movement and sensation or as a disposition to these two modifications. Each vital function was a mixture of movement and sensation. Haller's concept he described as a kind of misled variation of Montpellerian vitalism:

> ...one of the most distinguished philosophical physicians of this century, Haller... has considered the irritability of the parts of the living body as a general principle and has put it in the place of sensibility, which had been regarded as a general principle in the School of Montpellier before there was

> any question of irritability considered under this aspect.... Besides, these two opinions are very similar....[143]

Bordeu's description of Haller as a *médecin philosophe* who had established irritability as a 'general principle' and whose views were quite similar to his own certainly cannot be considered as a fundamental misunderstanding. It was a deliberate, although not very elegant move to get rid of a tedious debate, which Bordeu for whatever reason did not want to join.

Confrontation with Haller was not avoided in the *Encyclopédie*, however. We must distinguish between two groups of medical contributors. The majority of the anatomical and physiological articles were written by the Chevalier Louis de Jaucourt (1704–80) – a pupil of Boerhaave – and the physicians Pierre Tarin (1721–93?) and Arnulphe d'Aumont (1721–1800), who used Haller's edition of Boerhaave as one of their main sources.[144] They adhered to a rather traditional mechanism and their entries reflect neither Haller's critical approach to Boerhaave nor the newer developments emerging in the Netherlands. Their contributions are often disregarded, as their old-fashioned views do not fit with the general picture of the *Encyclopédie* as a motor of innovation. Here we are concerned with the second group of the vitalist authors, notably Ménuret de Chambaud and Fouquet.[145] Whereas Ménuret, wrote some eighty articles, Fouquet contributed only three, including the important entries *Sécretion* and *Sensibilité*. Together these articles form a more or less homogenous corpus of vitalist thought, mainly based on the works of La Caze and Bordeu. The opposition to Haller is manifested on the methodological and on the conceptual level. The experimental approach is rejected because it denaturalises the organic body and neglects its unity (see Chapter 4). Also, the single experimental findings are disputed, notably the insensibility of tendons and the lack of the irritability of the arteries.[146] Haller's concept is, of course, implicitly rejected throughout the exposition of vitalist thought, but explicitly refuted and discussed in detail only in the entries *Irritabilité* and *Sensibilité*. The former consists in the shortest possible judgement on Haller's theory:

> IRRITABILITY (Physiology) term invented by Glisson, and renewed in our time by the famous Mr Haller, to designate a particular mode of a more general faculty of the organic parts of animals, which will be treated under the name of sensibility.[147]

The detailed refutation of Haller is, in fact, to be found in Fouquet's article *Sensibilité*, which essentially repeats the notion, also supported by Ménuret, that irritability required a sensible principle and was nothing but a specific product of sensibility. This is not the place to discuss the details of

Fouquet's and Ménuret's positions. It is important, though, to realise that the corpus of vitalist thought displayed in the *Encyclopédie* helped foster an independent French vitalist tradition that considered La Caze and Bordeu as its leaders and Haller as an opponent.

Paul-Joseph Barthez (1734–1806) established a new manifestation of this tradition, which broke with several of its earlier elements. In 1761, he argued in the manner of Whytt and stated that irritability as the motive faculty of the muscles depended on a sentient principle or sensibility, which itself could not be reduced to the presence and activity of nerves.[148] Barthez thus already in his early work rejected the 'mechanical' correlation between material substrata (tension and irritation of nerves) and property (sensibility) still present in Bordeu. In his main works of the 1770s he went a step further with the postulation of a vital principle co-existing with the living body but to be distinguished from it and from the soul as well. Haller's remark that he assumed a principle of unknown nature in the manner of Gaub, is not without foundation, but Barthez stressed the radical difference of his concept from all previous ones.[149] Although the vital principle is endowed with both sensitive and motor forces and thus responsible for the vital actions, it has to be envisaged as a cause of a higher order than that of simple organs and their functions.[150] It is not a property but a principle – the name of which is of no importance – that correlates all these organic processes by means of sympathetic nervous connections, in order to maintain the unity of the body and life itself.

Just as in the case of the Netherlands, the foundations of French vitalism (La Caze, Bordeu) were laid before Haller's new theory appeared. The repetition and (mainly) refutation of his experiments – concurrent with the devaluation of the experimental approach as such – allowed the establishment of a system that emphasised the unity of the body under the guidance of the nervous system. Whereas Haller has often been credited as an important source for vitalist thought, in the case of the French and Dutch vitalists it seems rather to have been the opposite. Haller's late notions of an unconscious, involuntary, modulatory role of the nerves and his acknowledgment of a direct communication between nerves as an explanation for the phenomena of sympathy may have its roots in the works of his opponents.

The German countries

An abundance of physiological concepts was developed in Germany in the second half and especially the last quarter of the eighteenth century.[151] Many of the authors may be called vitalists only in the very broadest sense as they rejected the reduction of vital powers to the actions of mechanics or of the soul. The ontological character of these powers oscillated between the two extremes of Johann Friedrich Medicus, who considered the vital force as a third entity besides the body and the soul, and Johann Christian Reil, who searched for its material structure.[152] In contrast to the cases of the Netherlands and France, there was no proper vitalist or semi-vitalist tradition predating Haller's work, and the theory of irritability and sensibility therefore was often considered as a point of departure. The discussion centred primarily on two domains of animal economy: the neuro–muscular system and the questions of generation.

Until the 1760s, appropriations of the Hallerian model as presented in the first section of this chapter seem to have prevailed. Within a general non-reductionist mechanist frame, irritability was conceived as a property of the muscles which, despite its partial dependence on the nerves, retained its chief importance in the execution of vital actions. The role of the nerve in these concepts was envisaged primarily on a material level, either as anatomical component of the muscular fibre or as deliverer of nervous juice, which was a prerequisite for muscular contraction. In the last third of the century, however, the nerves were increasingly seen as actors in the vital processes and the notion of irritability was often, as in France, subsumed under the heading of nervous actions. A major spokesman of this trend was Johann August Unzer (1727–99), a well-known physician and medical author from Hamburg.[153] Unzer thought that Haller's work represented the actual state of knowledge in physiological matters but maintained that it nevertheless contained defects and errors.[154] On the structural level he went a step further than his predecessors and regarded the muscular fibres as extensions of the nerves. Although irritability manifested itself only in the muscles it was not an original property of these fibres but depended on the nerves. Unzer distinguished between external and internal sensual impressions (*äusserer und innerer sinnlicher Eindruck*) as the two forms of nervous power (*Nervenkraft*), which transcend the boundaries of mechanical laws. The former, the lower nervous powers, were responsible for perceptions upon irritation, which did not reach the brain and caused muscular contractions, as for instance in the heart. In this respect he talked also of a 'nervous irritability' (*Nervenreizbarkeit*). The latter impressions were transmitted to the brain and only these produced conscious sensations. The whole animal acted in coordination and sympathy thanks to the connection of the nerves on the levels of bifurcations, plexus, ganglia, and the brain. The lower intersections functioned as points of reflection (*Reflexionspunkte*) of sensual impressions without the brain being involved.

This short description of Unzer's views stands for the general tendency in this period to consider the nervous system as the primary vital organ of the human body.[155] Meanwhile, another fundamental physiological question had become a focus of attention. Generation had always been one of the great themes of physiology but with Trembley's discovery of the regeneration of polyps (1741), epigenetic models, which envisaged a continuous organisation of a previously unstructured mass, (re-)gained credibility.[156] The idea that specific powers directed this development (Maupertuis, Buffon, Needham) spread at the same time as the notion of particular forces or properties controlling corporeal movements gained acceptance. A certain reciprocal transfer of explanatory models took place. The Hallerians Pagani and Bonioli argued that the observations of Buffon and Needham had demonstrated motive powers in the human seed, which they identified with Hallerian irritability.[157] The French vitalist Henri Fouquet described the process of generation of the human embryo as a development originating from a first nervous cylinder and guided by the principle of sensibility.[158] The embryologist Needham, on the other hand, considered the vegetative force, which he had regarded as the guiding principle in generation, as a universal power responsible for all vital activity.[159] But, to a large extent, the questions of generation and motive powers were treated separately until the last quarter of the century. As to Haller, the investigations into irritability seem to have shaped his perception of the importance, but at the same time limited variability, of reaction of corporeal forces. This might, to some extent, have fostered his support for preformism because he realised that simple forces could not direct the complex embryological development.[160]

Haller's preformist theory was countered by the young and brilliant physician Caspar Friedrich Wolff (1734–1794) in two important works of 1759 and 1764.[161] Wolff considered Haller's and all other previous embryological concepts as mere descriptions, which could not be called theories or explanations because they did not furnish principles or laws that accounted for the processes in question. He thought he had identified such a principle, which he called 'essential force' (*vis essentialis*). In his extensive and detailed investigations he had observed that the embryo in its early stage is nourished through a system of gradually developing channels before the existence of the heart, vessels or muscular structures. There had to be a force which pushed the nourishing juice through the unstructured matter of the yolk towards the embryo, thereby creating these channels. The same essential force would transform the embryo from an amorphous liquid mass into an organic structure and direct also the further development and the vegetative functions of nutrition and growth. The force, therefore, preceded the structures and could not be derived therefrom. Circulation, respiration and excretion as mechanical processes could be explained merely by the structure of the organs. But these were not truly vital actions and not directly responsible for life. Only the vegetative actions of generation, nutrition, and growth as well as voluntary motions, sensation and mental processes as activities of the soul could be considered as real vital actions. Physiology, for Wolff, was the doctrine of the functions of the body, but the theory of generation was the science of the body itself as it explained the formation of its structures.

The description of the *vis essentialis* remained vague, and Wolff did not explain how it could steer the complex process of generation. Although he had disclosed several weak points of Haller's concept he could not prevail over the preformationist triumvirate of Haller, Bonnet and Spallanzani. To some of the younger generation, though, he had demonstrated the vulnerability of the structuro–functional model and the potentials of a theory that used organic powers to explain the formation of living structure. Blumenbach especially seems to have been stimulated by the Haller–Wolff controversy.

Johann Friedrich Blumenbach (1752–1840), professor of medicine in Göttingen, initially supported Haller's preformationist theory. In 1781, though, he presented the principles of a new epigenetic model in his treatise *Über den Bildungstrieb und das Zeugungsgeschäfte*, which he elaborated in later works.[162] Impressed by the regenerative capacities of the polyps but also of the human body after injuries, he concluded that there had to be a specific formative power (*Bildungstrieb, nisus formativus*), which strived to impart to all animated creatures a specific shape and to restore it if it was destroyed. Generation, nutrition and restoration were only modifications of the same force, which was activated on different occasions. It had to be distinguished from elasticity, irritability, and sensibility. Whereas irritability was described more or less in Haller's terms, sensibility included unconscious reactivity on the level of the spinal marrow or isolated nerves. A fourth power, called *vita propria*, was responsible for a few specific actions of some organs. Although the exact relation between these forces was not made very clear, the *Bildungstrieb* as the force which appeared first and was active in almost all parts of the body was less specific than the others and was geared to the more general aim of life and preservation. Blumenbach acknowledged the existence of Wolff's essential force but this power would only be 'a prerequisite *to* the formative force – but by no means the formative force itself.'[163] Wolff's force initiated the first formation of the organism but only Blumenbach's *Bildungstrieb* could determine the direction of further development. It is obvious that in this model, the transition from inorganic to organic matter preceded the *Bildungstrieb* and was indispensable for its action.

Kant recognised in Blumenbach's postulation of a *Bildungstrieb* a method which met his own demands on natural history. He stated in the *Critique of Judgement* (1790, § 80–81) that a purely mechanical description of organic bodies did not enable us to establish a systematic connection between the single discoveries. The intentionality (*Zweckmässigkeit*) of nature had to be introduced as a heuristic principle. The mechanical, empirical fundamentals of science had to be united with a teleological interpretation, and this is what Blumenbach, in his view, had done. Of course, Kant pointed out, only a subjective and regulative principle was thereby given, which did not say anything about the real conditions of nature, but this was the only feasible method for a systematic science. Blumenbach readily accepted this teleo–mechanical model because it agreed with his own views and adapted his later works accordingly. Nevertheless, his vital forces often did not appear as purely heuristic principles but rather as truly constitutive agents, and the same is true for many of his followers.[164]

Blumenbach was a highly influential figure and a very successful teacher in Göttingen for over fifty years. His concept of the *Bildungstrieb* in its association with Kantian teleology met with a very broad and mostly favourable reception. It was the first important alternative to Haller's structuro-functional model in the German countries. With regard to irritability, the intimate relation between structure and function was, as in the case of Wolff, maintained. But the process of unconscious co-ordination under the guiding principle of sensibility was not reducible to such a model. Even less could the nature of the *Bildungstrieb* be grasped by this concept. Most of the later authors agreed in this general assessment of vital forces. A few however, stressed their material basis.[165] Johann Christian Reil, for instance, thought that the organisation of matter and its chemical affinities would be sufficient to produce vital actions.[166] But such postulations lacked the intelligibility of Haller's model as Reil could not demonstrate the correlation between – more or less – visible structures and functions in the manner the Göttingen professor had. In contrast to Haller, who investigated organic structures and detected their specific functions, Reil and his contemporaries identified specific organic forces and tried to establish their relations. Whether they thought these forces emerged from specific organic structures or not, did not considerably change their approach. But they differed in their identifications of these forces as well as in the postulation of their relations. J.F. Kielmeyer (1793), J.D. Brandis, H.F. Link and C.W. Hufeland (all of them 1795) considered both generative powers and those of irritability and sensibility as the major vital forces. C. Girtanner (1791), partly adopting the popular ideas of John Brown, deemed irritability to be the leading power under which sensibility could be subsumed. J.U.C.

Schäffer (1793) simply reversed Girtanner's theory and regarded irritability as reducible to sensibility. Johann Daniel Metzger (1739–1805) was probably the author of the 1790s closest to Haller, as he strictly maintained the distinction between irritability and sensibility and the independence of the former from the latter. But he did not restrict irritability to the muscular fibre and conceived an unconscious activity of the nervous system. He was in fact more indebted to Albinus and Gaub, and considered the former as the greatest physiologist, and the latter as the greatest pathologist of his era.[167]

In all these models, from Wolff and Unzer, to Blumenbach and Metzger, Haller's notions of irritability and sensibility turned up continuously, but always transformed. Generally the qualities were less rigidly defined, and both were either considered as ubiquitous or as surpassing the restricted realm of visible contraction and conscious sensation. The structuro–functional model was thereby abandoned or at least damaged. The qualities of irritability and sensibility were regarded as vital principles upon which many, most or all organic processes depended. Other forces, notably those of generation, might also be judged as necessary for life, and often sensibility and irritability were considered only as modifications of one and the same force. We have to ask whether Haller's late formulations of his concept fostered this development. With his separation of cellular tissue (elasticity), muscular fibres (irritability) and nervous system (sensibility) he had created a 'new division' of the parts of the body. Besides this general model, in his later works, and especially after the 1765 edition of his physiological textbook, he envisaged another tripartite division of forces, now within the realm of contraction. He considered elasticity, irritability and the *vis nervosa* as a *triplex vis* acting in the muscle. The *vis nervosa* indicated the power of the muscles as activated through the nerves and therefore had clearly to be distinguished from sensibility involving consciousness. But the postulation of two tripartite divisions, in both of which the first two powers were elasticity and irritability might have produced some confusion, especially in the German countries where *vis nervosa* was translated with *Nervenkraft*, a term that several authors equally used to designate sensibility. Wolfgang Pross, for instance, has argued that Haller's description of the muscular triple force – as described in the *Encyclopédie* – was the model for Herder to postulate the unity of one vital force, including irritability and sensibility and replacing in this way the *vis nervosa* with sensibility.[168] This might well have been the case for Herder but in general, I think, the sequence of theories was different. The distinction of irritability and sensibility was the cornerstone of Haller's concept and everybody was well aware of it. Professional medical authors did not confound the *vis nervosa* with sensibility, and presumably also Herder's move was a deliberate choice. That Haller, of all authors, should have encouraged the unification of these two properties is rather unlikely, despite his late acknowledgement of a certain modulatory, unconscious role of the nerves in the *Encyclopédie*. It was not necessary to 'misread' Haller in order to form the idea of a unified nervous and muscular system, as this was part of the Boerhaavian legacy. It was turned from a mechanist into a more or less vitalist concept in the Netherlands and France before Haller had formulated his new theory, and it was partly preserved by non-reductionist mechanists whom we have described as 'semi-Hallerians'. In fact, it was the main common denominator of those who deviated from the Göttingen professor. Haller, no doubt, was very important as the central figure who rejected both Stahlian animism and Boerhaavian mechanism. But, as Unzer said, his move had only been the first step.[169] The younger authors like Wolff, Unzer, Blumenbach, and Kielmeyer were indebted to Haller but described their own theories in opposition to him. Haller's *themes* dominated German physiology of the second half of the eighteenth century, but not his concepts.

Scotland and England

Boerhaavian mechanism dominated the medical curriculum in Edinburgh until the mid-eighteenth century.[170] All four professors of the newly founded medical school (1726) had been pupils of the Leiden master. Their replacement in the 1740s and 1750s marked not only the advent of a new generation but also of new ideas. With Robert Whytt's installation in 1747 came the first reservations upon the mechanical model expressed by a professor, reservations which earlier on had already been voiced by Whytt's preceptor, George Young. Although Whytt still used Boerhaave's textbook – which he replaced only in 1762 with Gaub's *Pathology* – he must have presented his critical attitude and stressed the specific features of vital actions and the importance of the nerves. In 1754 one of his pupils, Alexander Monro secundus (1733–1817), was appointed as professor of anatomy, physiology, and surgery. Generally critical of the Boerhaavian heritage, he was open to the new developments in physiology. In his earlier works he maintained, like Haller, a certain independence of muscular action of any nervous control, but later – in contrast to the Göttingen professor – he rejected the existence of a *vis insita* as distinct from the *vis nervea* and considered the nervous system as the basic motor of life.[171] In this respect, he followed his colleague Cullen, whose physiological theory dominated the Edinburgh curriculum.

The physiology of William Cullen (1710–90) is an eclectic amalgam of contemporary thought, derived in great parts from the works of Whytt,

Haller and Gaub.[172] Cullen was at no pains to be particularly original, as he was primarily concerned with the practice of medicine. But, like Gaub, he thought that medicine had to be founded on a proper systematic understanding of physiology and therefore wrote a textbook that served as an introduction to the study of pathology and therapeutics.[173] He considered the muscular fibres as continuations of the nerves, motive extremities of the nervous system that itself represented the vital part of the solids of the body.[174] The muscles themselves had a particular organisation that rendered them irritable, ie. able to contract upon irritation. This quality might after Gaub also be called 'vital force'. But, as it seemed to be independent of the nerves, Cullen named it 'innate force' following Haller. It might be, as Haller argued in his late works, just an increase of elasticity, the mechanical power of the inanimate solid parts of the body. Nevertheless, experiments showed that contraction was also caused by the nerves or by a nervous power. Cullen tried to unite these two causes and declared that both innate and nervous power were of more or less the same nature and that in the case of living bodies both might depend on animal power, ie. the soul. The whole body, and especially the nervous system, was intimately connected with the soul. Just as thought could provoke movements, motions in the nervous system could produce thoughts. The state of the brain, the nervous spirits and the whole nervous system was defined by its degree of excitement or collapse. Life consisted in the excitement of the nervous system and especially of the brain, which united the different parts and formed a whole. This concept served as the basis of Cullen's influential pathological model, which explained diseases mainly as a result of the different states of excitement and collapse of the nervous system including the muscles as its motive extremities. In the whole exposition, Cullen was very careful not to make apodictic statements and often presented his ideas as probable hypotheses. The ambiguity and, to a certain extent, vagueness of his position was heightened by his attempt to link different models of explanation. Whereas he connected the innate force with a purely mechanical phenomenon, elasticity, he coupled the nervous force with the immaterial powers of the soul. Yet both these forces, the innate and the nervous force, ought to be of the same nature. His lingering between mechanist and animist positions did not result in the articulation of a decisive formulation of a vitalist alternative. Quite typically, like the vitalists, he stressed the importance of sympathy between the organs, but located their connection not in the periphery but in the centre of the nervous system, in the brain.

John Hunter (1728–93), the famous London surgeon, was partly a pupil of Monro and Cullen.[175] The most extensive exposition of his physiology is to be found in his *Lectures on the Principles of Surgery*, delivered in 1786–7 but published as late as 1835.[176] His *Treatise on Blood, Inflammation, and Gun-Shot Wounds*, too, in which Hunter demonstrated the practical application and significance of this theoretical model, was only edited posthumously in 1794.[177] Despite their late diffusion we have to touch upon some main points of his ideas, as they represent a significant shift in English physiological thought. Hunter broke with the traditional belief, still present in Monro and Cullen, that there is a connection between the organisation of matter and life. In Hunter's view, there had to be a power, a vital principle superadded to the material structure that accounted for vital processes. The presence of this principle was not only visible in the solids but also in the fluids of the body.[178] In case of injuries, the blood coagulated to form new solid matter. This had to be conceived as an active, purposeful process of repair, it was an 'operation of life'.[179] The vital principle was the 'immediate cause of action in every part' of the body, 'every individual particle of the animal matter, then, is possessed of life.'[180] The specific arrangement of these single particles into a structure like the muscular fibre was responsible for the phenomena of irritability, but the arrangement itself was not able to maintain the action. Even when the principle no longer existed, the arrangement remained, but life was lost. Just as in France with Barthez's*principe vital* and in Germany with Blumenbach's *Bildungstrieb*, English vitalism with Hunter's *vital principle* had definitely abandoned its connections with mechanical explanations or with structuro–functional models as proposed by Haller.

Conclusion

In our survey, a wide variety of mechanist, animist and vitalist theories has been presented. But only a few of these might be called 'Hallerian'. A considerable number of surgeons and physicians confirmed Haller's experiments and thereby his theory but these were mostly 'minor' authors, Caldani and Fontana being the only figures of European reputation. The case of Fontana shows that initial support did not exclude later deviations from Haller's concept. Another group has been identified as the 'semi-Hallerian' mechanists, who accepted the distinction of irritability and sensibility but rejected the idea that the former property was wholly independent of the nervous system. As another faction, we have recognised authors such as Tissot who equally maintained the distinction of the two corporeal qualities, but allowed for unconscious sensations and extended the realm of irritability. These two latter groups may be placed between the strict Hallerians and the remaining majority of physiological authors who adopted and developed a broad range of non-Hallerian theories of either mechanist, animist or vitalist colour. These 'non-Hallerians' were not necessarily 'anti-

Hallerians', since several of them, for instance Monro and Unzer, considered Haller as their main source of instruction. But they must be considered opponents of his theory in so far as they, even more than the 'semi-Hallerians', rejected one or more of the central tenets of the new concept. A prerequisite for the disapproval of Haller's notions was the refutation of his experimental results, or at least of his procedure and interpretation (see Chapter 4). This allowed, on the one hand, the conception of irritability as a purely mechanical phenomenon or as a vital faculty extended beyond the muscular fibres. On the other hand, sensibility could be conceived as an unconscious activity on a local or central level. As a result, most of these authors did not regard the muscular and the nervous system as entirely independent and neatly demarcated territories. Whereas mechanists and animists rejected the notion of specific 'biological' principles or forces and thus invoked the powers of the soul, vitalists stressed their existence and interaction. In principle, these were not entirely new positions, developed in order to refute Haller. They were either indebted to the Boerhaavian fibre theory or to animist and vitalist theories as established in the 1740s and early 1750s in France, Scotland, and the Netherlands (Sauvages, Whytt, Bordeu, Albinus, etc.). The vitalist theories in particular, however, were further elaborated, and here certain reconceptualisations in view of Haller's new findings – adaptations of his notions as well as explicit rejections – are visible. This is the case for some first-generation vitalists of the second half of the eighteenth century, notably Gaub, Unzer and Cullen, who were still partly bound to mechanical concepts and envisaged more or less strict relations between structures and functions. Ultimately, with Barthez, Blumenbach and Hunter these notions, dear to Haller, were definitely abandoned. Now, as earlier on already in France, vitalist theories were usually considered as implicit or explicit rejections of Haller's whole approach. Barthez, at the end of the century, was pleased to see that Blumenbach had adopted his own general views of animal economy, and thus had turned against Hallerianism, which still seemed to haunt the Göttingen professor.[181]

Thus, the second half of the eighteenth century produced a relatively small number of faithful Hallerians but a large quantity of new theories that rejected rather than supported the ideas of the Swiss physiologist. How, then, may the assessment of John Pringle (1707–82), the famous military physician, be explained, who wrote to Haller in 1764, that,

> Your discoveries concerning Sensibility & Irritability will always be placed next to that of the Circulation; & possibly in the next age, when the emulation among the now living shall cease, the question will be, whether Harvey or Haller was the greatest man?[182]

Were these the standard flatteries between correspondents or was Pringle just one of the few enthusiastic Hallerians? How do we account for Condorcet's eulogy in which he stated that Haller's work on irritability and sensibility 'were the beginning of a revolution in anatomy'? Condorcet offered the following explanation of his judgement:

> We have learnt that there exists a particular power in the living bodies which we could consider as the immediate principle of their movements, as the power which is spread out in the organs and makes each of them perform its proper function. Physiology, which was based on metaphysical and uncertain principles for too long a time, could finally have a general and experimentally proven fact as its basis.[183]

Condorcet emphasised two elements of Haller's theory: the detection of an independent corporeal force and the experimental proof of its existence. It was the combination of these two elements that impressed. The postulation of bodily powers in itself was not new, it was already an accepted principle in Dutch physiology. With their experimental verification, however, this concept gained precision and credibility. Unzer stated that authors before Haller 'had almost no concept' of specific vital powers and Christoph Heinrich Pfaff (1773–1852) in the 1790s considered the work on irritability and sensibility as the advent of a new era in physiology before which there had been mere ideas and often only dreams.[184] But although the experimental approach was considered as decisive by many, this did not result in its general acceptance and propagation. This must partly be explained by the lack of standards of procedure and thus the production of contradictory results (see Chapter 4). Also, the other element of Haller's theory was not adopted without reservations. It was the very general idea of innate forces or specific principles which gained increasing recognition but not necessarily Haller's specific and restrictive notions of these properties. The position of Antoine Louis (1723–92), professor at the *Ecole de Chirurgie* and secretary of the *Académie de Chirurgie*, is quite typical. Although he criticised Haller's experimental approach, rejected his limitation of the realm of sensibility and disputed the novelty of his remarks on irritability, he nevertheless acknowledged the importance of his work in so far as it had attracted attention to these qualities and thus had helped to discredit the mechanical and hydraulic principles of Boerhaave.[185] But, if neither the experimental approach nor the main tenets of his concept were accepted, little was left over of Haller's specific physiological contribution.

Theories, of course, are always adapted and transformed by following generations, but in Haller's case the transformations went immediately to the core of the concept. Tissot's description of irritability as 'the sum total of the powers of the vital principle', was just such an assault, although considered as the support of a devoted Hallerian. Similarly Vicq d'Azyr, who thought that Haller 'has laid the foundations of a science which has nothing in common with its predecessor but the name', and who asked to show respect for this great man 'by adopting his method and by trying to follow in his footsteps', conceived a general tension of the reticular tissue of nerves and arteries which depended on irritability, and thus violated the central principle of Hallerian physiology, the strict relation between structure and function.[186] There was a tendency to equate all concepts of irritability with Hallerianism. Pfaff generalised that those who considered irritability as the main vital power were Hallerians, and viewed Metzger as one of them.[187] Haller himself remarked already in 1764 to Tissot: 'Irritability begins to build a sect. This is not my fault.'[188] What did he mean to say? Was it Tissot's fault? In this case, Haller should have been more cautious, as he had himself approved Tissot's preface and fostered its dissemination. Furthermore, his own treatise, due to its descriptive character and its lack of theoretical speculation, demanded further interpretation. In the second half of the eighteenth century there was a strong temptation to maximise rather than to minimise the importance of qualities such as those described by Haller. There are three main reasons for this. First, the phenomena described by Bonnet, Trembley, and Haller heightened the general awareness of processes that seemed not to be reducible to the powers of mechanics or the soul. The recognition of the limitations of the animist and mechanist model highlighted the explanatory power of the new model of specific vital forces and thus advocated its application in all parts of physiology. Secondly, once the specificity of life was fully acknowledged, Newtonianism in its broadest sense fostered the conception of 'biological' principles as analogues of gravity.[189] The epistemological and ontological description of these forces did not necessarily equate with Newton's, but the notion of an ubiquitous force acting in all parts of living matter was suggestive. Haller's limitation of his properties to specific structures was therefore often rejected. Thirdly, the concept of active guiding principles was very well suited to the construction of a new pathological and therapeutic model. Authors like Gaub, Cullen and later Brown based their concept of disease on the notion of increase or decrease of the vital forces or principles, often identified as irritability. This enabled them to deliver a coherent account of both physiology and pathology, an account which was getting more and more complex and difficult to make clear in the mechanical concept. Haller's sophisticated physiology, in which innumerable processes had nothing to do with irritability and sensibility, could not furnish such simple explanations (see next chapter).

To sum up, Haller's work played a pivotal role in the development of a new understanding of physiological properties that rejected both animism and iatromechanism. It encouraged the creation of a wide variety of new theories of vitalist or non-reductionist mechanist colour. But, although Haller was an acknowledged authority of the first rank and a major point of reference in all physiological questions, remarkably few authors maintained a position that kept to the essentials of his concept. His views were defined to such a degree by a methodological approach and a conceptual framework particular to him, that they could not easily be adopted by others. Furthermore, they occupied a place of transition, a somewhat unstable position between mechanism and vitalism where it was difficult to keep the balance. Haller's own changing views on irritability and sensibility, although remaining within the limits of his concept, reflect this situation.

Notes

1. J. Weise, *De Irritabilitate Morborum Genitrice... Praeside Ernesto Godofredo Baldinger* (M.D. thesis, Jena: Heller, 1772), preface.
2. F.C. Medicus, *Von der Lebenskraft* (Mannheim: Hof- und akademische Buchdruckerei, 1774).
3. E. Platner, 'Über einige Schwierigkeiten des Hallerischen Systems', in E. Platner (ed.), *Anton von Haens... Heilungsmethode in dem Kaiserlichen Krankenhause zu Wien*, vol. 3 (Leipzig: Weygand, 1781), 373–87.
4. See especially the works of F. Duchesneau and M.T. Monti in the bibliography. Only R. Toellner still considers Haller as a genuine mechanist: 'Principles and Forces of Life in Haller', in G. Cimino and F. Duchesneau (eds), *Vitalisms from Haller to the Cell Theory* (Florence: Olschki, 1997), 31–7.
5. E.G. Baldinger, *Vindiciae Irritabilitatis Hallerianae* (Göttingen: Dietrich, 1775).
6. Azzoguidi to Haller, 13 July 1773 (*Haller Papers*).
7. This approach has been employed in particular by L. Brockliss and C. Jones, *The Medical World of Early Modern France* (Oxford: Oxford University Press, 1997).
8. *Bibliothèque Raisonnée*, 33 (1744), 46. Van Swieten actually was a Catholic.
9. For further details on the debate see E. Lesky, 'Albrecht von Haller, Gerard van Swieten und Boerhaavens Erbe', *Gesnerus* 15 (1958), 120–40 and E. Lesky, 'Neue Dokumente zum Streit Haller – van Swieten', *Clio Medica*, 7 (1972), 120–7.
10. G. Fè, 'Saggio Critico... in Difesa di Ermanno Boerhaave...', *Opuscoli*, ii, 141–77.
11. A. de Haen, *Lettre... à un de ses Amis, au Sujet de la Lettre de Mr. Tyssot à Mr.*

Hirzel (Vienna: Krüchten, 1763), 80.

12. Although de Haen published several hundred pages on the topic he never explicitly discussed this problem.
13. A. de Haen, *Vindiciae Difficultatum circa Modernorum Systema de Sensibilitate et Irritabilitate Humani Corporis, contra... Alberti v. Haller ad Easdem Difficultates Apologiam* (Vienna: Krüchten, 1762), 202.
14. De Haen, *op. cit.* (note 13), 234.
15. A. de Haen, *Difficultates circa Modernorum Systema de Sensibilitate et Irritabilitate Humani Corporis* (Vienna: Krüchten, 1761), 108–10.
16. 'De Insensibilitate, atque Irritabilitate Halleriana', *De Bononiensi Scientiarum et Artium Instituto atque Academia Commentarii*, 4 (1757), 208–17: 216.
17. Similar positions were maintained by Arrigoni (1756), Bianchi (1755, 1757, 1759), Brandt (1754), Girard (1757, 1763), Gnauck (1772), Goldhagen (1765), Kergel (1777), Krause (1755, 1756), Lorry (1756–57), Lotteri (1757) and Petrioli (1757), Struve (1765) (see bibliography).
18. For Le Cat's notions on movement and sensation see his *Dissertation... sur le Principe de l'Action des Muscles* (Berlin: Haude und Spener, 1753); 'Dissertation sur la Sensibilité de la Dure-Mere, de la Pie-Mere, des Membranes, des Ligamens, des Tendons, &c. sur l'Insensibilité du Cerveau & sur l'Irritabilité Hallerienne', *Journal Encyclopédique*, 1757, IV/iii, 129–41; V/i, 128–40; V/ii, 131–43; V/iii, 129–41; VI/i, 132–41; *Traité de l'Existance, de la Nature et des Propriétés du Fluide des Nerfs, et Principalement de son Action dans le Mouvement Musculaire... Suivie des Dissertations sur la Sensibilité des Meninges, des Tendons &c., l'Insensibilité du Cerveau, la Structure des Nerfs, l'Irritabilité Hallérienne, &c.* (Berlin: n.p, 1765).
19. Le Cat, *Dissertation 1753* (note 18), 20–1, 56–8.
20. *Ibid.*, 56–9.
21. Le Cat, *Traité* (note 18), 307.
22. *Ibid.*, 298.
23. *Ibid.*, 314–15.
24. See eg. the review in *Jenaische Gelehrte Anzeigen* 1755, 235–40.
25. J.P. Eberhard, *Conspectus Medicinae Theoreticae... Pars Prima: Physiologia et Diaetetica* (Halle: Renger, 1757); G. Azzoguidi, *Institutiones Medicae* (2 vols, Bologna: Saxius, 1775).
26. Caldani's textbook are the *Institutiones Physiologicae* (Padua: Comino, 1773). For his other publications on irritability and sensibility see the Bibliography.
27. For biographical details see H.R. Abe and W. Kaiser, 'Beiträge zum Leben und Werk von Andreas Elias Büchner (1701–1769)', *Beiträge zur Geschichte der Universität Erfurt (1392–1816): Zur Erfurter Medizin- und Hochschulgeschichte*, 18 (1975–1978), 91–138.
28. K.W. Schultz, *De Differentia Sensationis et Irritationis... Praeside... Andrea*

Elia Büchnero... (M.D. thesis, Halle: Hilliger, 1755).

29. G.V. Köhler, *De Destructa Partium Irritabilitate per Incongruum Adstringentium Usum... Praeside... Andrea Elia Büchnero...* (M.D. thesis, Halle: Hendel, 1763).
30. J.A. Pakendorf, *De Legibus Irritabilitatis Generalioribus... Praeside... Andrea Elia Büchnero...* (M.D. thesis, Halle: Hendel, 1765).
31. Haller did not totally refute the term *vis vitalis*, and in some cases used it himself, but he preferred the designation *vis insita* (see Chapter 3).
32. G.C. Busse, *De Actione Cordis Quatenus a Nervis Pendet... Praeside... Andrea Elia Büchnero...* (M.D. thesis, Halle: Litteris Stephanianis, 1769).
33. See for instance the works of Helwig (1760), Rehfeld (1771), Isenflamm (1774), and Niderhueber (1777). Some non-German authors maintained similar positions, eg. Smith (1767) and Fabre (1770) (see Bibliography).
34. J.G. Zimmermann, *De Principio in Homine Impetum Faciente Dissertatio*, fol. 4r–6v (*Zimmermann Papers*, MS XLII, 1933, B 6e).
35. See Tissot to Haller, 21 May 1754. *Corr. Tissot*, 26, 28.
36. See S.-A. Tissot, *Vie de Zimmermann* (Lausanne: Fischer & Vincent, 1797), 11–13.
37. *Mémoires*, i, xlv.
38. *Irritable Parts*, 4.
39. *Mémoires*, i, xlvii. On this passage see also F. Duchesneau, *La Physiologie des Lumières: Empirisme, Modèles, Théories* (The Hague: Nijhoff, 1982), 160–1.
40. To some extent this analogy had already been suggested by Haller who had declared that the cause of irritability was unknown just as the cause of gravity and attraction. *De Partibus*, 154.
41. *Johannes Franciscus Cigna... Publice Disputabat... MDCCLVII. Die XIV. Aprilis* (Turin: Typographia Regia, 1757); *Sull' Insensibilità e Irritabilità di Alcune Parti degli Animali: Dissertazioni de' Signori Haller, Zimmerman e Castell; Trasportate in Lingua Ital. dal Gian Vincenzo Petrini... ; Colle Lettere del Urbano Tosetti, sullo stesso Argomento* (Rome: Zempel, 1755), preface, ix–xxix.
42. Weise, *op. cit.* (note 1).
43. J.-L. Moreau de la Sarthe (ed.), *Œuvres de Vicq d'Azyr* (6 vols, Paris: Duprat-Duverger, 1805), iv, 80–1, 15–16.
44. *Ibid.*, v, 32–3 and iv, 233.
45. F. Fontana, *De Irritabilitatis Legibus* (Lucca: Riccomini, 1767); Fontana, *Ricerche Filosofiche sopra la Fisica Animale* (Florence: Cambiagi, 1775). It has to be stressed, though, that Fontana's works owed much to the critical remarks of Caldani. See: R.G. Mazzolini and G. Ongaro (eds), *Epistolario di Felice Fontana: I. Carteggio con Leopoldo Marc'Antonio Caldani 1758–1794* (Trento: Società di Studi Trentini di Scienze Storiche, 1980).

46. F. Fontana, *Traité sur le Venin de la Vipère* (Florence: n.p., 1781), 320–1.
47. On Fontana's late views see W. Bernardi, *I Fluidi della Vita. Alle Origni della Controversia sull'Elettricità Animale* (Florence: Olschki, 1992), 245–61.
48. T. de Bordeu, 'Recherches Anatomiques sur la Position des Glandes et sur leur Action' [Paris, 1751], *Œuvres Complètes de Bordeu* (2 vols, Paris: Caille and Ravier, 1818), i, 45–208: 203–6.
49. F. Boissier de Sauvages, 'A Messieurs les Journalistes de la Bibliothèque Raisonnée', *Recueil Périodique d'Observations de Médecine, Chirurgie, Pharmacie &c.*, Vol. 3 (1755), 83–116: 85. R. Whytt, *An Essay on the Vital and other Involuntary Motions of Animals* (Edinburgh: Hamilton, Balfour and Neill, 1751), 277.
50. R. Whytt, *Physiological Essays* (2nd edn, Edinburgh: Hamilton, Balfour and Neill, 1761), 222; Sauvages, *op. cit.* (note 49), 85; H.F. Delius, *Animadversiones in Doctrinam de Irritabilitate, Tono, Sensatione et Motu Corporis Humani* (Erlangen: Camerarius, 1752), vii.
51. 'Antwort' [answer to a letter of Sauvages], *Fränkische Sammlungen*, 5 (1760), 481–97: 488.
52. Whytt, *op. cit.* (note 49), 267.
53. R. French, 'Ether and Physiology', in G.N. Cantor and M.S. Hodge (eds), *Conceptions of Ether: Studies in the History of Ether Theories 1740–1900* (Cambridge: Cambridge University Press, 1981), 110–34: 130.
54. On Sauvages's physiology see R. French, 'Sauvages, Whytt and the Motion of the Heart: Aspects of Eighteenth Century Animism', *Clio Medica*, 7 (1972), 35–54; R. French, 'Sickness and the Soul: Stahl, Hoffmann and Sauvages on Pathology'; J. Martin, 'Sauvages's Nosology. Medical Enlightenment in Montpellier'. The latter two articles appeared in A. Cunningham and R. French (eds), *The Medical Enlightenment of the Eighteenth Century* (Cambridge: Cambridge University Press, 1990), 88–110 and 111–37.
55. F. Labroquere 'Dissertatio Medica de Motum Vitalium Causa... Praeside D.D. Francisco De Sauvages', [Montpellier, 1740], in A. Haller (ed.), *Disputationum Anatomicarum Selectarum Volumen iiii* (Göttingen: Vandenhoeck, 1749), 481–515.
56. Martin, *op. cit.* (note 54).
57. Labroquere, *op. cit.* (note 55), 504–8.
58. *Ibid.*, 500.
59. Sauvages to Haller, 12 November 1751 (*Haller Papers*). Following quote also from this source.
60. Sauvages to Haller, 10 May 1753 (*Haller Papers*).
61. Haller to Sauvages, 15 October 1762 (*Haller Papers*).
62. F. Boissier de Sauvages, 'Von der Reitzbarkeit der Theile des Menschlichen Körpers: Auszug eines Schreibens: Aus dem Lateinischen', *Fränkische*

Sammlungen, 1 (1756) 146–50.

63. J.-J. Dupont, *Dissertatio Medica Opposita Argumentis Celeberrimi Eberhardi de Animae Imperio in cor… Praeside Francisco de Sauvages* (Avignon: Tilan, 1760), 19.
64. Sauvages, *op. cit.* (note 49), 103; Sauvages, *op. cit.* (note 62).
65. F. Boissier de Sauvages, *Nosologie Méthodique* (10 vols, Lyons: Bruyset, 1772), i, 255–6.
66. The importance of Borelli has also been stressed by French, *Sauvages* (note 54).
67. In particular, he gained the conviction that muscular parts detached from the body are still governed by the soul independently of Whytt. He only became aware of Whytt through Haller's letter of 21 May 1752. See his answer of 15 August 1752 (*Haller Papers*).
68. Letter to Whytt, 1 May 1765, edited in French, *Sauvages* (note 54), 35.
69. *Relationes de Libris Novis*, 3 (1752), 155.
70. R. French, *Robert Whytt, the Soul, and Medicine* (London: The Wellcome Institute of the History of Medicine, 1969); Duchesneau, *op. cit.* (note 39), Chapter 6; J.P. Wright, 'Metaphysics and Physiology: Mind, Body, and the Animal Economy in Eighteenth-Century Scotland', in M.A. Stewart (ed.), *Studies in the Philosophy of Scottish Enlightenment* (Oxford: Clarendon Press, 1990), 251–302; A.-H. Maehle, *Drugs on Trial: Experimental Pharmacology and Therapeutic Innovation in the Eighteenth Century* (Amsterdam: Rodopi, 1999), Chapter 3; the dissertations of M.R. Barclay (Edinburgh, 1922), G. Miller (Johns Hopkins, 1939), and H. Hürzeler (Zurich, 1973), all of them devoted to Whytt, are not worth consulting.
71. Whytt, *op. cit.* (note 49), preface.
72. *Ibid.*, 268.
73. *Ibid.*, 239–40.
74. As we have seen in Chapter 3, for Haller, on the contrary, activity of matter agreed perfectly well with Newtonian philosophy.
75. W. Porterfield, 'Essay Concerning the Motions of Our Eyes', *Medical Essays and Observations*, iii (1735), 160–261, iv (1737), 124–294.
76. *The Works of Robert Whytt* (Edinburgh: Balfour, 1768), 154–62.
77. Whytt, *op. cit.* (note 49), 289.
78. French, *op. cit.* (note 70), 78.
79. W. Porterfield, *A Treatise on the Eye, the Manner and Phaenomena of Vision* (2 vols, Edinburgh: Miller, 1759), ii, 162.
80. Whytt, Essay *op. cit.* (note 49), 240, 244–5, 262–5.
81. *Ibid.*, 250–51.
82. R. Whytt, *Physiological Essays Containing Observations on the Sensibility and Irritability of the Parts of Men and other Animals, Occasioned by Dr. Haller's*

Late Treatise on these Subjects (Edinburgh: Hamilton, Balfour and Neill, 1755), 173–4.
83. *Ibid.*, 107–8; Whytt, *op. cit.* (note 50), 254–5.
84. A. von Haller, *Ad Nuperum Scriptum Roberti Whyttij Apologia; Leopoldi M. Antonij Caldani ad Albertum Allerum [sic] Epistola* (Yverdon: n.p., 1764).
85. Whytt, *op. cit.* (note 49), 229; Whytt, *op. cit.* (note 50), 268–9.
86. Whytt, *op. cit.* (note 82), 164–78.
87. Whytt, *op. cit.* (note 49), 241.
88. Whytt, *op. cit.* (note 82), 177.
89. On Delius see W. Kaiser, 'Heinrich Friedrich Delius (1720–1791)', *Harz-Zeitschrift*, 31 (1979), 65–82.
90. *Fränkische Sammlungen von Anmerkungen aus der Naturlehre, Arzneygelahrtheit, Oekonomie und den damit Verwandten Wissenschaften* (Nuremberg: Monath, 1756–68).
91. H.F. Delius, *Theoria et Foecundus in Medicina Usus Principii: Sensationem Sequitur Motus Sensatio Proportionatus* (Erlangen: Becker, 1749). I have used Delius's own German translation in the *Hamburgisches Magazin*, 16 (1756), 191–217.
92. Delius, *op. cit.* (note 50).
93. E. Bergmann, *Die Satiren des Herrn Maschine: Ein Beitrag zur Philosophie- und Kulturgeschichte des 18 Jahrhunderts* (Leipzig: Wiegandt, 1913); R. de Saussure, 'Haller and La Mettrie', *Journal of the History of Medicine*, 4 (1949), 431–49; *La Mettrie's L'Homme Machine: A Study in the Origins of an Idea,* Critical Edition with an Introductory Monograph and Notes by A. Vartanian (Princeton: Princeton University Press, 1960); E. Hintzsche, 'Neue Funde zum Thema 'L'Homme Machine' und Albrecht Haller', *Gesnerus*, 25 (1968), 135–66; U.P. Jauch, *Jenseits der Maschine: Philosophie, Ironie und Ästhetik bei Julien Offray de La Mettrie (1709–1751)* (Munich, Vienna: Hanser, 1998), Chapter 8.
94. H. Boerhaave, *Institutions de Médecine... Seconde Edition, avec un Commentaire. Par M. de la Mettrie* (8 vols, Paris: Huart, 1743–50).
95. For details see T. Verbeek, *Le Traité de l'Ame de la Mettrie. Edition Critique du Texte avec une Introduction et un Commentaire Historiques* (2 vols, Utrecht: OMI-Grafisch Bedrijf, 1988), ii, 13–26; A. Thomson, 'La Mettrie, Lecteur et Traducteur de Boerhaave', *Dix-huitième Siècle*, 23 (1991), 23–29; K. Wellman, *La Mettrie: Medicine, Philosophy, and Enlightenment* (Durham: Duke University Press, 1992), Chapter 5.
96. Certain materialist tendencies are already present in La Mettrie's own remarks in the Boerhaave-edition. See Thomson, *op. cit.* (note 95).
97. See Haller's review in *GGA*, 1747, 413–15.
98. J.O. de La Mettrie, *Œuvres Philosophiques* (London: Nourse, 1751), 145,

derived from *Praelectiones*, iv, 616.

99. La Mettrie, *op. cit.* (note 93), 90–3; Hintzsche, *op. cit.* (note 93), 159. It is not certain, however, whether La Mettrie met Winter before the publication of his book, as the latter presumably arrived only in August 1747, the same month in which La Mettrie may have completed the *Homme Machine.*
100. La Mettrie, *op. cit.* (note 93), 181–2, 186.
101. *Ibid.*, 180.
102. *Ibid.*, 186–97.
103. La Mettrie's thoughts were in no way 'a genial anticipation' of Haller's concept of irritability, as R. Toellner, 'Anima et Irritabilitas: Hallers Abwehr von Animismus und Materialismus', *Sudhoffs Archiv*, 51 (1967), 129–44: 138) and others have argued.
104. Cf. La Mettrie, *op. cit.* (note 93), 15–17; Jauch, *op. cit.* (note 93), passim.
105. See the dissertation of his pupik F.A. Scheid, *De Irritabilitate a Materialismo Vindicata* (M.D. thesis, Leiden: Hoogeveen and Heyligert, 1772).
106. The dedication was due less to the fact that La Mettrie based his materialist system on notions particular to Haller, and more because the latter had already criticised the Breton earlier and because he was one of the most distinguished authors who represented the unity of science and religion. Besides, his lack of humour made him an easy target.
107. *Irritable Parts*, 67 and *GGA* 1752, 459.
108. A. von Haller, *Briefe über einige Einwürfe Nochlebender Freygeister wider die Offenbarung* (3 vols, Bern: Typographische Gesellschaft, 1775–7), ii: 212–3.
109. See *GGA* 1770, 699–700.
110. *Elementa*, v, 529.
111. These sub-microscopic motions of the nervous fibres should not be confounded with the visible movements of the muscular fibres.
112. *Elementa*, v, 551–2.
113. Haller, *op. cit.* (note 108), ii, 97–8.
114. *GGA* 1762, 997–8.
115. Delius, *op. cit.* (note 51), 483–7.
116. Platner, *op. cit.* (note 3), 377–8.
117. *Ibid.*, 375.
118. Cf. G. Barsanti, 'Les Phénomènes 'Étranges' et 'Paradoxaux' aux Origines de la Première Révolution Biologique (1740–1810)', in Cimino and Duchesneau, *op. cit.* (note 4), 67–82.
119. R. Wittern (ed.), *Die Professoren und Dozenten der Friedrich-Alexander-Universität Erlangen: 1743–1960* (Erlangen: Universitäts-Bibliothek, 1993), 321–2.
120. Haller listed the supporters of animism in *Elementa*, iv, 522–3.
121. R. Toellner, 'Mechanismus – Vitalismus: Ein Paradigmawechsel? Testfall

Haller', in A. Diemer (ed.), *Die Struktur Wissenschaftlicher Revolutionen und die Geschichte der Wissenschaften* (Meisenheim: Hain, 1977), 61–72.

122. T.S. Kuhn, *The Structure of Scientific Revolutions* (Chicago: University of Chicago Press, 1962; 3rd edn 1996).
123. H. Driesch, *Der Vitalismus als Geschichte und als Lehre* (Leipzig: Barth, 1905), 1.
124. F. Duchesneau, 'Vitalism in Late 18th-Century Physiology: The Cases of Barthez, Blumenbach, and John Hunter', in W.F. Bynum and R. Porter (eds), *William Hunter an the Eighteenth Century Medical World* (Cambridge: Cambridge University Press, 1985), 259–95; F. Duchesneau, 'Territoires et Frontières du Vitalisme (1750–1850)', in Cimino and Duchesneau, *op. cit.* (note 4), 297–349.
125. H.D. Gaub, *Sermo Academicus de Regimine Mentis quod Medicorum Est* (Leiden: Van der Aa, 1747). The text is translated and commented by L.J. Rather, *Mind and Body in Eighteenth-Century Medicine: A Study Based on Jerome Gaub's De Regime Mentis* (London: The Wellcome Medical Historical Library, 1965).
126. Gaub, *op. cit.* (note 125), 41; Rather, *op. cit.* (note 125), 64.
127. Some of these are discussed by G. van der Waa, *'De Irritabilitate': Een Onderzoek naar de Betekenis van het Irritabiliteitsbegrip in de Geschiedenis van de Achttiende-Eeuwse Nederlandse Fysiologie* (Rotterdam: Erasmus, 1992).
128. H.D. Gaub, *Institutiones Pathologiae Medicinalis* (Leiden: Luchtmans, 1758). The later editions (Leiden 1763, 1781) contain only slight changes.
129. Gaub's pathology is discussed by P. Pogány-Wnendt, *Das Mechanistische Denken in der Modernen Medizin im Spiegel ihrer Geschichtlichen Entwicklung: Hieronimus David Gaub (1705–1780)* (Frankfurt am Main: Lang, 1991). The author fails, however, to recognise the importance of vital forces in Gaub's concept.
130. Gaub, *op. cit.* (note 128), § 169–88.
131. This aspect is especially developed in his *Sermo Academicus alter de Regimine Mentis quod Medicorum Est* (Leiden: Van der Aa, 1763), also translated and commented by Rather, *op. cit.* (note 125).
132. Gaub, *op. cit.* (note 128), § 647–9.
133. *Ibid.*, § 189–95.
134. W.F. Verschuir, *De Arteriarum et Venarum Vi Irritabili* (M.D. thesis, Groningen: Bolt, 1766); Verschuir, *Oratio Inauguralis de Recentiorum Medicorum, Inprimis Belgarum, Meritis* (Groningen: Spandaw, 1781).
135. See *GGA* 1756, 698–9 and 1758, 889.
136. German vitalism, or rather, the different theories of organic forces in the late-eighteenth century and Naturphilosophie, have also been studied in detail, but the development of these ideas in the early decades of the second

half of the century is much less understood than in the French case.

137. The most important work is R. Rey, *Naissance et Développement du Vitalisme en France de la Deuxième Moitié du 18e Siècle à la Fin du Premier Empire* (Oxford: Voltaire Foundation, 2000). See also her article 'Vitalism, Disease and Society', in R. Porter (ed.), *Medicine in the Enlightenment* (Amsterdam: Rodopi, 1995), 274–88. Further essential studies are Duchesneau, *op. cit.* (note 39); E. Haigh, *Xavier Bichat and the Medical Theory of the Eighteenth Century* (London: Wellcome Institute for the History of Medicine, 1984); E.A. Williams, *The Physical and the Moral: Anthropology, Physiology and Philosophical Medicine in France, 1750–1850* (Cambridge. Cambridge University Press, 1994); E.A. Williams, *A Cultural History of Medical Vitalism in Enlightenment Montpellier* (Aldershot: Ashgate, 2003).
138. Bordeu, *op. cit.* (note 48), 144.
139. *Ibid.*, 163–4.
140. *Ibid.*, 200–1.
141. *Ibid.*, 187.
142. 'Recherches sur le Pouls, par Rapport aux Crises' [Paris, 1756], in Bordeu, *op. cit.* (note 48), i, 253–421: 420.
143. Bordeu, *op. cit.* (note 48), ii, 668.
144. For biographical information on the *Encyclopédie* authors see F.A. and S.L. Kafker (eds), *The Encyclopedists as Individuals: A Biographical Dictionary of the Authors of the Encyclopédie* (Oxford: Voltaire Foundation, 1988).
145. The only other vitalist authors were Bordeu, who only wrote one, albeit important article ('Crise'), and Barthez who contributed eighteen entries, which are of no particular importance.
146. See Ménuret's articles 'Inflammation', 'Paraphrénésie', 'Pouls' and 'Sciatique' in the *Encyclopédie.*
147. *Encyclopédie*, viii, 909. This article is signed by Gabriel-François Venel (1723–1775), a colleague of Ménuret and close friend of Diderot.
148. P.-J. Barthez, *Quaestiones Medicae Duodecim* (Montpellier: Martel, 1761), 40–1.
149. See Barthez to Haller, 6 October 1773 (*Haller Papers*) and Haller to Barthez, 16 October 1773. The latter letter is edited by J. Lordat, *Exposition de la Doctrine Médicale de P.-J. Barthez, et Mémoires sur la Vie de ce Médecin* (Paris: Gabon, 1818), 111.
150. Barthez presented his new ideas first in an oration of 1772 – the one to which Haller objected – and developed them especially in his *Nouveaux Éléments de la Science de l'Homme* (Montpellier: Martel, 1778, 2nd edn Paris: Gujon and Brunot, 1806).
151. For short descriptions of many of these theories see J. Jantzen, 'Physiologische Theorien', in J. Jantzen, M. Durner and F. Moiso (eds),

Wissenschaftshistorischer Bericht zu Schellings Naturphilosophischen Schriften 1797–1800 (Stuttgart: Frommann-Holzboog, 1994), 375–668.

152. Medicus, *op. cit.* (note 2); J.C. Reil, 'Über die Lebenskraft', *Archiv für die Physiologie*, 1 (1796), 8–162 (Reprint Leipzig: Barth, 1910)
153. Unzer was especially famous as editor of the journal *Der Arzt.* See M. Reiber, *Anatomie eines Bestsellers: Johann August Unzers Wochenschrift 'Der Arzt' (1759–1764)* (Göttingen: Wallstein, 1999). Some aspects of his physiology are discussed in G. Canguilhem, *La Formation du Concept de Réflexe aux XVIIe et XVIIIe siècles* (2nd edn, Paris: Vrin, 1977), 108–12.
154. The general thoughts of Unzer's physiology are laid out in his *Erste Gründe einer Physiologie der Eigentlichen Thierischen Natur Thierischer Körper* (Leipzig: Weidmann, 1771). In answer to the reviews of this book – the most important of which was Haller's – he published a further account, *Physiologische Untersuchungen* (Leipzig: Weidmann, 1773).
155. The most notable works are A.D. Thaer, *De Actione Systematis Nervosi in Febribus* (M.D. thesis, Göttingen: Dietrich, 1774) and G. Prochaska, *Adnotationum Academicarum Fasciculus Tertius: De Functionibus Systematis Nervosi Commentatio* (Prague: Gerle, 1784).
156. On eighteenth-century embryology see J. Roger, *Les Sciences de la Vie dans la Pensée Française au XVIIIème Siècle: La Génération des Animaux de Descartes à l'Encyclopédie* (Paris: Collin, 1963); R.G. Mazzolini and S.A. Roe, *Science Against the Unbelievers: The Correspondence of Bonnet and Needham, 1760–1780* (Oxford: Voltaire Foundation, 1986); W. Bernardi, *Le Metafisiche dell'Embrione: Scienze della Vita e Filosofia da Malpighi a Spallanzani (1672–1793)* (Florence: Olschki, 1986).
157. O.M. Pagani and C. Bonioli, *Delle Parti Insensibili, ed Irritabili degli Animali Discorso Teorico-Pratico* (Venice: Occhi, 1757), 99–100.
158. 'Sensibilité', *Encyclopédie*, xv: 40–1.
159. J.T. Needham, *Nouvelles Observations Microscopiques* (Paris: Ganeau, 1750), esp. 221.
160. Haller's embryology cannot be discussed here. The best description is A. von Haller, *Commentarius de Formatione Cordis in Ovo Incubato*, M.T. Monti (ed.), (Basel: Schwabe, 2000), Introduction.
161. C.F. Wolff, *Theoria Generationis* (M.D. thesis, Halle: Hendel, 1759); Wolff, *Theorie von der Generation* (Berlin: Birnstiel, 1764). On Wolff see S.A. Roe, *Matter, Life and Generation: Eighteenth-Century Embryology and the Haller-Wolff Debate* (Cambridge: Cambridge University Press, 1981); R. Mocek, 'Caspar Friedrich Wolffs Epigenesis-Konzept – Ein Problem im Wandel der Zeit', *Biologisches Zentralblatt*, 114 (1995), 179–90.
162. J.F. Blumenbach, *Über den Bildungstrieb und das Zeugungsgeschäfte* (Göttingen: Dieterich, 1781); Blumenbach, *Institutiones Physiologicae*

(Göttingen: Dieterich, 1786). On Blumenbach see – besides the already mentioned works of Duchesneau – T. Lenoir, *The Strategy of Life: Teleology and Mechanics in Nineteenth Century German Biology* (Dordrecht and London: Reidel, 1982); P. McLaughlin, 'Blumenbach und der Bildungstrieb: Zum Verhältnis von Epigenetischer Embryologie und Typologischem Artbegriff', *Medizinhistorisches Journal*, 17 (1982), 357–72; R.J. Richard, 'Kant and Blumenbach on the 'Bildungstrieb': A Historical Misunderstanding', *Studies in the History and Philosophy of Biological and Biomedical Sciences*, 31 (2000), 11–32.

163. Blumenbach, *Bildungstrieb* (note 161), (2nd edn, Göttingen: Dieterich, 1789), 31.
164. On this aspect see J.L. Larson, 'Vital Forces: Regulative Principles or Constitutive Agents? A Strategy in German Physiology, 1786–1802', *Isis*, 70 (1979), 235–49 and B. Lohff, 'The Concept of Vital Force as a Research-Program', in Cimino and Duchesneau, *op. cit.* (note 4), 127–42.
165. The authors mentioned below are all discussed in Jantzen, *op. cit.* (note 151), *ad indicem.*
166. Reil, *op. cit.* (note 152).
167. J.D. Metzger, *Ueber Irritabilität und Sensibilität als Lebensprincipien in der organisierten Natur* (Königsberg: Hartung, 1794), iii–iv, 1.
168. Wolfgang Pross, '"Ein Reich unsichtbarer Kräfte". Was kritisiert Kant an Herder ?', *Scientia Poetica*, 1 (1997), 62–119: 83; J.G. Herder, *Ideen zur Philosophie Geschichte der Menschheit*, W. Pross (ed.), (2 vols, München: Hanser, 2002), ii, 184–91.
169. Unzer, *Physiologie* (note 154), preface.
170. For a general survey see C.J. Lawrence, 'The Nervous System and Society in the Scottish Enlightenment', in B. Barnes and S. Shapin (eds), *Natural Order. Historical Studies of Scientific Culture* (Beverly Hills: Sage, 1979), 19–40; C.J. Lawrence, 'Ornate Physicians and Learned Artisans: Edinburgh Medical Men, 1727–1776', in W.F. Bynum and R. Porter (eds), *William Hunter and the Eighteenth-Century Medical World* (Cambridge: Cambridge University Press, 1985), 153–76; Wright, *op. cit.* (note 70).
171. See his experimental essay in the *Essays and Observations Physical and Literary*, 3 (1771), 282–365 and his *Observations on the Structure and Functions of the Nervous System* (Edinburgh: Creech, 1783), 90–104.
172. On Cullen's physiology see W.F. Bynum, 'Cullen and the Nervous System', in A. Doig *et al.* (eds), *William Cullen and the Eighteenth Century Medical World* (Edinburgh: Edinburgh University Press, 1993), 152–62; and Wright, *op. cit.* (note 70).
173. The first edition from 1772 seems to be very rare, Haller didn't know it. I have used the third edition: *Institutions of Medicine: Part I, Physiology*

(Edinburgh: Elliot, 1785).

174. This short exposition is based on Cullen's textbook, especially § 29–144.
175. On Hunter's physiology see S.J. Cross, 'John Hunter, the Animal Oeconomy and Late Eighteenth Century Physiological Discourse', *Studies in History of Biology*, 5 (1981), 1–110; Duchesneau, *op. cit.* (note 124); and S. Jacyna, 'Physiological Principles in the Surgical Writings of John Hunter', in C. Lawrence (ed.), *Medical Theory, Surgical Practice: Studies in the History of Surgery* (London: Routledge, 1992), 135–52.
176. J.F. Palmer (ed.), *The Works of John Hunter* (4 vols, London: Longman *et al.*, 1835), i: 199–643.
177. J. Hunter, *A Treatise on Blood, Inflammation, and Gun-Shot Wounds* (London: Richardson, 1794).
178. *Ibid.*, 76.
179. *Ibid.*, 26.
180. Hunter, *op. cit.* (note 175), i, 223.
181. P.-J. Barthez, *Nouvelle Méchanique des Mouvements de l'Homme et des Animaux* (Carcassonne: Polere, An VI, 1798), v.
182. Pringle to Haller, late winter or early spring 1764, in: O. Sonntag (ed.), *John Pringle's Correspondence with Albrecht von Haller*, (Basel: Schwabe, 1999), 58.
183. M.-J.-A.-N. Caritat de Condorcet, 'Éloge de M. de Haller', *Histoire de l'Academie Royale des Sciences*, année 1777 (Paris: Impr. Royale, 1780), 127–54: 139–40.
184. Unzer, *Physiologie* (note 154), preface; C.H. Pfaff, *Über Thierische Elektricität und Reizbarkeit: Ein Beytrag zu den Neuesten Entdeckungen über diese Gegenstände* (Leipzig: Crusius 1795), 236.
185. 'Éloge de Haller, Lu dans la Séance Publique du 30 Avril 1778', in E.-F. Dubois (ed.), *Éloges Lus dans les Séances Publiques de l'Académie Royale de Chirurgie de 1750 à 1792 par A. Louis* (Paris: Barillière, 1859), 265–81: 274–5.
186. Vicq d'Azyr, *op. cit.* (note 43), iv, 12.
187. Pfaff, *op. cit.* (note 184), 237–40.
188. Haller to Tissot, 17 September 1764. *Corr. Tissot*, 192.
189. A general survey is given by T.S. Hall, 'On Biological Analogs of Newtonian Paradigms', *Philosophy of Science*, 35 (1968), 6–27. For an often diverging assessment of Newtonianism in several physiologists see Duchesneau, *op. cit.* (note 39), *ad indicem*.

6

The Debate and the Medical and Public Sphere

Medical theory in the eighteenth century was not the exclusive domain of learned physicians. It had always been part of the wider discourse of natural philosophy but in the ambitious enterprise of the Enlightenment it received heightened attention. It could be pursued as a 'science of man' which linked the physical with the moral and thus played an important role in the ideas of many enlightened thinkers, especially in France. As the sought-after rationale for the explanation and therapy of diseases it was essential for those who considered medicine to be not only an empirical art but a theoretical science. The confirmation and use of medical theory in clinical practice was a touchstone of the project of the Enlightenment, which wanted to be not only empirical but also rational. Medicine and medical theory were discussed in books and journals for a general reading public as never before.

The medical concept *par excellence*, which established such a link between the physical and the moral, formulated a theoretical model of practical relevance and affected the larger public by medical books and even more by journals and novels, was the concept of sensibility as it was developed especially in France. In the first half of the century *sensibilité* was mainly used to describe the faculty of reacting immediately and without reflection to certain impressions.[1] As a semi-conscious act between reason and passion it reflected the good nature of a man. From the 1750s onwards sensibility was increasingly discussed also as a physiological quality. The model for this discourse was Bordeu's concept of secretion as presented in his *Recherches Anatomiques sur la Position des Glandes et leur Action* (1751).[2] In an exemplary manner Bordeu demonstrated the inadequacy of purely mechanical theories to explain secretion and other physiological processes and introduced sensibility as a guiding principle and as the general vital property of all parts of the body. He did not himself develop the moral implications of such a doctrine, but the intricate relation between physical and moral was to become a main element of French physiological thought from La Caze's *Idées de l'Homme Physique et Moral* (1755) to Cabanis's *Rapports du Physique et du Moral de l'Homme* (1802).[3] The French discussion was part of a much wider European movement in which sensibility and the nerves dominated the medical, moral and literary discourse.

Several authors have tried to link this very broad development with Hallerian physiology. But, in my view, their line of argument is not convincing.[4] Karl Figlio, for instance, has maintained that Haller's notion of sensibility as a physiological rather than a mechanical quality, and his concept of the *sensorium commune* as the place of the activity of the soul, mitigated the reigning dualism, and was decisive in the development of future notions of the brain and the soul.[5] But this assessment overlooks the fact that, despite Haller's late acknowledgement of an unconscious, involuntary, modulatory role of the nerves, his concept of sensibility always retained the notion of a conscious process. There was, on the one hand, a physical level of perception which was explicable in purely mechanical terms and, on the other hand, there was the soul, the real place of sensible activity, which alone could produce reactions. Irritability, although not entirely reducible to a mechanical response, was totally independent of the soul. Haller maintained a strict dualism, he did not enter into speculations about the relation between body and soul and he showed no particular interest for such metaphysical questions.[6] To place Haller at the beginning of a development towards a monistic view of human nature dominated by the nervous system is, therefore, rather a forced construction.

Anne Vila, to take another example, has made out a specific connecting link between Haller and the French account of sensibility in Charles Bonnet's *Essai Analytique sur les Facultés de l'Âme* (1759).[7] She conceives Bonnet's detailed description of anatomical microstructures and their movements in the brain as an extension of Haller's model of the reactive, irritable muscular fibres into the realm of the nervous system. Bonnet would thereby have fostered a monistic conception of one vital force which rejected Haller's separation of motion and sensation and allowed only for a single property of reactibility. But Bonnet's account of moving nervous fibres was not new; it essentially followed the micromechanics of fibres which Boerhaave and Haller had described for all parts of the body, including the brain. Only the extent of imagined details that Bonnet furnished was new. Just as in Haller, the proportion between stimulation and movement of the nervous fibres was maintained. This had to be considered as a purely mechanical and passive process, the only place of activity was the soul, the rest was a 'play of fibres' (*jeu de fibres*).[8] The motion of the sensible fibres happened on the microscopic level and was fundamentally different from the macroscopic motion of heart, intestines, and muscles. Bonnet always maintained the neat separation between irritability and sensibility, and his work seems not to have served as a means to transform Haller's physiology into a concept of an overall *sensibilité physique*.

George Rousseau, our last example, has seen an even closer link between Haller and the discourse of sensibility. In his view, it was not the work of Haller, Whytt and others that led to the sensibility movement. The foundations of this development were laid much earlier with Thomas Willis' location of the soul in the brain (1664) and especially with Locke's sensualism.[9] But – according to Rousseau – Haller privileged sensibility over irritability, 'made sensibility the centrepiece of his physiology', considered the nerves as the basis for all human passion and reason and delivered thus a physiological doctrine that served as a strong impetus for an entire literary movement.[10] This judgement, however, is based on a misapprehension of Haller's physiology. Haller always maintained the radical distinction between the physical basis of physiological processes and the immaterial realm of ideas. Like Boerhaave, he considered the reduction of thoughts and ideas to feelings as the clearest way to distinguish between a Spinozist and a Christian.[11] Rousseau's mistake, though, is illustrative. A pioneer in the study of the relations between medicine and literature, he was one of the first to stress the parallels between concepts of sensibility in both disciplines, and it was quite natural to tie it to the figure who put sensibility on the agenda of experimental research. But Haller did not see any link between his physiological work and the sensibility movement in literature. He would have had ample opportunities to declare it. An ardent admirer of Richardson, he wrote an enthusiastic review of *Clarissa* – which seems to have appealed to English readers and appeared also in the *Gentleman's Magazine* – arranged for the German translation of the novel and even discussed it with his students during their work in the anatomical theatre.[12] Could we imagine a better occasion to establish a link than the discussion of moral sensibility at the dissection table on which animals were tortured in order to determine their physical sensibility? Yet Haller refused to do so, the realms of the physical and the moral remained separate for him. Rousseau, of course, quite rightly connects the physiological with the literary discussion, but this should not be associated with Hallerian physiology and his particular notions of irritability and sensibility. It is true, Haller's work was well-known and Cullen's edition of his textbook was published in Edinburgh up to 1801. However, it was not Haller's but rather Cullen's physiology that was here – as in America – a point of reference in the language of sensibility.[13] Something similar happened in France, and here the case of Diderot is typical. Certainly, he studied Haller's physiology, but Bordeu was his man and indeed the physician in the *Rêve de d'Alembert* (1769).[14] These were deliberate choices as the authors were well aware of Haller's concept and its particularities. The work of the Swiss physiologist had, of course, fostered the general idea of the reactive body, but this was due to his notion of irritability and not sensibility. The image of the sensible body guided by the nervous system was entirely opposed to his concept. Nobody had, in such a decisive manner as Haller, stressed the independence of main physiological properties from the nerves. The whole discourse on the nerves presupposed the *unity* of the body, whereas Haller's main contribution had been the *separation* of its different forces. It is not the case, as Rousseau has claimed, that '*all* Enlightenment physiology was 'nervous'.'[15] A great deal of physiological discussion had nothing to do with the culture of sensibility.

The debate on Hallerian irritability and sensibility was, however, not entirely limited to the sphere of professional physiological discourse. Admittedly, the cruel experiments themselves, though witnessed by a considerable number of people, were not an appropriate public amusement – as the fascinating presentation of electrical phenomena was – and the presumed insensibility of tendons led to therapeutical implications that were in no way as obvious and important as those of inoculation. Nevertheless, the question of whether certain parts of the body were sensible or not, was of some consequence for surgical procedures, and the whole concept of reactive fibres could serve as a new pathological model. Surgeons and physicians were consequently interested in the debate not only as theorists but also as practitioners. In the first section I would therefore like to focus on the reception of Haller's concept within the professional medical sphere. Beyond the specifically medical discussion, however, the controversy on irritability and sensibility also attracted the attention of a wider audience. Through books and especially periodicals it was presented to the Republic of Letters at large and, in theory, to the public as a whole. It provoked the statements of non-specialists and, like many other controversies, raised the question of who was able to judge the matter and how this should be done. In this respect, the review journals played a key role and they will be at the centre of attention in the second section on the culture of criticism.

Pathology and the practice of medicine and surgery

In 1974, in an article which is considered a classic of the sociology of medical history, N.D. Jewson raised a number of interesting points concerning the relation between theory and practice in eighteenth century medicine.[16] He argued that the physician's method of explaining and treating disease was profoundly shaped by the demands of a genteel clientele that dominated the practitioner–patient relationship. Medical knowledge thus revolved around therapy and prognosis rather than around theory. Pathology furnished simple models to describe the source and cure of the ailments. It dealt less with the proximate and localised causes of disease than with the overall constitution of the body, and its actual condition was usually considered as the main reason of illness. As most medical researchers were also practitioners they were unable to develop a disinterested position detached from the opinions of the patients, which would have been necessary for a 'scientific' approach. The result was the lack of interest in aetiology, pathological anatomy and a certain distaste for the study of anatomy and physiology which, according to Jewson, the eighteenth century displayed.

While Jewson – writing in a time when the history of medicine was much more concerned with doctors than with patients – was quite right to stress the importance of the clientele, he certainly overstated the neglect of theoretical medicine. To some extent, this may be due to his focus on the English medical system, which was more entrepreneurial than that of other countries. But we should not forget that extensive theoretical knowledge was the main element distinguishing the university-trained physician from his humbler colleagues. François Quesnay (1694–1774), famous as an economist but equally active as a surgeon and physician, emphasised that extensive medical practice ought not to be confounded with experience.[17] Physiological understanding, rooted in physical, chemical and anatomical studies, was the basis of the art of healing and allowed us to penetrate to the real causes of disease and to discover the mechanism of the action of drugs. Only with such a knowledge, coupled with practical expertise, could a physician be said to have real experience,[18] and this is what distinguished him from a mere empiric. This argumentation was commonplace in mid-eighteenth century medical literature. Such a position was not, in the view of most physicians, in strict opposition to an observational approach to medicine as favoured by Sydenham. Boerhaave, his pupil Haller, and many others, were all admirers of the 'English Hippocrates'. Of course, they disagreed with Sydenham's notion of medical progress as arising from a simple accumulation of careful practical observations, but they thought that his approach, if combined with theoretical studies, was essentially right. The link between practice and theory had to be established through a pathological model. Although deduced from physiological theory and part of theoretical medicine itself, such a model was of great practical relevance as it served the physician to explain the disease and the therapeutic strategies to the patient. Besides that, more than the practice of healing itself, the rational explanation of disease was important for the physician's self-awareness as a learned practitioner.[19] The problem that increasingly worried the physicians was not that theory and practice should not be reconcilable but that the *actual* theoretical model did not adequately explain disease. The result was not only a certain disinterest in theory and a detachment of practice therefrom – which Jewson was quite right to notice – but also the search for a *new* physiological and pathological model. Both these tendencies and their effects on the reception of Haller's new findings will be discussed successively.

Physiological models and medical practice

Ancient medical theory, based on the notion of health as an equilibrium of humours and their qualities of warmth, cold, wetness and dryness, and of illness as humoral imbalance, had been well suited to explaining the numerous symptoms of disease. Mechanical systems, which dominated medical theory in the first half of the eighteenth century, were built on the idea of movement of non-visible particles and fibres and were therefore far more abstract and removed from experience.[20] Christian Gottlieb Ludwig (1709–73), medical professor in Leipzig, who published a widely used pathological textbook based on Boerhaavian mechanics, pointed to the weakness of this system.[21] Haller, in a review of the book, confirmed this view:

> The author admits himself that what he tells us in a Boerhaavian manner about the diseases of the basic particles of the fibres and humours is only based on reasoning and not on experience. We have to imagine these particles as too densely, too loosely and too tensely arranged, and equally as taking up a too large or too small space... whence we then have to deduce all different kinds of effects in a philosophical manner. It is only a pity, as the author rightly reminds us, that we cannot arrive thereby at a clear understanding of diseases.[22]

The basic units that Ludwig used in his pathological descriptions, therefore, were not minute particles but compound fluids and solids, which were – as he said – closer to our sensual experience and could be investigated in certain experiments.[23] His account lost some of its mechanical character thereby, and the unity of medical science, which he was at pains to demonstrate, was somewhat undermined. Ludwig delivered a comprehensive account of medicine in four consecutive textbooks, of which the one on pathology (1754) linked the first on physiology (1752) with those on the principles of therapeutics (1754) and the practice of medicine (1758). Formally, the unity of these works was emphasised by a single paragraphic numbering running throughout all four volumes. On the conceptual level, the mechanical model furnished the common ground. Ludwig was no extreme systematist and accepted certain inconsistencies within his corpus. The fact that a purely mechanical model had apparent problems in explaining disease, did not lead him to replace the model – to which he saw no alternative – but it did cause a certain dissatisfaction.

Quesnay went a step further. In his view, theories such as Boerhaavian mechanics, which were true but imprecise, led to practical and therapeutic principles which were themselves correct but vague. These were as unreliable as erroneous dogmas. Imperfect knowledge was easily followed by the creation of rational, invalid and dangerous theories.[24] Quesnay, who principally favoured mechanical doctrines, doubted their application to the practice of medicine and tended towards a more empirically-based description and treatment of diseases. This view was also – if not explicitly – displayed in Morgagni's famous *De Sedibus et Causis Morborum* (1761). Although a mechanist himself, Morgagni made no effort to explain diseases in terms of a general mechanical model of pathology.[25] He was content to connect particular diseases with specific lesions observed in post-mortem dissection; he was concerned with special, not with general pathology. The mechanists' interest in specific pathology may be compared to the vitalists' or animists' interest in nosology. Instead of looking for the internal causes of disease they focused on their external symptoms. In the works of Boissier de Sauvages, the most important author on nosology, diseases were classified by their appearances and not according to some causal relationship.[26] Since Sauvages rejected the mechanical model, he did not describe ailments as disorders on a structuro–functional level. Diseases could be considered as constituted by symptoms and it was neither necessary nor revealing to search for systematic explanations.[27] This was an opinion shared by many vitalists. Paul Victor de Sèze, a typical vitalist physician from Bordeaux, asked for restraint. It had been Boerhaave's 'unfortunate ability to explain everything' which had generated a whole crowd of *médecins causeurs*, to which the *médecins observateurs* were definitely preferable.[28] Pathology and therapeutics had to be released from rationalisation.

How this approach conflicted with Haller's new concept can best be seen in his dispute with the Viennese clinician, Anton de Haen. De Haen had been a pupil of Boerhaave himself, but rather than developing the theoretical parts of Boerhaavian medicine he built on what he considered Boerhaave's 'Hippocratic' heritage.[29] He was therefore less worried about the theoretical aspects of Haller's doctrine than about their practical implications. The debate – which included some fifteen publications from both protagonists and from some of their adherents – was probably the best known of all the disputes within the whole controversy on irritability and sensibility, and it was conducted with considerable belligerence. De Haen justified his attacks on Haller on the grounds that it was his duty as a leading authority and member of the Viennese Medical Faculty to fight against all systems that might have perilous effects on practical medicine. He feared that the supposed insensibility of the tendons would render surgeons incautious and that the theory of irritability would lead to simplified systems of therapeutics that set aside the principles of Hippocrates and Boerhaave.[30] Vienna was on the point of becoming a leading centre for practical medicine and considered itself as the true heir to the Leiden model. In the view of van Swieten and de Haen there was no question that progress in medicine could hardly be achieved by physiological research but rather by the collection of practical observations. They both worked for decades on their lasting monuments of aphoristic medicine.[31] Hypotheses and medical systems could not be accepted and were banned as the 'pest of medicine'. This conviction was shared by the Medical Faculty, which in its instruction of 1775 stated: 'The main care of our school is that the students will not become prejudiced and belligerent nor infected by hypotheses, which in most cases are harmful to the practice of our art.'[32] It is this ideal of a pure, immaculate practice of medicine to which the Viennese Medical Faculty and Anton de Haen in particular were devoted.

The details of the debate cannot be discussed here.[33] Suffice it to say that de Haen rejected the validity of Haller's experiments because they were performed under unnatural circumstances and upon animals, which in their whole nature differed from man (see Chapter 4). Only observations on patients were therefore admissible and these – as regards the question of sensibility – testified against Haller. After some years of profound disagreement the debate arrived at an interesting point at which a rapprochement of the positions seemed to become possible. In 1766, Haller admitted that if ever any nerves were to be found in the tendons and membranes, then the latter would be sensible in a small degree because of these few and very subtle nerves.[34] For de Haen this was a withdrawal of the initial claim of the absolute insensibility of the tendons. He saw the possibility of a reconciliation and asked Bonnet to act as a mediator.[35] De Haen initiated a correspondence with Haller, in which both sides elucidated their standpoints. Finally, Haller formulated an official statement and asked de Haen to publish it in his well-known multi-volume work, *Ratio Medendi*. Here, Haller explicated his position once more: there were nerves furnishing the skin which passed over the tendons and membranes without entering their structure. Their irritation caused a sensation which was mistakenly attributed to the tendons themselves. In the state of inflammation, parts provided with nerves were even more sensible. But there was no way that parts deprived of nerves could be rendered sensible. Haller concluded: 'As to pathology, I never wanted to get involved with.'[36] For de Haen, this was decisive: if Haller had made such a clear statement in 1752, he said, he would never have felt obliged to intervene. What did it matter whether the tendons and membranes themselves were insensible or not? It was of no importance to the practice of medicine and surgery whether the lesion of these structures was painful due to their own nerves or due to those passing nearby. The main thing to keep in mind was that wounds to the tendons and membranes inflicted intense pain, often caused severe disorders and had to be treated accordingly. The old principles of pathology and therapy were therefore preserved. It had not been Haller's intention to change pathology but only to extend physiological knowledge. De Haen drew a neat line between physiology and pathology but, of course, not exactly in the manner that Haller had intended. Haller never left any doubt that he considered his research to be of great practical relevance. Already, in his orations of 1752, he described it as 'the source of a great many changes, both in physiology, pathology and surgery.'[37] His intention had been to furnish physiological principles, and he left it to others to draw the pathological conclusions. This was not tantamount to the denial of any pathological relevance to his research. In fact, in a reply to de Haen's *Ratio Medendi*, Caldani insisted on the practical significance of Haller's findings and pointed to the conclusions that Gaub had drawn in his pathological work.[38] In Caldani's view, it was not Haller who had given up his initial position but de Haen who, with his admission of the insensibility of the tendons, turned out to be a true Hallerian. Quite typically, although the topics of the debate had changed, a real reconciliation between the two parties was – as throughout the whole controversy – not feasible.

De Haen's attempt to separate pathology from physiology and practice from the theory of medicine was, as we have seen, part of a general current. This had, of course, considerable bearing on the methods with which the new findings had to be tested. For Haller, observations on patients served mainly as a confirmation of facts previously established by animal experiments. But for many others, the question of whether lesions of tendons caused pain and serious accidents had to be resolved entirely on clinical grounds. William Hunter (1718–83), the famous anatomist and obstetrician, for instance, endorsed this opinion. He agreed with Haller that the tendons were not very sensible and that whenever an extraordinary pain was caused during the procedure of bleeding, this was due to a puncture of the nerve and not of the tendon. But he thought that Haller had gone too far in denying the tendons any sense of feeling. Only observations on the living human body could inform us about more obtuse and gentle sensations. Furthermore, although the tendons were only slightly sensible, this should not lead us to the conclusion that their laceration was harmless. In fact, Hunter was convinced that this was a serious complication with a potentially lethal outcome, especially in joints where the want of room caused more painful inflammations and thus gave full vent to the disease.[39]

Jacques Farjon (1719–1800), physician in Montpellier, did not share Hunter's view on these accidents. But, like the English anatomist, he had been converted not by Haller's reports but by his own practical experience. In his letter to Haller he argued that observations on patients rather than the previous animal experiments would convince the sceptics.[40] Another, anonymous French author in the *Mercure de France* exclaimed that the whole discussion based on animal experiments was futile. These authors should have consulted the surgeons in order to ascertain the facts.[41] Thus, in the view of many of Haller's contemporaries, there was a more or less autonomous area of medical investigation within which the question of the sensibility of tendons and membranes had to be decided: the observation of wounded patients. For them, it was therefore mainly a surgical problem, and it will now be discussed as such.

The surgical debate

Eighteenth-century commentators agreed that in their own century surgery had made significant progress. Many – particularly surgeons – argued that it had developed from a craft to a science. This claim was made especially by the *Académie Royale de Chirurgie* and most clearly formulated by its secretary Quesnay in the preface to the first volume of its *Mémoires*. In his view, surgery was characterised by an inductive scientific method, which distinguished it from the more theoretical speculations of the physicians as well as from the more empirical rules of the barbers. Observation and experimental investigation (*physique expérimentale*) were its two modes of research. The former dealt with the sensible qualities of the body and the latter with its structures and driving forces. Both these methods had to serve as a corrective to the other. Experimental findings in particular had to be checked for their consequences for the practice of medicine. Harvey's discovery, for instance, led to disbelief in the efficacy of certain drugs, the use of which was only continued after the collection of new observational evidence. The Academy, therefore, had to be familiar with experimental science but its main duty was to publish observations.[42] A similar understanding of the science of surgery seems to have prevailed in the other European countries. In the *Philosophical Transactions* which, as an embodiment of a scientific approach and as the only English periodical with a significant number of surgical contributions, may be compared to the French *Mémoires*, an average of eighty per cent of these articles were descriptions of extraordinary observations. Some twelve per cent reported improvements of technique and seven per cent may be called surgical experiments, as they compared the efficiency of one surgical method with another. 'Scientific' surgery in England was mainly considered as a collection of observational data and its critical assessment.[43] In the German countries, the personification of learned surgery in the mid-eighteenth century was Lorenz Heister (1683–1758). In his view, mechanics furnished the rational basis for both theoretical medicine and surgery. All diseases originated from mechanical sources, therapy was a mechanical intervention, and surgery was the 'mechanic medicine' *par excellence*.[44] This portrayal, however, should not convey the impression that Heister deduced his surgical principles from a physiological system. His rhetoric is mainly directed against the Stahlians and is used to stress the importance of anatomical knowledge and the 'scientific' foundation of surgery. Heister's *Chirurgie* – the most comprehensive handbook of its time and the bible of German surgery in the eighteenth century – is not a systematic arrangement of surgical disorders according to mechanical principles.[45] Without any introduction to physiology or general pathology, it discusses a great number of specific symptoms. Mechanical explanations are included here and there but they appear as retrospective explanations and are not used as a rationale to prefer one treatment to the other. Heister's therapeutic principles are based on observational evidence, collected from his own and his predecessor's experience. Surgery in the eighteenth century was, in Germany as elsewhere, an 'observational science'.

Haller's experimental findings on the insensibility of membranes, and especially of the tendons, were tested on patients all over Europe. But since lesions of tendons were considered as potentially dangerous, many doubted the appropriateness of the procedure. Only in one case did the examiner presumably have fewer reservations. Jean-Baptiste Boyer (1693–1768), dean of the Paris Medical Faculty and private physician to Louis XV, examined the unfortunate Robert Damiens, who had failed in his attempted assassination of the King on 1 January 1757 and whose Achilles' tendon was denuded of its skin in subsequent torture.[46] Despite the risks which examinations raised, more than 100 observations were published up until 1770.[47] As in the case of the animal experiments, the greater part of these favoured Haller's opinion, but important figures in Austria (de Haen) and especially in France (Le Cat, and others) reported contrary results. Let us first look at the situation in France, where the *Académie Royale de Chirurgie*, an élitist body ruled by a few core members, formulated the official surgical doctrines.[48] The debate on irritability and sensibility was well known within the society. Haller was elected as one of its few *associés étrangers* in 1752. Between 1756 and 1774 some fourteen lectures or reports on the topic were presented in its meetings and in three instances this happened to be the annual public meeting (1761, 1762, 1764).[49] Several members published on the subject,[50] and some other members reportedly witnessed experiments on the sensibility of certain organs.[51] Yet, in the *Mémoires* edited by the Academy the issue is not mentioned at all. This is also the case for most of the French surgical textbooks, mainly written by members of the Academy. As blood-letting was still one of the most frequent treatments administered by surgeons, the serious complications following bleeding occupied considerable space in these handbooks. But Haller's findings and the debate on sensibility were not mentioned at all; until the surgical volumes of the *Encyclopédie Méthodique* (1790–2) appeared, the lesion of a tendon was – with one exception – unanimously considered as a serious and potentially fatal accident, which had to be treated accordingly, i.e. with the instillation of hot oil of turpentine.[52]

There seem to be two factors that led to such a disregard of the controversy. First, despite the many observations made on patients, the most prominent feature of the debate was the great number of animal experiments. In the view of the Academy, these were of limited validity. The experiments of the surgeon Jean-Jacques Hoin, for instance, which were read in a meeting in 1769, were considered as 'peu concluantes', and the subsequent discussion centred on the practical, therapeutic observations of the surgeons present without entering into a debate on basic principles.[53] Surgery remained virtually untouched by such discussions. Although physiology from 1750 onward was a constituent part of the lecture courses at the College of Surgery, the research done in this field had hardly any effect on surgical procedures. As in the case of Heister, the emphasis on the theoretical background of surgery had a lot to do with the status of surgery.

Secondly, the Academy considered it as its duty to furnish the surgeons with well-established facts and doctrines and therefore did not want to raise controversial issues. The judgement of Haller's works by Antoine Louis, the longstanding secretary of the Academy, is characteristic of this attitude. He had no appreciation of Haller's careful critical assessment of different theories and opinions as displayed in the *Elementa Physiologiae.* Louis considered Haller's *opus magnum* a laboured accumulation of scholarly knowledge and of much less use than his short textbooks. Hypotheses that were shown to be wrong should no longer be discussed.[54] Obviously, Louis expected from medical science definitive statements rather than elaborate presentations of controversial problems. This is also what the Academy wanted to deliver. The many hundreds of reports and observations sent to the Academy by local surgeons were all assessed by one of its core members before a selection of them could serve as illustrative material in the well-balanced surveys of the *Mémoires.* Thus, the authoritative doctrines established in Paris appeared as a joint product of the French surgeons under the guidance of the Academy. This monopoly on surgery was fiercely defended. When, in 1761, a society of provincial and foreign surgeons set up a surgical journal in Lille, the Academy successfully intervened and had the printing privilege – already conferred to the society – withdrawn.[55] The members themselves were also kept under control. If they wanted to use the title of 'Member of the Royal Academy of Surgery' in a book, they had to submit the book for inspection. Thus, the Academy worked like a filter of surgical knowledge. Although – as a French student tells us – Haller's ideas on sensibility and irritability were taught in some colleges of surgery in Paris, just as in Germany and England, this had no effect on the French surgical textbooks – and therefore on French surgery at large.[56]

In the mid-eighteenth century, the leading role of French surgery was unquestioned. The disregard of Haller's findings in France was therefore to some extent mirrored in other countries. Nevertheless, as surgery was not controlled by a single body in either England or Germany, there was less unanimity. Like William Hunter, his brother John, and Benjamin Gooch regarded the tendons as very sensible only in inflammation. In their view, Haller, although essentially right, underestimated the alterations under pathological circumstances. His research tended to mislead the surgeons, who ought to consider the puncture of tendons as a serious accident.[57] In Benjamin Bell's *System of Surgery*, the most authoritative multi-volume handbook of the late-eighteenth century, the ancient theory was even less transformed and lesions of tendons were considered as very painful even without inflammation, and as eminently dangerous.[58] The situation in Germany was similar. Following the great Heister, most authors described the fatal accidents following lesions of tendons in their handbooks.[59] Only one important author, Georg Heuermann (1722–68), the leading Danish surgeon, disagreed. A barber and autodidact who then studied medicine and obtained high medical offices, he was a particularly independent mind. Heuermann published a four-volume textbook of physiology (1751–5), performed animal experiments himself and – in his surgical textbook (1754–7) – recorded that lesions of tendons could be treated safely, like those of muscles and other tissues.[60] But this doctrine gained acceptance in the German countries only later in the century with August Gottlob Richter (1742–1812), the professor in Göttingen who dominated surgery in the late-eighteenth and early-nineteenth centuries and who gained a reputation almost equal to that of Heister. Richter, in his authoritative surgical textbook, agreed with the Hunter brothers that in inflammation tendons were rendered sensible, but he could not conceive how they could thereby become even more vulnerable than other inflamed organs, such as the muscles, which were more sensible than the nerves in their natural state. Furthermore, the serious complications arose only after a certain time and were often unaccompanied by pain. Neither a simple lesion of a tendon nor of a nerve could explain this. Richter concluded that the immediate cause of these accidents could not be determined with certainty.[61] The discussion was liberated from dogmatic conjectures and could be continued on a new level.

Erna Lesky has argued that Haller's doctrine of sensibility or rather insensibility had a great impact on the surgery of his era. Enlightened or academic surgeons, especially in France, would eagerly have incorporated the new findings in their therapeutical schemes but the ancient fear of wounded tendons would have survived for a long time in the minds of the barber–surgeons.[62] Lesky's assessment falls short in two respects. First, she consulted only a few surgical authors, and with preference those Haller mentioned as his supporters. Our survey demonstrates that most surgical textbooks were not affected by the new insights. Secondly, the distinction between a quick reception of new methods of treatment by the enlightened surgeons and a slow uptake by illiterate barbers is a somewhat rash conclusion. We have to agree that there was a great difference, in social and intellectual terms, between élite surgeons in larger cities and their rural fellow practitioners and barbers. On the level of surgical practice, the leading surgeons distinguished themselves especially by the performance of major operations such as lithotomy and the closure of anal fistulas, which they had perfected during the century.[63] But it is doubtful whether, in the handling of daily administrations such as phlebotomy and the general management of wounds, they differed greatly from their humbler colleagues. As our case suggests, the principles of these treatments did not change very fast. Although innovations were mainly introduced by the learned members of the profession, the local surgeons presumably caught up on these developments. The level of knowledge of the average surgeon seems often to be underestimated. At least, for approved surgeons and barbers, handbooks were an important source of information. As German book inventories show, they had their textbooks, even if only in the vernacular. The most widely used in Germany was the one of Lorenz Heister, which the medical regulations of Württemberg in 1755 explicitly recommended.[64] A survey of textbooks, I would argue, furnishes a fairly correct picture of the theory and practice of surgeons. If, as in our case, an overwhelming majority embraced a similar position, we may conclude that most trained practitioners followed this line. Whether untrained barbers and apothecaries, popular healers and quacks adopted the same practice is more difficult to say and can only be established on further evidence, which in our case is lacking. We can only assume that, as in other cases, orthodox and fringe medicine were not poles apart in theory and practice.[65]

Haller had pinned high hopes upon the surgeons. They would have more occasion, he said, to verify the findings on patients and therefore to help to establish the new truth. Their testimonies were at least as credible as those of physicians who were suspect of '*love of system*'.[66] In 1761, he noticed that his ideas were promoted more by surgeons and less by physicians,[67] and fifteen years later he ascertained that the insensibility of the tendons was 'quite generally the idea of modern surgeons.'[68] Although Haller could quote quite a few surgeons who supported his views, our survey shows that his assessment had more to do with wishful thinking than with reality. Only in the 1780s and 1790s did the fear of lesion of tendons slowly disappear from surgical textbooks. Surgery was 'systematic', too. Whereas in physiology the lack of a systematic exposition led to a disregard of Haller's specific notions as the basis of a new conceptual model, in surgery they were considered as a theoretical question. At the same time, the careful and rather slow evaluation of observational data – from recent *and* earlier times – fostered reservations as to the acknowledgement of the new findings. In striving for the creation of a solid and reliable 'science', surgery was inclined to ignore controversial topics, and the monopoly of the *Académie Royale de Chirurgie* in France amplified this tendency.

A new model for pathology

Haller's concept of irritability and sensibility was immediately conceived by many as a promising basis for a new pathological model. Tissot's preface provided a first framework, ready for further development:

> If the dependence of pathology on physiology was better known it would not be necessary to make us realise how great an influence the new discovery will have on the art of healing. But, unfortunately, we lack a work entitled *Application of Theory to Practice.* That's what determines me to venture some ideas on the practical benefits of irritability.[69]

For Tissot, irritability as the cause of muscular contraction and thus of circulation was the key to the explanation of fever, inflammation and many other diseases. An increase in its activity was the reason for the excessive motive reactions that we see in the case of vapours, hysteria and similar disorders. Tonics were therefore the only appropriate prescription and bleeding and purging had to be replaced by exercise and frictions. Further studies could furnish us with 'real rules of practice in many cases.'[70] In fact, 'irritability has come to open a new area of research, a new source of solutions.'[71] Haller's concept appeared as a solution to the impasse into which the mechanical model of pathology had run. Consecutively, a flood of new pathological concepts was developed. But just as the physiologists adapted and transformed Haller's notions, so did the pathologists. They did not adhere to his particular notions of irritability and sensibility but were captivated by the general idea of the body as an active and reactive organism.

Probably the best way to illustrate the changes in pathological models is to focus on the concept of inflammation. As inflammation was one of the most common symptoms of disease, the physicians had always offered explanations which were in line with the general outlook of their physiological and pathological theory. For Galen, inflammation was due to an increased flow of humours and their accumulation in the – in his view, almost bloodless – arteries.[72] Until the mid-eighteenth century it was generally accepted that the local stagnation of blood flow was the primary cause of the disorder. For the defenders of humoralism the specific form of inflammation varied according to the mixture of the humours. For the iatrochemists the stagnated blood entered into a process of effervescence or fermentation. The Stahlians thought that the soul tried to overcome the stagnation by increasing the action of the heart and the arteries; the consecutive inflammation was therefore a sign of the helping hand of nature. Finally, Boerhaave, who formulated the theory that was still dominant in the mid-eighteenth century, delivered a more sophisticated explanation. For him, the initial cause was an increased blood flow and/or a constriction of the arteries. The constriction caused a stagnation and the high pressure pushed the blood corpuscles into the smallest arteries, which normally contained only serous fluid. This led to a congestion of corpuscles and to a dilation and often a destruction of these vessels. Despite the obstruction, circulation was accelerated and the increased friction caused heat.

Haller rejected this model on the grounds of three microscopic observations.[73] First, he claimed that there were no such serous arteries. Boerhaave's famous example of the inflammation of the eye (conjunctivitis), in which previously invisible arteries would become apparent, was not admissible. According to his investigations these arteries contained blood corpuscles, even in their healthy state, although these were invisible to the naked eye. Secondly, he stressed the importance of extravasation. In most, if not in all cases of inflammation, the *globuli* seeped through the vessels, and, as the injection of wax apparently showed, this was not due to a rupture of the arteries but had to be explained by a dilation of their pores. The central place of inflammation was the cellular tissue rather than the lumen of the vessels. Thirdly, in his view the role of obstruction was overrated. The ligature of arteries simply effected a deviation of the blood flow. This he had observed repeatedly in living animals. In order to create an inflammation, an increased velocity and vigour of the circulation was prerequisite. This was mainly a local phenomenon. Haller rejected the common notion that a regional congestion would activate the heart and the whole circuit and thus create fever. This might happen in the event of large obstructions but not in the case of an inflammation of the finger or other local part, where no change of heart action, pulse, and general circulation was visible. The cause of the increased local blood flow, however, was difficult to determine. For Haller, it had to be due to a stimulation, a pain, a lesion of nervous parts or a forceful friction. Obviously, the question of inflammation fell into the domain of his new theory, but unfortunately it fell between the two concepts of irritability and sensibility. Haller considered the erection of the penis as a natural example of inflammation, in which the stimulation of nerves led to a local affluence of blood. But the accumulation of blood first happened in the large vessels of the shaft and not in the place of stimulation, the glans. It could not be explained by a simple constriction of local arteries. Furthermore, in animal experiments the small arteries had proved to be only very slightly irritable and could therefore not cause vehement contractions. Nevertheless, there had to be a certain vasoconstriction, because the sole increase of inflow would only result in an increase of outflow and not in a considerable congestion of blood. Haller had no explanation to hand. The only thing which seemed to be clear was that – in erection as well as in pathological inflammation – there was a stimulation that activated the nerves and an unconscious reaction which had to be explained by muscular contraction. But how these two elements should be linked was unclear. Having separated the two realms of nerves and muscles and having limited their interaction, Haller was not able to furnish a proper concept of inflammation; and he was well aware of this. In his own review of the second volume of the *Elementa*, which would have been the right place to discuss the problem, he noted: 'Inflammation is not treated, probably because [the author] did not dare to determine its causes.'[74] In case of uncertainty Haller preferred not to present any solution. What he did, though, was to open the path for a new model. He did not deny the existence of congestion, but he stressed the importance of extravasation, and rather than simply ascertaining a constriction of the vessels he looked for its cause and shifted the focus of attention from the congestion of blood towards local irritation. Thereby, he delivered the keyword for the next generation.

The idea of irritation as the primary cause of inflammation was not entirely new – it had already been put forward by van Helmont and later by the Dutch physician, Jan de Gorter.[75] But with Haller's concept of irritability and sensibility it received fresh attention. The new authors did not necessarily deliver more detailed explanations than Haller but they formulated their views with less reservation.[76] Gaub, for instance, stated in a very general manner that irritations caused vibrations, tensions, contractions, obstructions, and inflammations.[77] Similarly, the French physician Pierre Fabre explained that the sole cause of inflammation lay in the irritation of nervous fibres, which attracted the fluids.[78] In contrast to Haller, most of these authors had no difficulty in connecting processes of sensation with subsequent motion and in connecting stimulations in one place with reactions in another. Consequently, and despite the lack of visible evidence, they thought that stimulation of the nerves caused a contraction of the small arteries. This was also the position of the French vitalists, which was presented in great detail in Ménuret's article 'Inflammation' in the *Encyclopédie.* In their view, the principle of sensibility or irritability – which was the same – was sufficient to explain inflammation. There were two ways in which the disorder could arise. Either it originated in a nervous irritation, which led to a contraction of the vessel and consecutive stagnation, or an obstruction led to a stagnation, which irritated the nerves. The process had to be considered as a vicious circle that could start turning at any point. Such a law-like, rather uniform operation was replaced by later teleological concepts such as the one developed by John Hunter, the most important author on inflammation in the late-eighteenth century. For Hunter, inflammation was a sign of the stimulated vital principle, which attempted to restore the state of health.[79]

The case of inflammation was quite typical. Haller rejected the traditional model and furnished observational evidence as to why it could not be correct; he also pointed to some fundamental operations that had to take place, but he did not present a new model. Rather than using a physiological theory to create new pathological concepts he used pathological observations to gain insight into physiological processes. Inflammation was only one of the many disorders which were attributed to an alteration of the state of irritability or sensibility. The pathology of the nerves and the vessels, which pervaded the whole body and were thus connected with almost all forms of disease, received special attention. The most famous treatise on nervous diseases was Tissot's *Traité des Nerfs* (1778–80). Tissot described nervous affections mainly as a change in the receptive quality of the nervous system or as due to an alteration of its stimulation. The tension or laxity of the nerves and the sensibility of the *sensorium commune* would set the intensity of reaction to stimuli. The acridity of the animal spirits and of other humours determined the intensity of stimulation. The irritability of the muscles was a decisive factor in most nervous diseases as it regulated how the nerves acted upon the organs of movement.[80] Tissot maintained the old concept of the laxity of fibres and he also considered the obstruction of animal spirits as a possible cause of disorders, but the notions of irritability and sensibility – although not exactly in Haller's sense – dominated his concept of nervous and thus of a great range of diseases. For many of his contemporaries, however, the vessels played the key role in the balance between health and illness. Increased or decreased irritability and consecutive constriction or relaxation of the vessels would determine the condition of circulation as well as all the processes depending thereon. For Haller, who considered the heart the exclusive motor of circulation, and the arteries only very slightly irritable, this was not an acceptable theory. The Parisian student de La Motte, however, encouraged by Zimmermann's dissertation, already declared in 1752 that in the end all diseases would depend on the irritability of the vessels.[81] Most later authors did not exclude other causes, but many regarded arteries and veins as still the main pathological factors.[82] The most common and most important sign of disorder was, of course, fever, which was mainly considered as a symptom of increased blood flow. The relation between fever and irritability and sensibility, therefore, received special attention.[83] In most of these concepts, Haller's neat separation between the two faculties was abandoned and the nerves were believed to be responsible for or closely related with irritability as well. Quite understandably, the pathological models reflected the development in physiology.

Despite their pathogenic explanations the new concepts often remained rather vague. Irritability and – especially in France – sensibility were not linked with a precise pathological mechanism but rather used as general terms to denote a broad group of disorders. John Pringle wrote to Haller in 1763:

> Tho' our physicians here have no great turn for theory, yet they seem in general disposed to receive those lights that you have thrown upon the animal oeconomy; so that spasms & irritability have of late become common terms amongst us, when we reason upon the nature of a disease & the cause of the symptoms.[84]

Although physicians were not very interested in extensive theoretical expositions they still needed some general principles to explain diseases and their treatment. A simple description of ailments without any conceptual background would not suffice. William Cullen reproached Joseph Lieutaud (1703–80) for having published a textbook of practical medicine with a large number of facts but without any exposition of the causes of disease, based on principles of physiology and pathology.[85] For Tissot, it was especially Haller's handbook of physiology which helped the physician to establish principles of practice more simple, more certain, and more luminous than those found in most of the books of practitioners.[86] As author of the *Traité des Nerfs*, he had in mind especially Haller's notions of irritability and sensibility, which he interpreted, as we have seen, in his own way. Johannes Weise, however, stressed in 1772 that these notions had not yet been sufficiently applied to pathology.[87] Irritability and sensibility, he said, were the only two vital properties and they affected one another. As diseases were caused by an excess or lack of motion, they had to be explained by the force of irritability, the motor of life. But Weise was less original than he had probably tried to appear. Several authors, both minor and well-known, had presented general pathological models based on the two properties before him.[88] Gaub's pathology, for instance, was based on the notion of increase or decrease of one main vital force. The vitalist authors of the *Encyclopédie* considered illness as the uneven distribution or action of sensibility.[89] Even the Chevalier de Jaucourt, a pupil of Boerhaave, maintained that the state of illness depended on the balance between solid and fluid parts and on the irritability and sensibility of the organs.[90] After Weise, and especially in the 1780s, the amount of new pathological concepts based on the notion of reactive bodily properties increased rapidly. Most of the new authors considered nervous powers to be decisive factors in the development of diseases. Irritability was either considered as a manifestation of this or as a second essential power. Kurt Sprengel presented these various theories of a 'pathology of solids' (*Solidarpathologie*), as he called it, in his extensive history of medicine.[91] The principal impetus came from Edinburgh and especially from Cullen, who argued that research on diseases should start with the study of the motive powers.[92] From Scotland these ideas spread to the continent and found their adherents particularly in the German countries. The main general handbooks, which Sprengel mentioned, were those of David Macbride (1772), Samuel Musgrave (1776), James Gregory (1782), John Gardiner (1784), De la Roche (1778), Johann Ulrich Gottlieb Schäffer (1782), C.G. van den Heuvell (1787), Francesco Vaccà Berlinghieri (1787), and his own (1795). Many further treatises delivered pathological explanations for individual disorders. The conception of disease as increased or decreased activity of the nervous and muscular system was widespread in the last decades of the eighteenth century.

Haller's whole project of experimental physiology consisted in the creation of a new science with its own research agenda. He thereby supported the separation of theoretical from practical medicine. His descriptions of various physiological processes were far too complex to serve as models for pathology. But with his theory of irritability and sensibility he provided a concept which, in an adapted form, could serve as a link between the two strands of medicine. The new pathological models incorporated only parts of Haller's original ideas and made use equally of the notions of Whytt,

Bordeu, and others. They delivered powerful explanations for a wide variety of disorders and in some cases furnished detailed accounts of the underlying processes. The books of Gaub and Cullen were particularly successful representatives of this new pathology and were translated into all major European languages. The physiological notion of irritation and reactive forces was only successful because it could serve as a model to explain diseases. In the 1790s, and especially in Germany, this model evolved in two opposing directions. In the movement of *Naturphilosophie* the notion of vital forces was coupled with much more general ideas on the nature of life and lost its intimate connection with pathological processes.[93] In the system of the Scottish physician John Brown (1735–88), however, which caused considerable uproar in Germany, theoretical and practical medicine were as closely united as probably never before.[94] Brown considered health as a state of balance between a stimulus and irritable matter. Diseases were classed as sthenic or asthenic according to the increase or decrease of the external stimuli that caused them. Treatment was correspondingly simple as it consisted in the modulation of stimulation through exercise, diet or administration of simple drugs. These therapeutic principles spread widely among German practitioners, but the whole theory was largely rejected by academic medicine. Among the various factors responsible for this rejection was the radicality and simplicity of Brown's system, which was contrived and considered as an attack on traditional theory and medicine itself. A pathological model had to furnish some general principles explaining diseases and serving as guidelines for therapy. But it had also to reflect the large amount of scholarly work and research performed in the last decades in order to be accepted by physicians who considered medicine a science. This is what the models of Gaub and Cullen did, which were both original in their conception of certain pathological principles and eclectic in their incorporation of traditional knowledge. Haller neglected the demands of the practitioners and Brown those of the professors. Haller's physiology was too complicated, and Brown's too simple.

The review journals and the culture of criticism

The debate on irritability and sensibility was not conducted in some secluded corner of the Republic of Letters reserved for medical discussion. There was no such corner. Some professional journals such as the *Journal de Médecine* (1754–93), the *Giornale di Medicina* (1763–81) or the *Magazin für Aerzte* (1775–99) emerged in the second half of the eighteenth century, but although they were widely read by physicians they did not monopolise the medical discourse.[95] Nor were they only read by doctors. Medicine pervaded most of the journals of the period. Roy Porter has vividly described the wide range of medical topics discussed in the *Gentleman's Magazine*.[96] Both laymen and practitioners wrote to the periodicals, and the articles often displayed a considerable level of professional knowledge. Nevertheless, they dealt predominantly with questions of medical practice and not with theory. Medical theory was the domain of the transactions of the scientific societies such as those of London, Paris, Göttingen, and Bologna. Besides the professional medical journals, medical theory was also discussed in some German monthlies that were not exclusively published for physicians but for those with more general interests in natural history and physics.[97] However, the main general periodicals in which medical concepts were considered were the review journals. The period in question is the golden age of these general review journals; the increase of publications led to the founding of many of them between 1730 and 1760 and the very same reason led to their ending before the turn of the century as the literature became more specialised, the selection of titles more difficult, and the reader more puzzled. Despite their number and notoriety they have not been recognised as important elements of scientific debates. The role of review journals and criticism has been studied extensively in the history of art and literature but has received almost no attention in the history of science and medicine.[98] The following discussion will show, however, that they are a good means of locating medical discourse within the realm of the Republic of Letters and of addressing the question of professional and scientific judgement in the eighteenth century. As the ideal of how a review had to be written changed substantially during this period and as Haller contributed to this development, I will first enlarge upon the general background of review culture before discussing its part in the controversy on irritability and sensibility.

Ideals of criticism

We may distinguish three types of general review journal.[99] First, there were national or international ones like the *Journal des Savants* (Paris, 1665–), the *Acta Eruditorum* (Leipzig, 1682–1776), the *Monthly Review* (London, 1749–1845) or the *Allgemeine Deutsche Bibliothek* (Berlin, 1765–1806). They all reviewed predominantly national literature, even the *Journal des Savants*.[100] This priority was somewhat at odds with the ideals of the Republic of Letters and several, mostly French journals were explicitly published to reduce these shortcomings and reviewed exclusively foreign works. This is the second type of general review journal, and the best known of these were the *Bibliothèque Germanique* (Berlin, 1720–60), the *Bibliothèque Britannique* (The Hague, 1733–47) and the *Journal Britannique* (The Hague, 1750–57), the *Journal Étranger* (Paris, 1754–62), and the *Estratto della Letteratura Europea* (Bern and Yverdon, 1758–66). They were important elements of cultural transfer, especially to France.[101] But they all had only temporary success, and many foreign and particularly German and Latin books did not reach French readers. Horace Bénédict de Saussure informed his friend Haller in 1767 that although he read virtually every French journal he knew almost nothing about German books on natural history but what Haller had told him.[102] French books were much better known in other countries, especially in Germany where they were translated and then reviewed in great amounts.[103] Here, the review culture was even more complex. The editor and journalist Christoph Friedrich Nicolai (1733–1811) noted that in countries with a single cultural centre, information about new books spread easily and reviews were less important. But in Germany with its many distant towns, review journals were of considerable moment and in some areas entirely indispensable.[104] Several of the larger cities like Leipzig and Frankfurt, and especially the university towns like Halle, Erlangen or Tübingen, therefore, had their own periodical. This is the third type of the general review journal. They were mainly of local importance. The most widely read of these, however, the *Göttingische Gelehrte Anzeigen* (*GGA*) (1739–), enjoyed, with some 500–800 copies sold, a national reputation.[105] All these learned journals reflected the German notion of erudition which, more than in other countries, focused on the comprehensive collection and compilation of knowledge.[106] They contained therefore many more but shorter reviews or summaries than other journals. The *GGA*, for instance, discussed 700–900 titles on 1,400–2,000 pages annually. The greater number of articles led also to an increased appreciation of publications from abroad. This was especially the case for the *GGA*, which considered the advertisement of foreign books as their special merit and devoted half of their reviews to them. As in the *Journal des Savants*, roughly forty-five per cent of all articles were assigned to medicine and the natural sciences.[107] But, as the *GGA* was more international and discussed four-to-five times more books than the French journal – and, quite importantly, transactions and periodicals – it avoided the lacunae to be found in the Paris periodical. Especially as regards the medical sciences, there was virtually no book of any importance which was not reviewed in the Göttingen journal. In fact, the *GGA* was the only review journal that could rightly claim to deliver a comprehensive covering of the whole scientific production of their time.[108]

The *GGA*, and especially its medical part, was to a large extent the work of Haller.[109] It seems that, in particular, the controversies about the mechanism of respiration (see Chapter 4) and the heritage of Boerhaave (see Chapter 5), which both began in 1744 and which were also conducted through review journals, persuaded Haller of the crucial importance of institutionalised criticism in the Republic of Letters. He had written some reviews already in the 1730s and started to work for the *Bibliothèque Raisonnée* in 1742, but only in 1745 did he begin to write for the Göttingen journal.[110] In 1747, finally, he became chief editor of the *GGA*. He set up clear standards of reviewing – which will be discussed later – and guaranteed to furnish half of all the articles. Despite his return to Switzerland in 1753, he almost lived up to his promise and penned some 9,000 reviews before his death in 1777. Haller's correspondence gives abundant proof of the weight which authors and readers accorded to the reviews.[111] Legions of authors sent Haller their books, asked him to write a review and pointed to some aspects of their work that required special attention. They did not go as far as the booksellers who occasionally offered remuneration.[112] But in some cases, upon an unfavourable review, they did not accept his verdict and sent an apology to the journal – which was usually not published. More often, however, they appealed to other journals, asking them to publish a refutation or another review.[113] Evidently, authors and booksellers believed that the judgements of the journalists would have a great effect upon the opinions of their readers. This was especially the case for the arts and *belles lettres*. It is true, the 'criticks' were constantly attacked, especially in the early era of literary criticism in the 1750s and 1760s, but this should be considered as a reluctant acknowledgement and not as a denial of the power of the journalists.[114] The situation seems not to have been very different in the natural sciences. Charles Bonnet wrote to Haller:

> I well know that the enlightened public judges for itself. But I well know too that a quantity even of reasonable people follow the journalists in their judgement. This is not a small abuse the journals have introduced into the Republic of Letters. A journalist decides from the height of his tribunal without having the time nor the means to inform himself about the case before him, and the author is condemned without having been heard.[115]

Bonnet's remark points to a principal problem for criticism in the Republic of Letters: the 'enlightened public' was considered the appropriate judge of a book. In his preface to the *Encyclopédie*, d'Alembert stated that only the *savants* were allowed to judge a work, which was written by a *société de gens de lettres*.[116] *Savants* or members of the Republic of Letters were generally regarded as those who either knew Latin, had attended a university or published a book.[117] Although this excluded the great mass of the general public, it admitted a considerable number of learned persons as 'authorised' judges. A critic who pronounced an authoritative judgement was guilty of abuse of power if he seduced the readers into neglecting their own critical faculties. Rather than judging the book he should deliver a summary of its content, which would allow the reader to form his own idea. This view was consistent with a notion of the Republic of Letters as Anne Goldgar describes it.[118] In this republic, criticism should be avoided because author and reviewer were part of a community in which the behaviour of its members towards each other was of prime importance. Moderation and politeness were the key virtues and questions of conduct tended to displace aspects of content. By the mid-eighteenth century, however, according to Goldgar, the republic of scholars was in decay and increasingly replaced by a younger generation of *beaux-esprits* and *philosophes* who were less concerned with themselves and with questions of conduct than with the common ideas and goals to be achieved. The (French) Enlightenment succeeded the older Republic of Letters.

Goldgar's book is a vivid and illuminating portrayal of the community of scholars in the early-eighteenth century, but through its focus on the conduct of the *savants* and its reduction of the Enlightenment to the French *philosophes*, it tends to underrate the continuity of the Republic of Letters.[119] The conflicting notions about the function of reviews, to which Goldgar herself points, are a good example. Already in the second decade of the century, Bayle's ideal of a review as an unbiased account of the content of a book had become questionable. By the mid-1720s, the notion prevailed that not only should good books be praised but bad ones should be censured.[120] This does not mean, however, that the older practice of politely praising and excerpting rather than judging disappeared. Not at all. Both types of review continued to co-exist and the journals which offered critical judgements still felt obliged to defend their principles. The community of scholars, united in politeness and learning, did not disappear with the arrival of Voltaire and his friends. This is particularly true for the realms of medicine and natural history. The critical reviews in these branches of knowledge also reflect, however, the advent of a new model which, besides the ideas of the *philosophes*, competed with the older ideal of the community of scholars. It had less to do with the new understanding of man, society and religion than with a new conception of science and particularly of scientific research. In this model the members of the Republic of Letters were considered neither as erudite polymaths, able to judge all productions of their colleagues, nor as *beaux-esprits* who, while entertaining society, aimed at its enlightenment and transformation, but as researchers and specialists in a more or less concise field of knowledge or – to use our modern term – as scientists. The transition from the older notion of a polite to a newer concept of a scientific community happened within the general Republic of Letters. It was gradual, and both models existed throughout the eighteenth century. Moreover, as the reviews demonstrate quite clearly, many critics of science books adopted elements of both views. Haller is a good case for this.

Upon his appointment as chief editor of the *GGA* in 1747, Haller described the characteristics of a good review journal.[121] He defined three main areas: the resources of the editors, the knowledge and abilities of the reviewers, and the content and style of the review. As to the resources, he stressed the importance of the supply of books, which should be international, comprehensive and as rapid as possible. A large correspondence was essential as this not only furnished the reviewer with books but also with further news about scholars and their work. This helped him to arrive at a more appropriate and original judgement. In his second point, Haller outlined the learning and skills required for writing reviews. A journalist had to know as many sciences and languages as possible. But as the range of sciences was too big for a single person, several scholars had to work together, each of them dealing with the publications in his own realm of knowledge. The reviewer had to know its history in order to distinguish what was implausible, old, common, probable, new, and true. Thus he had to be in command of his science to such a degree that he was able to make 'a valid judgement on the value of things'. Haller, in 1747, also felt compelled to stress that a simple summary of the content, which would leave the reader to draw his own conclusions, was not sufficient, at least in many cases. To recount the subject of a poem without delivering an interpretation and a portrayal of its quality was useless. A simple account of the content of, let's say, Winslow's *Anatomy*, was nonsense as even the most wretched handbook would discuss the same topics. The reviewer had to declare that the detailed descriptions were based not on other authors but on Winslow's own research and that they corresponded to nature. Furthermore, he had to assert that the author has made several improvements in various respects and that his descriptions of the nerves and the arteries were the best we had. Such a statement furnished the reader with useful information. In Haller's view, the review had to encourage the reader to buy good books and to keep him from reading bad ones. What was required, was the verdict of a specialist. His demand for professional judgement corresponded to his demand for specialisation in scientific research (see Chapter 2). Haller, who was admired for his breadth of learning and who is often called the last universal scholar, did not consider universality as a quality of any value. In Haller's opinion, Leibniz had worked on a wide range of scientific fields at the expense of his accomplishments in its different parts. He would have contributed more to our knowledge had he cultivated only one science.[122] In line with this view, Haller gave up the annual encyclopaedic summary of the entire scientific production that used to precede each volume of the *GGA*. The scholars in his Republic of Letters were united less through shared knowledge than through a common goal – the advancement of science – to which each member contributed in his own way. In his third point – the declaration of how a review should be written – Haller did, however, adopt some of the older ideals of community and conduct. A reviewer should rather go too far with politeness than with censure, he said. Since we should promote the happiness of other people and since happiness – especially of scholars – depended heavily on fame, we should rather seek to increase than to diminish reputation. But here the parallels end. Haller's notion of politeness had a different foundation to that of the older community of obligation:

> I often hear this word [politeness] but it signifies almost always only a light veil to cover the desire to humiliate. One would like to enjoy simultaneously the honour of victory and the favour which moderation attracts. Politeness, for me, lies in the things. If we really want to adorn ourselves with this virtue, we have to know how to do justice to our adversary, to recognise in his work with pleasure the good, right and new things and to oppose only with reluctance to the hypothetical and erroneous elements these might be mixed with.[123]

To be polite was not to honour a person because he was a member of the Republic of Letters but to respect the 'things', the 'facts', which the critic had to know. The result was a review which exposed both the strengths and the flaws of a work. Leniency was only admissible as long as truth was not distorted. False theories were like contagious plagues: they spread quickly at the expense of truth, which had to be defended.[124] Truth was not, however, a fixed set of dogmas that should never be questioned. Haller defended John Shebbeare's *Practice of Physic* (1755), which had been condemned in other journals because of its attack upon generally accepted medical principles:

> We believe, however, that it is not harmful to science when such freethinkers emerge who doubt everything that has everywhere been considered as well-founded. Thereby an impartial way of thinking arises, experimental investigations into the causes of our own medical belief are carried out, the false is, like an unpunished malefactor, called to account and truth still maintains its eternal rights in the judgement of the experts [*der Kundigen*].[125]

Those authors, however, who had a high opinion of themselves but shallow learning, who considered their thoughts as novel and true although they were old, doubtful or wrong, who sought to gain reputation by running down others and who, by stealing the properties of their colleagues, strove for undue fame were the object of condemnation. Most contemptible, thus, were those who violated both the principles of science and those of the Republic of Letters. These were the enemies of truth and the readers had to be warned against their books. Like all journalists of his age, Haller stressed that the reviews were written for the use of the community of readers, the public. But he pointed to an even greater use, which indicates that he valued science itself above the Republic of Letters: a good review rewarded the efforts of a scholar and encouraged him to continue with diligence. Haller said:

> We are almost certain that just and sound criticism is an indispensable business in the learned world. It puts off the miserable scribblers from writing, forces the mediocre into self-criticism and warns the great not to slacken and not to deliver an imperfect or precipitate work. It disseminates taste throughout the countries.[126]

Also talking about the arts, Haller added that it was because of too lenient criticism that German poetry had fallen behind its neighbours. In this system, criticism was not a means of self-assurance, as in Goldgar's static community of obligation, but an essentially dynamic element, which aimed at the development of science. Not the prosperity of the community but the advancement of science was the objective of criticism.

The different ideals of criticism become most visible in the case of controversies, as it is here where the reviewer is tempted to favour a particular position and must argue on what grounds. If he remains neutral this also reveals a lot about his standards of criticism. For those who stressed the value and importance of community, criticism and debate could be regarded as a threat to their ideals. Haller's anonymous criticism of Boerhaave in 1744 (see Chapter 5) was such a case. What made it serious was that he had chosen not just an ordinary but one of the most prestigious members of the Republic of Letters. Boerhaave was not only venerated as a hero of medicine and science but also as a model of politeness and humanity. Master of an unequalled number of pupils, he was, in fact, as Haller himself called him, *communis Europae praeceptor* – 'the common teacher of Europe'.[127] To describe such a figure as an inexperienced anatomist and inventor of doubtful hypotheses and to blame his widely respected pupil van Swieten because he adopted without reservation the views of his master was considered by many as a 'very shocking behaviour which can not but catch the eye of respectable people.'[128] Haller's anonymous article was primarily condemned for its offence against common decency but its author was, of course, also accused of having attacked Boerhaave on insufficient grounds. Willem van Noortwyck, the anonymous author of the 'Defense of Mr Boerhaave' in the *Bibliothèque Britannique*, gave many examples that revealed Haller, and not Boerhaave, to be inexperienced in anatomy and susceptible to hypotheses. This seems to be what Haller had waited for. His first review was clearly designed as a provocation, but he had not yet dared to launch a concentrated attack on Boerhaave's authority. Now, as his general assessment of Boerhaave's anatomy and physiology was challenged, he had to back up his judgement. In his reply to Noortwyck, he demonstrated how Boerhaave's physiological theories lacked any anatomical foundation and revealed no less than twenty-one anatomical errors in the first third of Boerhaave's *Institutiones Medicae* (1734) alone.[129] Haller used his detailed knowledge as a means of liberation from the constraints of misconceived politeness and from the authority of Boerhaave. He wanted to familiarise the Republic of Letters with the fact that Boerhaave's system had great flaws and that science had to move beyond him and, more precisely, that it should follow Haller.

The difference between Haller's and van Swieten's dealings with the Boerhaavian legacy had less to do with their attitudes towards the great master, whom they both venerated throughout their lives, than with their general ideals of science and scientific community.[130] Whereas van Swieten refused to engage in a debate with Haller or any other scholar, the Göttingen professor considered controversies as a necessary and productive element of scientific research. Talking about his debate with Hamberger on the mechanism of respiration he said:

> Assuming that Boerhaave's and my own opinion on respiration had never been doubted, then I would have satisfied myself with one or two basic principles and would not have sought to strengthen my conviction. The extent of the sciences is immeasurable, one does not know where to start working in a field whose breadth and fertility are equally great. But controversy teaches us to select a portion of the field which we will cultivate more assiduously, and if it is disputed, we fence it in rigorously. I was compelled to carry out new experiments and to repeat them often, and I found not only the truth of what I defended, but also new foundations.[131]

He concluded: 'Disputing sects are like flint and steel, which indeed generate fire but also illuminating light thereby.' Haller had quite a down-to-earth opinion of scholars. Although he maintained, quite traditionally, that scientists searched for higher truths and sought to work for the benefit of mankind, he stressed – probably more than any other scholar of his generation – the importance of ambition, emulation, jealousy and the strive for fame as the driving forces behind research in general and controversies in particular.[132] Although these were, as such, reprehensible qualities, they ultimately fostered the advancement of science and thus of humanity: 'This is how the evil enters the plan of a better world.'[133] Haller was fully aware that he had these qualities to an eminent degree but he tended to suppress and deny them. He described it as his fate to attract each year new accusations and quarrels despite his constant praise of other scholars.[134] This is certainly not a very convincing explanation as to why Haller, besides the debate on irritability and sensibility, was engaged in numerous disputes in anatomy (with B.S. Albinus, G.D. Coschwitz, J.G. Günz, J.J. Huber, C.C. Schmidel, J.B. Senac), physiology (Hamberger), embryology (C.F. Wolff), botany (Linné, Schmidel), and religion and philosophy (La Mettrie, Voltaire). It was the combination of his ambitious nature, enormous urge for scientific progress, and exercise of open criticism which made him the focus of many controversies. In particular, his work as a journalist made him aware of the conflict between critical science, as he understood it, and friendly relations between scholars. In a letter to Christian Gottlieb Ludwig, the editor of the Leipzig review journal *Commentarii de Rebus in Scientia Naturali et Medicina Gestis*, he wrote:

> I had to smile sweetly, my dear Ludwig, when I called to my mind that you admonished me to refrain from controversies and that now, since you are the head of a [review] journal, you realise that you are approached in quarrels [*litibus implorari*]. It is like that, excellent man. Men do not want to hear the truth, they want to be flatterers, who must be judged. Now, I fear, you experience that it is almost impossible to fulfil the duty of a just critic and to preserve friendship with the authors as well.[135]

Criticism in the debate on irritability and sensibility

The review journals concerned with the debate on irritability and sensibility were, to varying degrees, committed to one of the two ideals of criticism described so far; they furnished benevolent summaries or critical, professional judgements. The *Acta Eruditorum* are the classic representative of the summary-type journal. Their Latin articles were reports of the content of books and often of such a detailed nature as to replace the reading of the book itself. They rarely ever contained a verdict. The *Commentarii* from Leipzig, focusing on works of medicine and natural history, were designed after this model. The determination with which their editor Ludwig refrained from any judgement is remarkable. Ludwig was, as we have seen, an undogmatic mechanist. A close correspondent of Haller, he was not enthusiastic about Haller's concept but he favoured the new experimental approach. In his reviews, however, a trace of his own convictions is very seldom visible. He presented the majority of all books without any judgement. Only in the case of a few works favouring animist theories could he not abstain from critical remarks. After an unbiased synopsis of fifteen pages of Whytt's *Essay on the Vital and Other Involuntary Motions* he added one critical sentence on Whytt's use of hypotheses. And in the review of Le

Cat's work on the action of the muscles he noted that he had doubts about Le Cat's system but that the principles of the journal would not allow him to present them and that therefore he left the judgement to the reader.[136] Some journals, like the *Tübingische Berichte von Gelehrten Sachen* were even more committed to the principles of politeness and praised, almost without exception, the efforts and skills of the authors. Most other periodicals, from the international *Journal des Savants* to the national *Frankfurter Gelehrte Anzeigen* and the local *Critische Nachrichten* from Greifswald, allowed for selective criticism, but this usually occupied only a modest place. A critical assessment was especially avoided in the case of a controversy between two distinguished authors. The *Monthly Review*, for instance, presented Haller's treatise on irritability and Whytt's *Physiological Essays* in immediate succession but favoured neither, arguing that 'the acknowledged abilities of both may serve to excuse our not interposing any judgement of our own on either side.'[137] Those journals like the *Bibliothèque Raisonnée* or the *Bibliothèque Germanique* which published essay-like reviews, usually discussed the works from a wider perspective and thus included also more general criticism. The most decisive judgements were, of course, expressed by reviewers who were directly involved in the debate. Hamberger, the editor of the *Jenaische Gelehrte Zeitung* and Delius, the editor of the *Erlanger Gelehrte Nachrichten*, plainly and constantly criticised Haller's works on respiration and irritability and were, in return, censured by Haller in the *GGA*. It was obvious to anyone even superficially informed about the debate that the impartiality of the journals in these cases had to be questioned. The *Jenaische Gelehrte Zeitung* argued that Haller misused the *GGA* for his own interests and tried to 'deceive the readers through all sorts of incomplete information and judgements regarding his own matters.'[138] Furthermore, it asserted that Haller had succeeded in winning other review journals over to his side. The Jena journal, on the other hand, would desist from any judgement. Their reviews of Haller's works were written not by Hamberger, but by his son, and completed by a friend of his because of fear of partiality.[139] It must be doubted, however, whether anybody considered this as a guarantee of impartiality. In these, as in many other cases, the reviews were constitutive and, at times, the main elements of the disputes.[140] They often discussed many different topics and a heap of unrelated details that were difficult or impossible for the reader to follow without a precise knowledge of the controversy. In his review of a work of Giambattista Bianchi, for instance, Haller wrote that the author 'acknowledges the experiment with the left heart', without mentioning what the experiment was about.[141] Haller levelled this kind of detail-criticism not only in controversies – it is a general characteristic of his review style. The aim of such critiques was less to inform the reader about the general content of a book than to pronounce a clear verdict on the reliability of its single results and on its general value. They were not very helpful for those who wanted to get an overview of the development of a branch of knowledge by reading journals. But they were particularly useful for those who intended to read or, even more, for those who had already read the book. As a matter of fact, they were most helpful for Haller himself as they contained the essential information he needed for his own work.[142] Thus, the reviews were primarily written for professional use. This was not only Haller's objective but that of the *GGA* in general, and it is expressed in the government's decree of 1745 to exempt the *GGA* from censorship because the reviews were exclusively written by professors of the university.[143] This decision was not only a mark of confidence in the academic servants of the state but should be considered as a sign of the awareness that the discussion of scientific works – with which the *GGA* were primarily concerned – had moved from the general public sphere into a more professional environment.

The *GGA*, like other review journals, was in a somewhat contradictory position. Although their single reviews were directed at a more or less specialist audience, the journal itself covered all branches of science and thus maintained the unity of the Republic of Letters. This reflects the general situation, that although different sciences increasingly developed specific social systems, these were not yet marked off and closed but open to the general public and especially the scientific community at large.[144] Professional discussions could always arouse the interest of a wider audience and could not simply be settled between a few specialists. Especially in noisy controversies, a whole range of persons who joined the discussion and expressed their opinion had to be taken into consideration. Haller wrote to Tissot regarding the dispute with de Haen: 'We should answer de Haen, not for the knowledgeable judges [*juges connoisseurs*], they are a small number. But we should not leave to the badly informed judges and journalists the glory to say that we have not answered.'[145] And to de Haen he wrote: 'Two *savants* in dispute expose themselves to the judgement of ignorants and *demi-savants*, and it is already a degradation to be judged by such a people.'[146] For Haller it was clear that only specialists were able to judge a complicated controversy such as the one on irritability and sensibility. He informed de Haen that only ten-to-twelve people were sufficiently qualified judges. De Haen answered in a letter to Zimmermann:

> I only mention that when Mr Haller maintains that of all mortals there are only 10 or 12 persons capable to judge our dispute, he seems to be mistaken.
>
> I hope to show that every physician with a second-rate understanding of the

history of medicine is in a position to decide.[147]

Non-physicians like Bonnet, however, were, in his view, not able to judge the matter.[148] Other authors adopted a more inclusive position. Tissot noted in a letter to de Haen about their controversy on inoculation:

> The process seems to me to be sufficiently prepared. Advocates and opponents, we both have said what we thought to be the best we could say. The files are before the eyes of the judges. Let us wait for the judgement of the public and of posterity.[149]

Compared with de Haen and Tissot, Haller's ideal of professional criticism appears particularly exclusive. But we should not neglect the common ground upon which their ideals were based. Both the Republic of Letters and scientific discourse were considered as an open, public sphere to which everybody had access.[150] They were restrictive only in so far as one had to acquire a certain knowledge and skill in order to be considered a valid member or judge. But, in principle, it was open to anyone to qualify himself. The disagreement resulted from different conceptions as to the qualifications required. Tissot thought that the disputed matter could be published in such a manner that all enlightened readers could judge it. De Haen, however, restricted the circle of judges to the physicians, as only they had the necessary knowledge of medical literature ('history of medicine'). Presumably, although he did not explicitly say so, he believed that some experience in practical medicine was prerequisite in order to assess the disputed facts. Haller went a step further. Some general medical experience, in his view, was not sufficient. Out of thousands of physicians, only ten-to-twelve people were competent arbiters. These were those who not only possessed extensive anatomical and physiological knowledge, but who had themselves performed experiments. In his debate on the mechanism of respiration Haller said that 'Mr Hamberger should know that one should never deny experiments without having performed them oneself, and, in fact, more often and with more care than the author whose trials one refutes.'[151] Haller limited the jury to the experts, to those who had done the research themselves. This did not mean that science was the exclusive domain of preselected members. Felice Fontana and Cesare Pozzi, although not physicians, had acquired the knowledge and skill to perform experiments and thus were accepted as experts. Everybody could follow their path, and anybody with sufficient qualification was invited to pronounce his judgement. The ideology of the public sphere as the arena of science was retained. It was, as Thomas Broman has stressed, the principle on which the authority of science was based.[152] This was the case for Tissot, de Haen and Haller. It is still the case today and the professionalisation of research has not changed that.

If we consider how the debate evolved we must agree that Tissot's and de Haen's ideals were closer to reality than Haller's. There was a great variety of descriptions, judgements and interpretations of the phenomena and although some figures enjoyed a greater reputation than others, there was no unquestioned group of experts and authorities. Haller's model of science, in which the specialists produced results, which others had to accept, was probably considered less as meritocratic than as aristocratic and therefore conflicted with the democratic values of the Republic of Letters. Haller's authoritarian character presumably added to this general impression. Zimmermann remarked in a letter to his friend Tissot on Haller's conduct as a scientist: 'Mr Haller appears to fear republican equality, he wants to command, as it seems. But he has only to become senator in Bern, and he will command.'[153] Unfortunately for Haller, he was not elected senator and thus could not command in politics either.

The various types of review journals and their differing opinions and ideals of criticism argue for the existence not of one homogenous sphere of scientific discourse but rather of several partially merging sub-spheres. The selection of books and the manner in which they were reviewed helped to create different literary cultures. Readers of the *Journal des Savants* received a more or less neutral description of Haller's concept of irritability and sensibility with the appended advice that the new theory had to be taken with a pinch of salt. They would have gained the impression that the alternative theories of La Caze and Bordeu should be preferred.[154] Besides that, they received little information about the ongoing controversy. In a similarly exclusive manner, the *Monthly Review* presented Haller's notions in comparison with those of Whytt but it did not pronounce any judgement. The only Italian journal containing medical reviews of any importance, the *Giornale di Medicina*, overtly adopted Haller's and Caldani's position. In addition, the Italians read the *Commentarii* from Leipzig, the *Journal des Savants* and some other French journals. The German readers were confronted with a wide variety of portrayals. The *GGA* delivered full coverage of the whole controversy and published some 120 reviews which clearly favoured Haller's position although they acknowledged the merits of some of his opponents. The *Commentarii* from Leipzig, with sixty reviews, were similarly comprehensive but furnished only summaries of the books in question. Most of the other journals devoted only five to twenty reviews to the subject, supporting one or the other side in varying degrees or leaving the judgement to the reader. The great variety of review- and other journals was not to everyone's taste. Tissot observed: 'It seems to me that we could work out a plan for the journals in such a manner that six would suffice for the whole of Europe. Probably eighty or a hundred are now published – what a misery. Quis leget illas?'[155] As Tissot's question suggests, nobody could read them all. Everybody had to make his own choice. The review journals, as a whole, fostered the establishment of neither overtly international nor explicitly national or regional realms of science and discourse.[156] In some German towns, however, such as Göttingen, Jena or Erlangen, the usually university-linked journals could convey a kind of corporate identity to the local intelligentsia which furnished most of the articles. But, for the learned community at large, the journals could not assume this function. Only Friedrich Nicolai's *Allgemeine Deutsche Bibliothek*, for which, in the period from 1765 to 1785, some 180 people from a great variety of places wrote reviews, may be considered as a relevant vehicle of integration within the German Republic of Letters.[157] But this did not result in a community of scholars with similar scientific interests and beliefs. Professional journals alone – for instance Lorenz Crell's *Chemische Annalen* (1784–1804) and Johann Friedrich Meckel's *Deutsches Archiv für die Physiologie* (1815ff.) – provided a well-defined forum that helped to transform the collectivity of scholars working in a specific field of research into a proper community.[158] The debate on irritability and sensibility was still considered as belonging to the wider profession of physicians and the Republic of Letters at large, and the contributions to it from various sides were therefore heterogeneous in quality and character.

Notes

1. F. Baasner, *Der Begriff 'Sensibilité' im 18. Jahrhundert: Aufstieg und Niedergang eines Ideals* (Heidelberg: Winter, 1988).
2. R. Rey, 'La Théorie de la Sécrétion chez Bordeu, Modèle de la Physiologie et de la Pathologie des Vitalistes', *Dix-Huitième Siècle*, 23 (1991), 45–58.
3. S. Moravia, *Il Pensiero degli Idéologues: Scienza e Filosofia in Francia (1780–1815)* (Florence: La Nuova Italia, 1974); E.A. Williams, *The Physical and the Moral: Anthropology, Physiology and Philosophical Medicine in France, 1750–1850* (Cambridge: Cambridge University Press, 1994); R. Rey, *Naissance et Développement du Vitalisme en France de la Deuxième Moitié du 18e Siècle à la Fin du Premier Empire* (Oxford: Voltaire Foundation, 2000).
4. The first to stress the link, was R.F. Brissenden in his *Virtue in Distress. Studies in the Novel of Sentiment from Richardson to Sade* (London: Macmillan, 1974). Many others followed. M.W. Dull, *Irritability and Fiction, 1814–1837* (Virginia Univ. D.Phil. thesis, 1997), is one of the first to criticise the focus on Haller's physiology. For her, not Haller's strict separation of conscious feeling and unconscious reaction served as a model in Britain, but Whytt's concept, which was less dividing.

5. K.M. Figlio, 'Theories of Perception and the Physiology of Mind in Late Eigtheenth Century', *History of Science*, 12 (1975), 177–212: 199–200.
6. *Elementa*, v, 535, 551–2.
7. A.C. Vila, *Enlightenment and Pathology: Sensibility in the Literature and Medicine of Eighteenth-Century France* (Baltimore: Johns Hopkins University Press, 1998), Chapter 1.
8. C. Bonnet, *Essai Analytique sur les Facultés de l'Ame* (Copenhague: Philibert, 1760; Reprint Hildesheim: Olms, 1973), 86–7.
9. G.S. Rousseau, 'Nerves, Spirits and Fibres: Towards Defining the Origins of Sensibility', *Studies in the Eighteenth Century*, 3 (1976), 137–57.
10. G.S. Rousseau, 'Discourses of the Nerve', in F. Amrine (ed.), *Literature and Science as Modes of Expression* (Dordrecht: Kluwer, 1989), 29–60: 44–6. Rousseau's judgement is partly based on a quote mistakenly attributed to Haller. See also his *Enlightenment Crossings: Pre- and Post-Modern Discourses – Anthropological* (Manchester: Manchester University Press, 1991), 188.
11. See eg. his review of Helvétius's *De l'Esprit* in the *GGA* 1759, 1034–41: 1036.
12. The original review appeared in the *Bibliothèque Raisonnée*, 42 (1749), 324–36, the English translation in the *Gentleman's Magazine* 1749, 245–6, 345–9. Discussions about Clarissa in the anatomical theatre are reported by a student of Haller in E. Bodemann (ed.), *Von und über Albrecht von Haller: Ungedruckte Briefe und Gedichte Hallers sowie Ungedruckte Briefe und Notizen über denselben* (Hanover: Meyer, 1885), 199.
13. Cf. C. Lawrence, 'The Nervous System and Society in the Scottish Enlightenment', in B. Barnes and S. Shapin (eds), *Natural Order: Historical Studies of Scientific Culture* (Beverly Hills: Sage, 1979), 19–40. S. Knott, *A Cultural History of Sensibility in the Era of the American Revolution* (Oxford Univ. D.Phil. thesis, 1999).
14. K. Ballstadt, *The Natural Philosophical Thought of Denis Diderot* (Oxford Univ. D.Phil. thesis, 2002).
15. Rousseau, *Discourses* (note 10), 40.
16. N.D. Jewson, 'Medical Knowledge and the Patronage System in Eighteenth-Century England', *Sociology*, 8 (1974), 369–85.
17. F. Quesnay, *Essai Physique sur l'Oeconomie Animale*. Seconde Édition (3 vols, Paris: Cavelier, 1747), xxix ('Discours Préliminaire sur l'Expérience & la Théorie en Médecine').
18. *Ibid.*, vii–viii.
19. Thomas Broman has argued – quite convincingly – that the claim to join medical theory and the practice of healing was part of the physician's professional ideology only in late- but not in mid-eighteenth century. I would like to stress that this was not tantamount to the denial of the

intimate link between theory and practice. Even if classical training and gentlemanly behaviour rather than the practice of healing at the bedside was the source of a physician's identity, the explanation of disease was indispensable for his self-awareness as a competent and professional healer. Cf. T. Broman, 'Rethinking Professionalization: Theory, Practice, and Professional Ideology in Eighteenth-Century German Medicine', *The Journal of Modern History*, 67 (1995), 835–72.

20. This point is also made by L.S. King, *The Philosophy of Medicine: The Early Eighteenth Century* (Cambridge, Mass.: Harvard University Press, 1978), 192–3 and by T. Broman, *The Transformation of German Academic Medicine, 1750–1820* (Cambridge: Cambridge University Press, 1996), 77–9.
21. C.G. Ludwig, *Institutiones Pathologiae Praelectionibus Academicis Accomodatae* (Leipzig: Gleditsch, 1754).
22. *GGA*, 1754, 784.
23. Ludwig, *op. cit.* (note 21), § 931: 12–13.
24. Quesnay, *op. cit.* (note 17), xciv–civ.
25. A thorough analysis of Morgagni's pathology is still lacking. For a good introduction see L. Belloni, 'L'Opera di Giambattista Morgagni – Dalla Strutturazione Meccanica dell'Organismo Vivente all'Anatomia Patologica', in L. Belloni, *Per la Storia della Medicina* (Bologna: Forni, 1980), 239–46.
26. Sauvages's main nosological works are *Pathologia Methodica* (Amsterdam: De Tournes, 1752); *Nosologia Methodica* (2 vols, Amsterdam: de Tournes, 1768); and the posthumous French version *Nosologie Méthodique* (10 vols, Lyons: Bruyset, 1772).
27. See R. French, 'Sickness and the Soul: Stahl, Hoffmann and Sauvages on Pathology', in A. Cunningham and R. French (eds), *The Medical Enlightenment of the Eighteenth Century* (Cambridge: Cambridge University Press, 1990), 88–110.
28. P.V. de Sèze, *Recherches Physiologiques et Philosophiques sur la Sensibilité ou la Vie Animale* (Paris: Prault, 1786), 5–6.
29. E. Lesky, 'Vom Hippokratismus Boerhaaves und De Haens', *Boerhaave and his Time* (Leiden: Brill, 1970), 123–43.
30. A. de Haen, 'Caput VII: Finis Quaestionis de Sensibilitate et Irritabilitate', in A. de Haen, *Pars Duodecima Rationis Medendi in Nosocomio Practico* (Vienna; Trattner, 1768), 263–74.
31. G. van Swieten, *Commentaria in Hermanni Boerhaave Aphorismos de Cognoscendis et Curandis Morbis* (5 vols, Leiden: Verbeek, 1742–72); A de Haen, *Ratio Medendi in Nosocomio Practico* (15 vols, Vienna: Trattner, 1757–74).
32. *Instituta Facultatis Medicae Vindobonensis* [curante Anton Störck] (Vienna: Trattner, 1775), 9.

33. For details see E. Lesky, 'Albrecht von Haller und Anton de Haen im Streit um die Lehre von der Sensibilität', *Gesnerus*, 16 (1959), 16–46 and U. Boschung, '"Multa pro Nostra Innocentia" – L'Implication de Tissot dans la Querelle Haller–de Haen', in V. Barras and M. Louis-Courvoisier (eds), *La Médecine des Lumières: Tout Autour de Tissot* (Chêne-Bourg: Georg, 2001), 113–47.
34. *Elementa*, viii, xxiii.
35. See de Haen's letters to Bonnet, especially 19 May and 27 December 1766 (Bibliothèque Publique et Universitaire de Genève, Ms. Bonnet 29.1–2).
36. 'Pour le Pathologique, je n'en ai jamais voulu m'en meler.' Haller to de Haen, 29 October 1770, edited in de Haen, *op. cit.* (note 31), Vol. 14, 1770, 122.
37. *Irritable Parts*, 3.
38. L.M. Caldani, *Esame del Capitolo Settimo Contenuto nella XII: Parte dell'Ultima Opera del... Antonio de Haen... Indiritto allo stesso Autore* (Padua: Comino, 1770), 28–9.
39. 'Dr. Hunter's Thoughts on the Sensibility of the Tendons, Periosteum, Ligaments, Dura and Pia Mater, &c. From his Medical Commentaries, &c.', *The Medical Museum*, 1 (1763), 214–21.
40. Farjon to Haller, 31 October 1755 (*Haller Papers*).
41. Anon., 'Esquisse sur l'Irritabilité & la Sensibilité des Parties dur Corps de l'Homme', *Mercure de France*, Avril 1757, 144–51.
42. F. Quesnay, 'Préface', *Mémoires de l'Académie Royale de Chirurgie*, 1 (1743), ix–xxxi. On Quesnay's preface see T. Gelfand, 'Empiricism and Eighteenth-Century French Surgery', *Bulletin of the History of Medicine*, 44 (1970), 40–53.
43. P. Wilson, 'An Enlightenment Science? Surgery and the Royal Society', in R. Porter (ed.), *Medicine in the Enlightenment* (Amsterdam: Rodopi, 1995), 360–86.
44. U. Boschung, 'Iatromechanik und Chirurgie bei Lorenz Heister', *Gesnerus*, 40 (1983), 31–41.
45. Six original German editions as well as Latin, Dutch, English, French, Italian and Spanish translations appeared between 1718 and 1770.
46. L.-M. Girard de Villars to Haller, 31 Mars 1757 (*Haller Papers*).
47. A list is given in the *Elementa*, viii, preface. Further evidence may be gathered from other publications listed in the bibliography.
48. For the Academy and the general situation of surgery in France see T. Gelfand, *Professionalizing Modern Medicine: Paris Surgeons and Medical Science and Institutions in the 18th Century* (London: Greenwood, 1980) and L. Brockliss and C. Jones, *The Medical World of Early Modern France* (Oxford: Oxford University Press, 1997), Chapter 9.

49. Cf. the Procès-verbaux in the Bibliothèque de l'Académie de Médecine, Paris, Mss. 18–26. In six cases this was due to the reading of articles from the *Journal de Médecine* or the *Mercure de France.*
50. Toussaint Bordenave (1757), Ferrand (1760), Pierre Fabre (1770) and the associated correspondents Le Cat (1753, 1757, 1765), Jacques Daviel (1762), and Jean Jacques Hoin (1769). For details see the bibliography.
51. Andouillé, Jean Nicolas Moreau, Jean-Joseph Sue.
52. See the textbooks of Jean-Joseph Sue (1755), Planque (1757, 1770), Dionis (1757, 1773), La Faye (1761), Portal (1768), Pierre Sue (1771, 1774, 1777), Louis (1772), Simon (1780), La Roche (1790–92), and Hévin (1793). The only handbook I found that considers the fear of lesions of tendons as exaggerated, is that of Hugues Ravaton (1768). Towards the end of the century, R.B. Sabatier (1796), and presumably other authors took a similar position. The titles of the books are given in the Bibliography.
53. Bibliothèque de l'Académie de Médecine, Paris, Ms. 23, 1769, fo. 6.
54. A. Louis, 'Éloge de Haller, Lu dans la Séance Publique du 30 Avril 1778', in E.-F. Dubois (ed.), *Éloges Lus dans les Séances Publiques de L'Académie Royale de Chirurgie de 1750 à 1792 par A. Louis* (Paris: Barillière, 1859), 265–81: 272–3.
55. Procès-verbaux, June to September 1761 (Ms. 22, fol. 54–83). Interestingly, the general assembly first accepted the dedication of the journal, for which the new society had asked, by a large majority. It seems to have considered it as a useful undertaking. Only two weeks later, upon intervention of the director of the Academy, was opposition mobilised. It was stated that the foreign society with its claim to 'un éspece de jugment Doctrinal' on surgical matters would enter into rivalry with the Royal Academy.
56. L.-M. Girard de Villars to Haller, 31 Mars 1757 (*Haller Papers*).
57. B. Gooch, *A Practical Treatise on Wounds and other Chirurgical Subjects* (2 vols., Norwich: Chase, 1767), i: 89–93; J. Hunter, *A Treatise on Blood, Inflammation, and Gun-Shot Wounds* (London: Longman *et al.*, 1794), 288.
58. B. Bell, *A System of Surgery* (6 vols, Edinburgh: Elliott, 1783–8), i: 105–34.
59. See the textbooks of Heister (1750, 1770), Platner (1757, 1770), Bilguer (1763), Schneider (1775), Pallas (1777), and Bernstein (1790). The problem is not discussed in the handbooks of Schmucker (1774) and Theden (1774). One textbook, published in Riga by Alix (1772), supported Haller. I have not consulted the work of Henckel (1770–76). For details see the Bibliography.
60. G. Heuermann, *Abhandlungen der Vornehmsten Chirurgischen Operationen* (3 vols., Copenhagen and Leipzig: Pelt, 1754–57), ii: 43–50.
61. A.G. Richter, *Anfangsgründe der Wundarzneykunst* (6 vols, 2nd edn, Frankenthal: Gegel, 1788–1800), i: § 396–403.

62. Lesky, *op. cit.* (note 33).
63. The distinction between two communities of surgeons in France is stressed by Brockliss and Jones, *op. cit.* (note 48), Chapter 9.
64. S. Sander, *Handwerkschirurgen: Sozialgeschichte einer Verdrängten Berufsgruppe* (Göttingen: Vandenhoeck & Ruprecht, 1989), 81–2.
65. W.F. Bynum and R. Porter (eds), *Medical Fringe & Medical Orthodoxy 1750–1850* (London: Croom Helm, 1987)
66. Letter to Somis, 18 January 1757, see E. Hintzsche (ed.), *Albrecht von Haller – Ignazio Somis. Briefwechsel 1754–1777* (Bern and Stuttgart: Huber, 1965), 35.
67. Letter to Caldani, 6 April 1761, *Corr. Caldani*, 90.
68. 'Ligamens', *Encyclopédie, Suppl.*, iii (1777): 744.
69. S.-A. Tissot, 'Discours Préliminaire du Traducteur', in A. von Haller, *Dissertation sur les Parties Irritables et Sensibles des Animaux* (Lausanne: Bousquet, 1755), xxii–xxiii.
70. *Ibid.*, xxviii.
71. *Ibid.*, xlix.
72. For a general survey see M. Wolman, *Entzündung: Studie zur Geschichte eines Biologischen Begriffs* (Heidelberg: Hüthig, 1962). Most helpful are contemporary accounts such as those by J.-J. Ménuret de Chambaud, 'Inflammation, Inflammatoires maladies (Méd.)', *Encyclopédie*, viii, 708–27 and E. Platner, *Zusätze zu seines Vaters Einleitung in die Chirurgie* (Leipzig: Dyck, 1776).
73. Haller's main work on haemodynamics are the *Deux Mémoires sur le Mouvement du Sang et sur les Effets de la Saignée, Fondés sur des Experiences Faites sur des Animaux* (Lausanne: Bousquet, 1756). More specific comments on inflammation are to be found in the *Elementa*, i, 35–8, 112–17, 436, 444–5; ii, 202–6, 213–14, 251–2 and in the article 'Inflammation', *Encyclopédie, Suppl.*, iii (1777), 599–601.
74. *GGA*, 1760, 84.
75. J. de Gorter, *Medicinae Compendium, in Usum Exercitationis Domesticae* (2 vols, Leiden: Van der Aa, 1737), i: 239.
76. Detailed expositions are given eg. by D. Magenise, *The Doctrine of Inflammations Founded upon Reason and Experience; and Intirely Cleared from the Contradictory Systems of Boerhaave, Van Swieten, and Others* (London: Owen, 1768); and A. Odendahl, *Inflammationis Ratio Dissertatione Inaugurali Praeside G. M. Gattenhof... Proposita* (M.D. thesis, Heidelberg: Haener, 1773).
77. H.D. Gaub, *Institutiones Pathologiae Medicinalis* (Leiden: Luchtmans, 1758), § 195.
78. P. Fabre, *Essais sur Différens Points de Physiologie, de Pathologie et de*

Thérapeutique (Paris: Didot, 1770), 101–8.

79. Hunter, *op. cit.* (note 57), 226–76.
80. S.-A. Tissot, *Traité des Nerfs et de leurs Maladies* (4 vols, Paris and Lausanne: Didot et al., 1778–80), ii, 270–310.
81. J.G. de La Motte, *Quaestio Medica... Discutienda in Scholis Medicorum Die Decimo-Tertio Januarii MDCCLII. Guillelmo De Magny Praeside. 'An a Vasorum Aucta aut Imminuta Irritabilitate Omnis Morbus?'* (M.D. thesis, Paris: Quillau, 1752).
82. See eg. W.F. Verschuir, *De Arteriarum et Venarum Vi Irritabili: ejusque in Vasis Excessu: et inde Oriunda Sanguinis Directione Abnormi* (M.D. thesis, Groningen: Bolt, 1766) and P.J. Gruber, *De Excessu Vis Vitalis Vasorum, Variisque inde in Machina Animali Pendentibus Phoenomenis* (M.D. thesis, Prague and Dresden: Walther, 1775).
83. See eg. A.D. Thaer, *De Actione Systematis Nervosi in Febribus* (M.D. thesis, Göttingen: Dieterich 1774) and M. Stoll, *Aphorismi de Cognoscendis et Curandis Febribus* (Vienna: Kurzbeck, 1786). See also W.F. Bynum, 'Cullen and the Study of Fevers in Britain, 1760–1820', in W.F. Bynum and V. Nutton (eds), *Theories of Fever from Antiquity to the Enlightenment* (London: Wellcome Institute for the History of Medicine, 1981), 135–47.
84. Pringle to Haller, 11 June 1763 in O. Sonntag (ed.), *John Pringle's Correspondence with Albrecht von Haller*, (Basel: Schwabe, 1999), 55.
85. W. Cullen, *First Lines of the Practice of Physic,* Corrected and Enlarged (4 vols, 4th edn, Edinburgh: Elliot, 1784), xxxiv–xlvi. Lieutaud's book is the *Précis de la Médecine Pratique.* Nouvelle Édition (2 vols, Paris: Barrois, 1781).
86. Tissot, *op. cit.* (note 80), i, xxv (préface).
87. J. Weise, *De Irritabilitate Morborum Genitrice... Praeside Ernesto Godofredo Baldinger* (M.D. thesis, Jena: Heller, 1772), 3–4.
88. As one of these 'minor' authors we may mention Carl Abraham Gerhard who published a *Triga Dissertationum Physico-Medicarum* (Berlin: Rudiger, 1763), which included a 'Specimen Irritabilitatis ad Pathologiam et Therapiam Applicatae', 7–64.
89. H. Fouquet, 'Sensibilité, Sentiment (Méd.)', *Encyclopédie*, xv (1765), 38–52.
90. L. de Jaucourt, 'Maladie', *Encyclopédie*, ix, 929–38: 931.
91. K. Sprengel, *Versuch einer Pragmatischen Geschichte der Arzneikunde* (5 vols, Halle: Gebauer, 1792–1803), v: 214–42.
92. Cullen, *op. cit.* (note 85), xlviii.
93. My analysis differs somewhat from Thomas Broman's, who argues that the search for vital forces quite generally drew physiology away from the practical portions of medicine. Focusing on the ideas of Naturphilosophie developed in the 1790s, he underrates the importance of vitalist and semi-

vitalist pathological models. In particular, he depicts Gaub's pathology as a traditional account rather than as a new model based on the notion of vital properties, as it was clearly conceived by his contemporaries. Broman, *op. cit.* (note 20), Chapter 3.

94. On Brown and the reception of his system in Germany see G.B. Risse, *The History of John Brown's Medical System in Germany During the Years 1790–1806* (Chicago Univ. D.Phil. thesis, 1971); T. Henkelmann, *Zur Geschichte des Pathophysiologischen Denkens: John Brown (1735–1788) und sein System der Medizin* (Berlin: Springer, 1981); and Broman, *op. cit.* (note 20), Chapter 5.
95. For these journals see J. Sgard (ed.), *Dictionnaire des Journaux, 1600–1789* (2 vols, Paris: Universitas, 1991), *sub voce*; G. Mann, 'Ernst Gottfried Baldinger und sein Magazin für Aerzte', *Sudhoffs Archiv*, 42 (1958), 312–18. The *Giornale di Medicina* has never been studied.
96. R. Porter, 'Laymen, Doctors and Medical Knowledge in the 18th Century: The Evidence of the Gentleman's Magazine', *Medical History*, 29 (1985), 138–68.
97. Eg. the successful *Hamburgisches Magazin, oder Gesammelte Schriften zum Unterricht und Vergnügen aus der Naturforschung und den Angenehmen Wissenschaften Überhaupt* (Hamburg, 1747–81) or the *Allgemeines Magazin der Natur, Kunst und Wissenschaften* (Leipzig, 1753–67) and the *Fränkische Sammlungen von Anmerkungen aus der Naturlehre, Arzneygelahrtheit, Oekonomie und den damit Verwandten Wissenschaften* (Nuremberg, 1756–68).
98. For a general assessment of literary criticism and its importance for the Republic of Letters and the public sphere see M. Fontius, 'Tendenzen der Literaturkritik in Frankreich und Deutschland im 18. Jahrhundert', in S. Jüttner and J. Schlobach (eds), *Europäische Aufklärung(en): Einheit und Nationale Vielfalt* (Hamburg: Meiner, 1992), 127–40. For art criticism see R. Wrigley, *The Origins of French Art Criticism: From the Ancien Régime to the Restoration* (Oxford: Clarendon Press, 1993).
99. There is no comprehensive account of the review journals and the review culture. The best analysis is A. Goldgar, *Impolite Learning: Conduct and Community in the Republic of Letters, 1680–1750* (New Haven and London: Yale University Press, 1995), esp. Chapter 2. It focuses, however, on Dutch and French journals of the early eighteenth century and, in my view, overstresses the aspects of the 'community of obligation' (see below). For Germany and the second half of the century see U. Schneider, *Friedrich Nicolais Allgemeine Deutsche Bibliothek als Integrationsmedium der Gelehrtenrepublik* (Wiesbaden: Harrassowitz, 1995) and H. Rowland and K.J. Fink (eds), *The Eighteenth Century German Book Review* (Heidelberg: Winter, 1995). For further literature see Sgard, *op. cit.* (note 95), and the

extensive bibliography of D. Kuhles, *Deutsche Literarische Zeitschriften von der Aufklärung bis zur Romantik* (2 vols., Munich: Saur, 1994). For England see F. Donoghue, 'Colonizing Readers: Review Criticism and the Formation of a Reading Public', in A. Bermingham and J. Brewer (eds), *The Consumption of Culture 1600-1800: Image, Object, Text* (London and New York: Routledge, 1995), 54-74. Some comparative work has been done by C.B. O'Keefe, *Contemporary Reactions to the Enlightenment (1728–1762): A Study of Three Critical Journals, the Jesuit Journal de Trévoux, the Jansensist Nouvelles Ecclésiastiques, and the Secular Journal des Savants* (Geneva: Slatkine, 1974) and by P.-E. Knabe, *Die Rezeption der Französischen Aufklärung in den 'Göttingischen Gelehrten Anzeigen' (1739–1779)* (Frankfurt am Main: Klostermann, 1978).

100. Seventy per cent of the reviews in the *Journal des Savants* were devoted to French publications and to those published in France. See Knabe, *op. cit.* (note 99), 52.
101. On this aspect see H. Bots (ed.), *La Diffusion et la Lecture des Journaux de Langue Française sous l'Ancien Régime* (Amsterdam and Maarssen: APA-Holland University Press, 1988).
102. Letter from 6 March 1767. O. Sonntag (ed.), *The Correspondence between Albrecht von Haller and Horace-Bénédict de Saussure* (Bern etc.: Huber, 1990), 334.
103. See R. Nohr, E. Papacek and A. Vetter, '"Das Richtige Urteil über den Zustand der Vaterländischen Literatur?": Zum Anteil des Rezensionswesens an der Französisch-Deutschen Kulturvermittlung im Zeitalter der Aufklärung', in H.-J. Lüsebrink *et al.* (eds), *Kulturtransfer als Epochenumbruch Frankreich – Deutschland: 1770 bis 1815* (Leipzig: Universitätsverlag, 1997), 499–535.
104. Fontius, *op. cit.* (note 98), 128.
105. G. Roethe, 'Göttingische Zeitungen von Gelehrten Sachen', *Festschrift zur Feier des Hundertfünfzigjährigen Bestehens der Königlichen Gesellschaft der Wissenschaften zu Göttingen* (Berlin: Weidmann, 1901), 569–688: 631, 656.
106. Cf. M. Gierl, 'Kompilation und die Produktion von Wissen im 18. Jahrhundert', in Helmut Zedelmaier and M. Mulsow (eds), *Die Praktiken der Gelehrsamkeit in der Frühen Neuzeit* (Tübingen: Niemeyer, 2001), 63–94.
107. Knabe, *op. cit.* (note 99), 11–52.
108. Haller himself stated: 'Unstreitig sind folglich diese Anzeigen das reichste Wochenblatt worden, das in Europa über gelehrte Sachen herauskömmt.' *Sammlung Kleiner Hallerischer Schriften* (3 vols, Bern: Emanuel Haller, 1772), i: 129.
109. On Haller and the *GGA* see Roethe, *op. cit.* (note 105); K.S. Guthke, *Haller und die Literatur* (Göttingen: Vandenhoeck & Ruprecht, 1962) and U. Enke

(ed.), *Samuel Thomas Soemmering: Rezensionen für die Göttingischen Gelehrten Anzeigen* (2 vols, Stuttgart: Fischer, 1995), introduction.

110. K.S. Guthke, 'Haller und die Bibliothèque Raisonnée', *Jahrbuch des Freien Deutschen Hochstifts* (1973), 1–13.
111. Martin Stuber furnishes many examples and classifies the different requests and reactions of the readers to the *GGA* in his paper 'Journal and Letter: The Interaction Between two Communication Media in the Correspondence of Albrecht von Haller', *Studies on Voltaire and the Eighteenth Century* (2004), 114–41.
112. Cf. eg. the letters of the librarians G. Lichtensteger from 16 May 1746 and F.C. Pelt from 12 May 1752 (*Haller Papers*).
113. Cf. eg. the letters of G.M. Bose from 30 March 1747, of C.F. Helwing from 23 November 1750 and the whole correspondence of G.A. Müller (*Haller Papers*).
114. See A. Forster, ' "The Self-Impannelled Jury": The Reception of the Review Journals, 1749–1760', *Studies in Newspaper and Periodical History*, 1993, 27–51.
115. Letter from 13 March 1762, *Corr. Bonnet*, 268.
116. J. le Rond d'Alembert, 'Discours Préliminiaire', *Encyclopédie*, i: i.
117. H. Bosse, 'Die Gelehrte Republik', in H.-W. Jäger (ed.), *Öffentlichkeit im 18. Jahrhundert* (Göttingen: Wallstein, 1997), 51–76: 61.
118. Goldgar, *op. cit.* (note 99).
119. The continuity of the Republic of Letters and the fact that the Enlightenment was part of it is stressed by L. Brockliss, *Calvet's Web. Enlightenment and the Republic of Letters in Eighteenth-Century France* (Oxford: Oxford University Press, 2002).
120. H. Mattauch, *Die Literarische Kritik der Frühen Französischen Zeitschriften (1665–1748)* (Munich: Hueber, 1968), 70–1.
121. See Haller's prefaces to the volumes of 1747 and 1748.
122. See his review of Leibniz's correspondence with Johannes Bernoulli in the *Bibliothèque Raisonnée*, 37 (1746), 178–93; and O. Sonntag, 'Albrecht von Haller on Academies and the Advancement of Science: The Case of Göttingen', *Annals of Science*, 32 (1975), 379–91. It must be noted that Haller actually lived up to this idea. Although his erudition covered almost all parts of knowledge (except jurisprudence), he performed active research only in the medical sciences, botany, natural history, and agriculture. He also reviewed books from other branches – including literature, history, philosophy, theology, mathematics, physics and chemistry – but to what extent has yet to be ascertained.
123. A. von Haller, 'Mémoire sur Plusieurs Phénomènes Importants de la Respiration: Fondé sur les Expériences', in A. von Haller, *Sur la Formation*

du Coeur dans le Poulet, sur l'Oeil, sur la Structure du Jaune &c. (2 vols, Lausanne: Bousquet & Comp., 1758), ii, 197–366: 206–7.

124. *GGA*, 1748, 502.
125. *GGA*, 1756, 689–90.
126. Preface to the *GGA* 1748.
127. A. von Haller, *Bibliotheca Anatomica* (2 vols., Zurich: Orell, Gessner, Fuessli et Socc., 1774–77), i, 756.
128. W. van Noortwyck, 'Defense de Mr. Boerhaave', *Bibliothèque Britannique*, 23 (1746), 336–67 and 24 (1746), 187–216: 199.
129. 'Lettre de l'Auteur d'un Extrait de la Bibliothèque Raisonnée Attaqué dans la Bibliothèque Britannique', *Bibliothèque Raisonnée*, 40 (1748), 205–26, 454–68.
130. See E. Lesky, 'Albrecht von Haller, Gerard van Swieten und Boerhaavens Erbe', *Gesnerus*, 15 (1958), 120–40: 135–7.
131. 'The Haller Preface to the German Translation of the Histoire naturelle (1750)', in J. Lyon and P.R. Sloan (eds), *From Natural History to the History of Nature: Readings from Buffon and his Critics* (Notre Dame, Ind.: Notre Dame University Press, 1981), 295–310: 304–5. The original text is the preface to *Allgemeine Historie der Natur nach allen ihren Besondern Theilen Abgehandelt, nebst einer Beschreibung der Naturalienkammer.... Erster Theil* (Hamburg and Leipzig: Grund and Holle, 1750), ix–xxii. I have corrected Sloan's translation.
132. See O. Sonntag, 'The Motivations of the Scientist: The Self-Image of Albrecht von Haller', *Isis*, 65 (1974), 336–51.
133. A. Haller, 'Mémoire sur une Controverse au Sujet de la Respiration', *Nouvelle Bibliothèque Germanique*, 1748, 412–28: 428.
134. *Elementa*, v (1763), preface, iii.
135. Letter from 11 March 1753 (*Haller Papers*). How Haller tried to keep to his commitment of professional criticism without violating the principles of conduct that the Republic of Letters expected from its members, is best visible in the dispute about his review of a work of Casimir Christoph Schmiedel. See H. Steinke (ed.), *Der Nützliche Brief: Die Korrespondenz zwischen Albrecht von Haller und Christoph Jakob Trew, 1733–1763*, (Basel: Schwabe, 1999), introduction.
136. *Commentarii de Rebus in Scientia Naturali et Medicina Gestis*, i (1752), 601 and iii (1754), 420–1.
137. *Monthly Review*, 14 (1756), 140
138. *Jenaische Gelehrte Anzeigen*, 1750, 23
139. *Ibid.*, 1750, 228.
140. A review often served as an incentive for or even as the immediate topic of a new publication. Cf. eg. H.F. Delius, 'Abgenöthigte Erläuterung der

Göttingischen Beurtheilung einiger Aufsätze im Dritten Bande der Fränkischen Sammlungen', *Fränkische Sammlungen*, 5 (1760), 99–144; J.A. Unzer, *Physiologische Untersuchungen: Auf Veranlassung der Göttingischen, Frankfurter, Leipziger und Hallischen Recensionen seiner Physiologie der Thierischen Natur* (Leipzig: Weidmann, 1773).

141. *GGA*, 1755, 1436.
142. To facilitate the access to this information he compiled an index of all his published and unpublished reviews.
143. Roethe, *op. cit.* (note 105), 606.
144. See R. Stichweh, *Zur Entstehung des Modernen Systems Wissenschaftlicher Disziplinen: Physik in Deutschland, 1740–1890* (Frankfurt am Main: Suhrkamp, 1984).
145. Letter from 20 March 1761, *Corr. Tissot*, 115
146. Letter from 29 October 1770, edited in de Haen, *op. cit.* (note 31), xiv: 122
147. Zimmermann quoting de Haen in his letter to Haller from 24 February 1762. 'J.G Zimmermanns Briefe an Haller nach dem Manuskript der Stadtbibliothek', in R. Ischer (ed.), *Neues Berner Taschenbuch auf das Jahr 1904–1912* (Bern, 1903–1911), 1910: 175–6.
148. De Haen to Bonnet, 19 May 1766 (Bibliothèque Publique et Universitaire de Genève, Ms. Bonnet VI.1).
149. Edited in de Haen, *op. cit.* (note 31), xii: 275–76. Boschung, *op. cit.* (note 33), 143 points also to the different positions on criticism of Haller, de Haen and Tissot.
150. This point has been stressed by T. Broman, 'The Habermasian Public Sphere and "Science in the Enlightenment"', *History of Science*, 36 (1998), 123–49. It has to be noted, however that, in Broman's view, the public sphere for the older Republic of Letters was only an imagined ideal whereas in the Enlightenment it was reality.
151. *GGA*, 1748, 908.
152. Broman, *op. cit.* (note 150).
153. Letter from 27 August 1763 (Burgerbibliothek Bern, MS. hist. helv. XVIII 71).
154. *Journal des Savants*, 1753, 550–5; 1755, 226–33 and 771–4; 1757, 73–82 and 290–302; 1759, 138–40.
155. Letter to Zimmermann, October 1758 (Niedersächsische Landesbibliothek, Hanover, *Zimmermann Papers*, A II, 96).
156. This assessment is in line with more recent studies on the Enlightenment that do not presuppose the unity of science but try to reveal the interaction between local, separating and international, unifying discourses and practices. See W. Clark, J. Golinski and S. Schaffer (eds), *The Sciences in Enlightened Europe* (Chicago and London: Chicago University Press, 1999),

introduction.

157. Schneider, *op. cit.* (note 99).

158. See K. Hufbauer, *The Formation of the German Chemical Community (1720–1795)* (Berkeley: University of California Press, 1982) and Broman, *op. cit.* (note 20), 188. An earlier example in the transitional period towards the specialised periodical is Jean François Rozier's *Journal de Physique.* See J.E. McClellan, 'The Scientific Press in Transition: Rozier's Journal and the Scientific Societies in the 1770s', *Annals of Science*, 36 (1979), 425–49.

Conclusion

This book has been primarily concerned with Haller's concept of irritability and sensibility and its reception in the second half of the eighteenth century. It discussed also the research, theories and thoughts of several of Haller's contemporaries but always in their relation to the specific notions of the Swiss physiologist. It is thus not a general history of the idea of irritation and irritability, as Owsei Temkin intended to write it.[1] Such a history has still to be written. This study may be described as an intellectual or social rather than as a traditional history of ideas in the manner of Arthur Lovejoy.[2] Of course, I have tried to understand the originality of Haller's concept and the theoretical and philosophical background from which it emerged but, even more, I have stressed and sought to identify the variety of 'internal' and 'external' factors that characterised and determined his research and the debate on irritability and sensibility. The controversy was about facts, theories and methods. But in none of these a settlement could be reached. The controversy was neither resolved by an agreement of the Republic of Letters nor closed by any external power, it just withered away.[3] It was not, however, without effect. Although it did not provoke a revolution of physiological methodology, it helped to transform the face of physiological and pathological concepts. The various models of the body as an active and reactive organism replaced the older mechanical ones. Having been the major figure to initiate this change, Haller was hailed as one of the – if not the – greatest physiologists of the eighteenth century. But his particular ideals of physiology – on both the methodological and the conceptual level – were, in fact, largely rejected. Haller's successors seem not to have realised to what extent they contradicted his fundamental beliefs.

The rejection of Haller's specific notions – which has hitherto been underestimated – has to be explained by the particularities of his theoretical position and the general state of physiological research. On the one hand, Haller's descriptive account of irritability and his separation of two specific faculties with clearly defined and limited functions was rejected because it did not answer the needs for a comprehensive explanation of all physiological processes and the fundamentals of life. On the other hand – and this aspect I have particularly considered – Haller did not succeed because his ideals of physiological research and scientific discourse were generally not accepted. The principal method of physiological investigation, in his view, was animal experimentation. The experiments had to be performed with great care and according to certain standards, which Haller had tried to establish in regard to the trials on irritability and sensibility. The researcher should select a small and well-defined area of exploration, which had to be examined scrupulously. All details had to be taken into consideration. Commenting on his embryological research, Haller said: 'This is a minute detail, but truth is based on details and error arises from a superficial knowledge of things.'[4] The science of physiology consisted of many different areas of research with a countless number of details and its complexity had to be acknowledged. Physiological knowledge was necessary to understand malfunctions of the body and thus to develop sound principles of pathology and medical practice. But it was not an ancillary science. It had to be considered a science on its own and to be pursued as such, regardless of the consequences a physiological finding might have for the practice of medicine. Finally, the results thus achieved had to be judged by professionals or, more exactly, by specialists, who performed physiological, ie. experimental research themselves.

All these Hallerian ideals of research and scientific discourse conflicted with reality. There were no established criteria how to perform animal experiments and these were not used as a standard physiological research method. In Haller's view, physiology was therefore still a 'very imperfect science'.[5] In fact, as late as 1776 he stated that 'the career [of physiology] has, indeed, only just begun'.[6] Despite some encouraging developments, physiology was still a rational rather than an experimental science. It was based on broad concepts rather than on specialist studies and the need for a rational explanation of disorders and diseases had much bearing on the acceptance or rejection of a new notion. Last but not least, there were no accepted professional experts who would have decided matters of dispute. Every physician, even if inexperienced in physiological research, could claim to be a professional judge of all medical questions.

A hundred years later, when the question of muscular irritability was again put on the agenda of experimental research, these parameters had changed radically. In the early 1850s, Claude Bernard (1813–78) performed several series of animal experiments with curare. These experiments showed that curare destroyed the capacity of the nerves to act upon the muscles. Irritations of nerves could not provoke a muscular movement anymore, but the muscle itself retained its ability of contraction. This ability was entirely independent of any nervous action. Bernard concluded that the separation of the nervous from the muscular system and the independence of Hallerian irritability, which had been doubted since Haller's time and which now was called contractility, was finally proven.[7] Albert Kölliker (1817–1905) performed similar experiments with a similar outcome independently of Bernard, and although some objections were raised against Kölliker's and Bernard's research, their results were quickly accepted as standard physiological knowledge.[8] The reason for this acceptance has to be located in the same set of parameters which had impeded the approval of Haller's findings.[9] In Haller's time, the postulation of an independent power inherent in the muscles had touched basic physiological problems, and it had been expected that any explanation of the phenomena of motion and sensation had also to serve as a model explaining other physiological processes and the fundamentals of life. In the mid-nineteenth century, however, it was common to acknowledge that each organ had its specific mode of action and reaction, an idea that had been put forward by Bordeu and established notably by Johannes Müller and his authoritative *Handbuch der Physiologie* in the 1830s.[10] A new explanation of muscular or nervous action, therefore, did not change the whole outlook of physiology. When Bernard and Kölliker confirmed Haller's postulation of an independent muscular power, they made a contribution to the specific area of muscular, and not to general, physiology. Even less were their results conceived as tools to construct a new model of pathology. Physiology was installed as a more or less independent branch of science, and animal experimentation was one of its main methods. Claude Bernard stated somewhat later, in the 1870s: 'Nowadays all the physiologists are without exception experimentalists.'[11] As the standards of experimentation were much more firmly established than in Haller's time, Bernard did not have to publish the protocols of his trials with curare. His results were judged not by the whole community of physicians but by experimental physiologists who worked in his own or similar research areas. Haller's ideals of scientific research seemed finally to be realised. It has to be doubted, however, that Haller would have felt comfortable in the laboratories of Bernard in Paris or of Kölliker in Würzburg. The transformation of scholars into scientists and of the Republic of Letters into a Scientific Community was part of a much wider social, political, and religious revolution, which Haller – conservative in these respects – would not have welcomed.

Notes

1. O. Temkin, *The Double Face of Janus and Other Essays in the History of Medicine* (Baltimore and London: Johns Hopkins University Press, 1977), 31.
2. A.O. Lovejoy, *The Great Chain of Being: The Study of the History of an Idea* (Cambridge, Mass.: Harvard University Press, 1936). For a recent assessment

of these different approaches and their importance for the historiography of medicine see C. auf der Horst, 'Vorstellungen, Ideen, Begriffe: Intellectual History in der Medizingeschichtsschreibung am Beispiel des Naturbegriffs', in N. Paul and T. Schlich (eds), *Medizingeschichte: Aufgaben, Probleme, Perspektiven* (Frankfurt am Main and New York: Campus, 1998), 186–215.

3. On the various ways in which scientific controversies end cf. H.T. Engelhardt and A.L. Caplan (eds), *Scientific Controversies: Case Studies in the Resolution and Closure of Disputes in Science and Technology* (Cambridge: Cambridge University Press, 1987).
4. A. von Haller, *Sur la Formation du Coeur dans le Poulet, sur l'Oeil, sur la Structure du Jaune &c.* (2 vols, Lausanne: Bousquet & Comp., 1758), i: iv (pages not numbered).
5. *GGA*, 1760, 606–7.
6. E. Hintzsche (ed.), 'A.v. Hallers "Prospectus d'un Dictionnaire Universel de Médecine" [1776]', *Gesnerus*, 23 (1966), 48–54: 49.
7. C. Bernard, 'Analyse Physiologique des Propriétés des Systèmes Musculaire et Nerveux au Moyen du Curare', *Compte Rendu des Séances de l'Académie des Sciences*, 43 (1856), 825–9. Bernard, Rudolf Virchow (1821–1902) and others used the term 'irritability' in a much broader sense than Haller; similar to Glisson, they regarded it as the most simple and most general property of interaction, which all living organic substances possessed. See F. Mondella, 'L'Irritabilità ed il Programma di Ricerca nella Fisiologia di C. Bernard', in M. Di Giandomenico (ed.), *Claude Bernard: Scienza, Filosofia, Letteratura* (Verona: Bertani, 1982), 177–88; R. Virchow, 'Reizung und Reizbarkeit', *Archiv für Pathologische Anatomie und Physiologie und für Klinische Medicin*, 14 (1858), 1–63.
8. A. Kölliker, 'Physiologische Untersuchungen über die Wirkung einiger Gifte', *Archiv für Pathologische Anatomie und Physiologie und für Klinische Medicin*, 10 (1856), 3–77, 235–96; A. Kölliker, 'Note sur l'Action du Curare sur le Système Nerveux', *Compte Rendu des Séances de l'Académie des Sciences*, 43 (1856), 791–2.
9. I would like to stress that my aim is not to show that the 'Great Haller' was right. It just happens to be that he was right in his main postulation regarding the general independence of both properties. From our modern point of view, he was wrong in his underestimation of subconscious nervous action and of the sensibility of certain organs.
10. J. Müller, *Handbuch der Physiologie des Menschen für Vorlesungen*, (3rd edn, 2 vols, Coblenz: Hölscher, 1838–40), i, 54–7. The handbook was also translated into English and French.
11. C. Bernard, *Principes de Médecine Expérimentale*, L. Delhoume (ed.), (Paris: Presses Universitaires de France, 1947), 191.

Bibliography

Manuscripts

BURGERBIBLIOTHEK BERN

Nachlass Albrecht von Haller (Haller papers)

Correspondence (Letters to Haller from G. Azzoguidi, E.G. Baldinger, P.J. Barthez, A. Beurer, G.M. Bose, J.H. Brunn, J. Farjon, A. Ferrein, J.-E. Gilibert, L.-M. Girard de Villars, J.F. Gmelin, F.M. Grimm, C.F. Helwing, J.P. Housset, D.R. Ith, C.-N. Le Cat, G. Lichtensteger, G.A. Müller, A. Monro, F.C. Pelt, J.S. Pütter, J.C. Ramspeck, J.G. Runge, F. Boissier de Sauvages, F. Thierry, C.-A. Vandermonde, P.G. Werlhof and J.G. Zinn; letters from Haller to C.G. Ludwig, R.A.F. de Réaumur and F. Boissier de Sauvages).

Mss. 23–31 (Observationes Anatomicae).

Mss. 32–55 (Iudicia Librorum).

Other collections

Ms. hist. helv. XVIII 71 (Letters from J.G. Zimmermann to S.-A. Tissot).

BIBLIOTHÈQUE PUBLIQUE ET UNIVERSITAIRE, GENEVA

Ms. Bonnet 29.1–2 (Letters from A. de Haen to Ch. Bonnet).

BIBLIOTECA NAZIONALE BRAIDENSE, MILAN

A.D. XI. 6 (Haller, Albrecht, 'Excepta de Usu Partium harumque Fabrica Subtili, ex Ore Clar. Viri D. Bernhard Siegfried Albini').

A. XIV. 2432/29 (N.N., 'Experimenta a Studiosis Capta...'; 10 October 1751 to 12 March 1752).

University Library, Leiden

Ms. Leningrad XIII. 32 (Herman Boerhaave, 'Praelectiones Publicae de Corde', 1735–37; microfilm of the original manuscript in the Military Academy in St. Petersburg).

Niedersächsische Landesbibliothek, Hanover

Zimmermann papers (MS XLII 1933)

AII 83 (Letters fom Samuel Schmid von Auenstein), A II 96 (Letters from S.-A. Tissot).

B1 ('De Irritabilitate'), B 6a ('Von den Temperamenten'), B 6b ('Oratio de Temperamentis'), B 6c ('De Viribus Naturae Insitis ad Ingeniossissimos et Eruditissimos Italiae Medicos et Philosophos Dissertatio Epistolica'), B 6d ('De Viribus Naturae Insitis Dissertatio'), B 63 ('De Principio in Homine Impetum Faciente Dissertatio'), B 19c ('Obs. Medicae et Physicae').

Archives de l'Académie des Sciences, Paris

Letter from Haller to R.A.F. de Réaumur, 23 January 1753.

Bibliothèque de l'Académie de Médecine, Paris

Archives de l'Académie Royale de Chirurgie (1731–93)

Mss. 18–26 (Procès-Verbaux des Séances).

Printed Sources

Primary Sources

Review journals

The titles of the journals often changed slightly over the years.

Latin

Acta Eruditorum (Leipzig, 1682–1776), *Relationes de Libris Novis* (Göttingen, 1752–6), *Commentarii de Rebus in Scientia Naturali et Medicina Gestis* (Leipzig, 1752–98).

French

Journal des Savants (Paris, 1665–), *Bibliothèque Germanique* (Berlin, 1720–60), *Bibliothèque Raisonnée* (Amsterdam, 1728–53), *Bibliothèque Britannique* (The Hague, 1733–47), *Journal Britannique* (The Hague, 1750–7), *Journal Étranger* (Paris, 1754–62).

English

Monthly Review (London, 1749–1845).

Italian

Estratto della Letteratura Europea (Bern, Yverdon, 1758–66).

German

Frankfurter Gelehrte Anzeigen (Frankfurt am Main, 1736–90), *Göttingische Gelehrte Anzeigen* (Göttingen, 1739–), *Erlangische Gelehrte Anzeigen* (Erlangen, 1746–87), *Jenaische Gelehrte Zeitungen* (Jena, 1749–57, 1765–82), *Critische Nachrichten* (Greifswald, 1750–4, 65–74), *Tübingische Berichte von Gelehrten Sachen* (Tübingen, 1752–63), *Allgemeine Deutsche Bibliothek* (Berlin, 1765–1806).

Other sources

In order to furnish a comprehensive bibliography on irritability and sensibility, I have added some fifty titles that are not mentioned in the text.

Anon., 'Lettre de l'Auteur d'un Extrait de la Bibliothèque Raisonnée Attaqué dans la Bibliothèque Britannique', *Bibliothèque Raisonnée*, 40 (1748), 205–26, 454–68.

Anon., 'Esquisse sur l'Irritabilité & la Sensibilité des Parties du Corps de l'Homme', *Mercure de France*, Avril 1757, 144–51.

——'Lettre a Messieurs les Auteurs du Journal des Sçavans, sur la Cause des Mouvemens Musculaires', *Journal des Sçavans*, Avril 1766, 221–36.

——'Lettera Critica, che Favorisce la Sentenza Halleriana sull'Irritabilità', *Giornale di Medicina*, 30 (1770), 233–8.

Alberghetti, Domenico Agostino, *De Naturali quorumdam Corporis Humani Partium Anaisthasia, nec non Irritabilitate de Peculiari Cruris Amputatione..., cum Diagrammate Dynastae Alberto de Haller* (Venice: Recurti, 1770).

Alembert, Jean le Rond d', 'Discours Préliminiaire', *Encyclopédie* (q.v.), i (1751), i–xlv.

——'Expérimental (Philosophie nat.)', *Encyclopédie* (q.v.), vi (1756), 298–301.

Alix, Matthaeus Franciscus, *Anweisung zur Lehrenden und Ausübenden Wundarzeneykunst: Nach dem Lehrgebüude der Neuern zum Nutzen Angehender Wundärzte, besonders derer auf dem Lande Verfasst* (Riga and Leipzig: Hartknoch, 1772).

Andreae, Jakob Eberhard, *De Irritabilitate Animali... Praeside... Philippo Friderico Gmelin* (M.D. thesis, Tübingen: Erhard, 1758).

Arrigoni, Antonio, 'Dissertazione Epistolare Indiritta al Signor Giacinto Fabri dell' Irritabilità e Sensazione delle Parti nei Viventi [17.12.1756]', in Giacinto Bartolomeo Fabri (ed.), *Sulla Insensitività ed Irritabilità Halleriana: Opuscoli di vari Autorii* (4 vols, Bologna: Corciolani, Colli, 1757–59), ii, 131–40.

Arthaud, Charles, *Dissertations sur la Dilatation des Artères, et sur la Sensibilité; Apuyées de plusieurs Expériences, Faites sur les Animaux Vivans, Auxquelles on a Joint Deux Observations sur l'Hydropisie du Péritoine* (M.D. thesis, Paris: Cavelier, 1771).

Ascanius, Peter, *De Irritabilitate... Praeside Georgio Christiano Oeder* (M.D. thesis, Copenhagen: Berling, 1752).

Azzoguidi, Germano, *Institutiones Medicae* (2 vols, Bologna: Saxius, 1775).

Baldinger, Ernst Gottfried, *Vindiciae Irritabilitatis Hallerianae* (Göttingen: Dieterich, 1775).

——... *Tertio Indicit Octo Candidatis Doctissimis Gradum Doctoris*

Medicinae Contulisse Ordinem Medicorum. Praefatio Illustrat Malignitatem in Morbis, ex Mente Hippocratis, per Recentiorum Irritabilitatem et Sensibilitatem (Göttingen: Dieterich, 1775).

——*Vestigia Irritabilitatis in Veterum Monumentis* (Göttingen: Dieterich, 1775).

Bange, Paulus, *De Sanatione Epilepsiae per Imminutum Vitium Irritabilitatis et Mobilitatis Nervorum... Praeside Eberhardo Rosen* (M.D. thesis, Lund: [n.p.], 1771).

Barbiellini, Camillo, *Riflessioni Anatomiche Intorno alla Moderna Dissertazione del Sig. Alb. Haller, dal quale si Vogliono Sostenere per Insensibili Molte Parti del Corpo Umano contro l'Esperienze Anatomiche, e contro le Autorità di Gravissimi Scrittori di Medicina, e Chirurgia* (Rome: Barbiellini, 1755).

Barthez, Paul-Joseph, *Quaestiones Medicae Duodecim* (Montpellier: Martel, 1761).

——*Nouvelle Méchanique des Mouvements de l'Homme et des Animaux* (Carcassonne: Polere, An VI, 1798).

——*Nouveaux Éléments de la Science de l'Homme* (Montpellier: Martel, 1778, 2nd edn, Paris: Gujon and Brunot, 1806).

Bayer, Thaddäus, *De Animi Affectibus* (M.D. thesis, Vienna: Kurzbeck, 1760).

Beek, Elbartus, *De Partium Humani Corporis Irritabilitate* (M.D. thesis, Harderwyck, 1770). [not consulted]

Bell, Benjamin, *A System of Surgery* (6 vols, Edinburgh: Elliott, 1783–8).

Bernard, Claude, 'Analyse Physiologique des Propriétés des Systèmes Musculaire et Nerveux au Moyen du Curare', *Compte Rendu des Séances de l'Académie des Sciences*, 43 (1856), 825–9.

——*Principes de Médecine Expérimentale*, Léon Delhoume (ed.), (Paris: Presses Universitaires de France, 1947).

Bernstein, Johann Gottlob, *Praktisches Handbuch für Wundärzte und Geburtshelfer*, Neue durchaus Umgearbeitete Auflage (2 vols, Leipzig: Schwickert, 1790).

Bertossi, Giuseppe, *Della Sensibilità, ed Irritabilità delle Parti del Corpo Umano. Lettera... in Risposta a Richiesta Fattagli da un Amico sopra tale Argomento* (Padua: Conzatti, 1756).

Beyckert, Philipp Jakob, *De Nervis Durae Matris... Praeside Johanne Friderico Lobstein* (M.D. thesis, Strasburg: Heitz, 1772).

Bianchi, Giovanni Battista, *Lettera... sulla Sensibilità, ed Irritabilità delle Parti negli Uomini, e ne' Bruti: In Risposta d'altra Lettera Scrittali dal Signor Dottor Giambattista Bassani* (Turin: Campana, 1755).

——'Lettera Seconda... alli... Signori Lettori Pubblici dell' Università di Bologna' [14.9.1756], in Giacinto Bartolomeo Fabri (ed.), *Sulla Insensitività ed Irritabilità Halleriana: Opuscoli di vari Autorii* (4 vols, Bologna: Corciolani, Colli, 1757–59), ii, 25–52.

Bikker, Lambertus, *Specimen Physiologico-Medicum Inaugurale de Natura Humana, quae Medicorum est* (M.D. thesis, Leiden: n.p., 1757).

Bilguer, Johann Ulrich, *Anweisung zur Ausübenden Wundarzneykunst in Feldlazarethen* (Glogau and Leipzig: Günther, 1763).

[Bittermann], *Epistola quam de Vindiciis Haenii ad... Albertum de Haller Mittunt Lippi et Tonsores Haenii* (Hanover: n.p., 1762).

Blasche, Johann Christian, *Das Leben des Herrn Hofraths und Professors Georg Erhard Hambergers, nebst einer Nachricht von seinen Schriften, und Gelehrten Streitigkeiten* (Jena: Güth, 1758).

Blumenbach, Johann Friedrich, *Über den Bildungstrieb und das Zeugungsgeschäfte* (Göttingen: Dieterich, 1781; reprint Stuttgart, 1971).

——*Institutiones Physiologicae* (Göttingen: Dieterich, 1786).

Boerhaave, Herman, *Institutiones Medicae in Usus Annuae Exercitationis Domesticos* (5th edn, Leiden: Theodor Haak *et al.*, 1734).

——*Institutions de Médecine.... Seconde Edition, avec un Commentaire. Par M. de la Mettrie* (8 vols, Paris: Huart, 1743–50).

——*A New Method of Chemistry.* Transl. by Peter Shaw (2nd edn, 2 vols., London: Longman, 1741).

——*Hermanni Boerhaave Praelectiones de Morbis Nervorum 1730–35*, B.P.M. Schulte (ed.), (Leiden: Brill, 1959; Analecta Boerhaaviana, 2).

——*Boerhaave's Orations.* Transl. with Introductions and Notes by E. Kegel-Brinkgreve and A.M. Luyendijk-Elshout (Leiden: Brill, 1983).

Bohl, Johann Cristoph, *Von der Nöthigen Vorsichtigkeit bey denen in Lebendigen Geschöpfen Anzustellenden Erfahrungen, von der*

Unempfindlichkeit der Sehnen: Erfahrungsmässig Angezeiget (Königsberg: Hartung, 1764; enlarged version 1766).

Bonnet, Charles, *Essai Analytique sur les Facultés de l'Ame* (Copenhague: Philibert, 1760; Reprint Hildesheim: Olms, 1973).

Bordenave, Toussaint, 'Remarques sur l'Insensibilité de Quelques Parties, Établie par la Pratique', *Mercure de France*, Juin 1757, 137–54.

Bordeu, François, *De Sensibilitate et Contractibilitate Partium in Corpore Sano. Theses aliquot... Praeside D. Francisco de la Mure* (Montpellier: Martel, 1757).

Bordeu, Théophile de, *Oeuvres Complètes de Bordeu* (2 vols, Paris: Caille and Ravier, 1818).

Borelli, Giovanni Alfonso, *De Motu Animalium* (2nd edn, Leiden: Gaesbeek, 1685; engl. transl. by Paul Maquet, Berlin: Springer, 1989).

Bos, Iman Jacob van den, *Specimen Physiologico-Medicum Inaugurale de Vivis Humani Corporis Solidis* (M.D. thesis, Leiden: Luchtmans, Moens, 1757).

Brandt, Michael, *De Differentia Elateris, Toni, Contractionis Vitalis, Voluntariae, Sensibilitatis et Irritabilitatis... Praeside Io. Gottlob Krügero* (M.D. thesis, Helmstedt: Schnorr, 1754).

Brocklesby, Richard, 'An Account of some Experments on the Sensibility and Irritability of the several Parts of Animals; in a Letter ... to the Reverend Thomas Birch', *Philosophical Transactions*, 49, 1 (1755), 240–5.

Brunn, Johann Heinrich von, *Dissertatio Inauguralis Medica Sistens Experimenta quaedam circa Ligaturas Nervorum in Vivis Animalibus* (M.D. thesis, Göttingen: Vandenhoeck, 1753).

Burkhard, Hans Rudolf, 'Bestätigung des Hallerischen Lehrgebäudes von der Unempfindlichkeit Verschiedener Theile des Menschlichen Cörpers, besonders der Sehnen, durch einige Chirurgische Beobachtungen und Versuche', *Abhandlungen der Naturforschenden Gesellschaft in Zürich*, 1 (Zürich: Heidegger, 1761), 321–48.

Busse, Gottlieb Christian, *De Actione Cordis quatenus a Nervis Pendet... Praeside... Andrea Elia Büchnero...* (M.D. thesis, Halle: Litteris Stephanianis, 1769).

Caldani, Leopoldo Marcantonio, 'Sur l'Insensibilité et l'Irritabilité de Mr. Haller: Seconde Lettre... à Mr de Haller', in Albrecht von Haller (ed.), *Mémoires sur la Nature Sensible et Irritable des Parties du Corps Animal* (4 vols, Lausanne: Bousquet, Arnet, 1756–60), iii, 343–485.

——'Sull'Insensitività, ed Irritabilità di Alcune Parti degli Animali: Lettera Scritta al... Signore Alberto Haller [25.11.1756]', in Giacinto Bartolomeo Fabri (ed.), *Sulla Insensitività ed Irritabilità Halleriana: Opuscoli di vari Autorii* (4 vols, Bologna: Corciolani, Colli, 1757–59), i, 269–336.

——*Riflessioni Fisiologiche... sopra Due Dissertazioni del Signor Claudio Nicola Le Cat* (Venice: Pasquali, 1767).

——*Esame del Capitolo Settimo Contenuto nella XII: Parte dell'ultima Opera del... Antonio de Haen... Indiritto allo stesso Autore* (Padua: Comino, 1770).

——*Institutiones Physiologicae* (Padua: Comino, 1773).

Carrère, Joseph, *De Vitali Corporis et Animae Foedere... pro Baccalaureatus Gradu* (Perpignan: Reynier, 1758).

Chomel, Amabilis, *An Vitae Exercitium a Fibrarum Sensibilitate? ...Praeside Petro Ludovico-Maria Maloet* (Paris, 1752).

Cigna, Giovanni Francesco, *Johannes Franciscus Cigna... Publice Disputabat... MDCCLVII. Die XIV. Aprilis* (Turin: Typographia Regia, 1757).

Collomb] *Lettre de M. Collomb Étudiant en Médecine, en l'Université de Paris, à M. Collomb, Médecin à Lyon; sur un Cours de Physiologie Expérimentale, Fait cette Année 1771, au Collége de France, par M. Portal* (Paris: Didot, 1771).

Condorcet, Marie-Jean-Antoine-Nicolas Caritat de, 'Éloge de M. de Haller', *Histoire de l'Academie Royale des Sciences,* année 1777 (Paris: Impr. Royale, 1780), 127–154.

Coutavoz, Joannes-Augustinus, *Quaestio Medica Quodlibetariis Disputationibus Mane Discutienda in Scholis Medicorum Die Vigesima-Sexta Februarii MDCCLXVII. Henrico Raymond Praeside. 'Num Arteriae ut Sensibiles sic Irritabiles?'* (M.D. thesis, Paris: Quillau, 1767).

Crantz, Heinrich Johann Nepomuk von, *Solutiones Difficultam circa Cordis Irritabilitatem* (Vienna: Kraus, 1761).

——*Lettre... à Mr. Tyssot au Sujet de sa Dispute avec Mr. de Haen* (Vienna: Kraus, 1763).

Cullen, William, *Observations on the Structure and Functions of the Nervous System* (Edinburgh: Creech, 1783).

——*First Lines of the Practice of Physic* (4th edn, corrected and enlarged, 4 vols, Edinburgh: Elliot, 1784).

——*Institutions of Medicine. Part I. Physiology* (3rd edn, Edinburgh: Elliot, 1785).

Daviel, Jacques, 'Lettre de M. Daviel... Écrite au... M. le Baron de Halle[r]', *Mercure de France*, Janvier 1762, 146–58.

Delius, Heinrich Friedrich, *Theoria et Foecundus in Medicina Usus Principii: Sensationem Sequitur Motus Sensatio Proportionatus* (Erlangen: Becker, 1749).

——*Animadversiones in Doctrinam de Irritabilitate, Tono, Sensatione et Motu Corporis Humani* (Erlangen: Camerarius, 1752).

——'Theoria et Foecundus in Medicina Usus Principii: Sensationem Sequitur Motus Sensatio Proportionatus. d.i. Theorie und Nützlicher Gebrauch des Satzes in der Arzneykunst; daß nämlich auf eine Empfindung eine Bewegung Folge, so der Empfindung Proportional ist', *Hamburgisches Magazin*, 16 (1756), 191–217.

——'Physiologische Anmerkungen', *Fränkische Sammlungen von Anmerckungen aus der Naturlehre, Arzneygelahrheit, Oekonomie und den damit Verwandten Wissenschaften*, Vol. 3 (Nuremberg: Monath, 1758), 406–26.

——'Abgenöthigte Erläuterung der Göttingischen Beurtheilung einiger Aufsätze im Dritten Bande der Fränkischen Sammlungen', *Fränkische Sammlungen von Anmerckungen aus der Naturlehre, Arzneygelahrheit, Oekonomie und den damit Verwandten Wissenschaften*, Vol. 5 (Nuremberg: Monath, 1760), 99–144.

——'Antwort' [answer to a letter of Sauvages], *Fränkische Sammlungen von Anmerckungen aus der Naturlehre, Arzneygelahrheit, Oekonomie und den damit Verwandten Wissenschaften*, Vol. 5 (Nuremberg: Monath, 1760), 481–97.

Dennison, Richard, *Disputatio Physiologica Inauguralis, Arterias Omnes, et Venarum Partem, Irritabilitate Praeditas Esse* (M.D. thesis, Edinburgh: Smellie, 1775).

Dionis, Pierre, *Cours d'Opérations de Chirurgie, Démontrées au Jardin Royal*, Georges de La Faye (ed.), (5th edn, Paris: D'Houry, 1757; 7th edn, *ibid.*, 1773).

Doeveren, Wouter van, *De Imprudenti Ratiocinio ex Observationibus et Experimentis Medicis: Sermo Adacemicus* (Groningen: Crans, 1754).

——*Specimen Observationum Academicarum ad Monstrorum Historiam, Anatomen, Pathologiam, et Artem Obstetriciam, praecipue Spectantium* (Groningen and Leiden: Bolt, Luchtmans, 1765) [On 208–54 there is an account of experiments on irritability and sensibility, which was translated into German and published seperately in Leipzig: Krause, 1767].

Dupont, Jean-Jacques, *Dissertatio Medica Opposita Argumentis Celeberrimi Eberhardi de Animae Imperio in Cor... Praeside Francisco De Sauvages pro Baccalaureatus Gradu* (Avignon: Tilan, 1760).

Eberhard, Johann Peter, *Conspectus Medicinae Theoreticae... Pars Prima: Physiologia et Diaetetica* (Halle: Renger, 1757).

Eloy, Nicolas-François-Joseph, *Dictionnaire Historique de la Médecine Ancienne et Moderne* (4 vols, Mons: Hoyois, 1778).

Encyclopédie, ou Dictionnaire Raisonné des Sciences, des Arts et des Métiers, Denis Diderot and Jean le Rond d'Alembert (eds), (17 vols, Paris: Briasson etc., 1751–65).

[Encyclopédie] Supplément à l'Encyclopédie, ou Dictionnaire Raisonné des Sciences, des Arts et des Métiers, par une Société de Gens de Lettres (4 vols, Amsterdam: Rey, 1776–7).

Encyclopédie, ou Dictionnaire Universel Raisonné des Connoissances Humaines. Mis en Ordre par M. de Félice (42 vols, 6 vols of suppls, Yverdon: [n.p.], 1770–9).

Fabre, Pierre, *Essais sur Différens Points de Physiologie, de Pathologie et de Thérapeutique* (Paris: Didot, 1770).

Fabri, Giacinto Bartolomeo (ed.), *Sulla Insensitività ed Irritabilità Halleriana: Opuscoli di vari Autori* (4 vols, Bologna: Corciolani, Colli, 1757–9).

Fè, Giambattista, 'Saggio Critico... in Difesa di Ermanno Boerhaave...', in Giacinto Bartolomeo Fabri (ed.), *Sulla Insensitività ed Irritabilità Halleriana: Opuscoli di vari Autorii* (4 vols, Bologna: Corciolani, Colli, 1757–9), ii, 141–177.

Ferrand, 'Lettre [du 4.9.1760] à M. Lami... sur la Sensibilité des Parties du Corps Animal, en Réponse à une Lettre d'un Chirurgien de Turin, Insérée dans le Mercure de France, Août 1760', *Mercure de France*, Juillet 1761, 124–48.

Fontana, Felice, 'Dissertation Épistolaire de Mr. L'Abbé Felice Fontana... Adressée au R.P. Urbain Tosetti [23.5.1757]', in Albrecht von Haller (ed.), *Mémoires sur la Nature Sensible et Irritable des Parties du Corps Animal* (4 vols, Lausanne: Bousquet, Arnet, 1756–60), iii, 157–243.

——*De Irritabilitatis Legibus, nunc Primum Sancitis, et de Spirituum Animalium in Movendis Musculis Inefficacia* (Lucca: Riccomini, 1767; Engl. transl. by Felix Marchand and Hebbel Edward Hoff published in the *Journal of the History of Medicine and Allied Sciences*, 10 (1955), 197–206, 302–26, 399–420).

——*Ricerche Filosofiche sopra la Fisica Animale* [Florence, 1775]. Introduzione e Cura di Giulio Barsanti (Florence: Giunti, 1996; Biblioteca della scienza italiana, 13).

——*Traité sur le Venin de la Vipère, sur les Poisons Américains, sur le Laurier-cerise et sur quelques autres Poisons Végétaux* (2 vols, Florence: n.p., 1781).

——*Epistolario di Felice Fontana, I. Carteggio con Leopoldo Marc'Antonio Caldani 1758–1794* , Renato G. Mazzolini and Giuseppe Ongaro (eds) (Trent: Società di Studi Trentini di Scienze Storiche, 1980).

Fouquet, Henri, 'Sensibilité, Sentiment (Méd.)', *Encyclopédie* (q.v.), xv (1765), 38–52.

Galen on Anatomical Procedures: De Anatomicis Administrationibus. Translation of the Surviving Books with Introduction and Notes by Charles Singer (Oxford: Oxford University Press, 1956).

Gaub, Hieronymus David, *Sermo Academicus de Regimine Mentis quod Medicorum Est* (Leiden: Van der Aa, 1747).

——*Institutiones Pathologiae Medicinalis* (Leiden: Luchtmans, 1758).

——*Sermo Academicus alter de Regimine Mentis quod Medicorum Est* (Leiden: Van der Aa, 1763).

Gerhard, Carl Abraham, *Triga Dissertationum Physico-Medicarum* (Berlin: Rudiger, 1763; includes a *Specimen Irritabilitatis ad Pathologiam et Therapiam Applicatae*, 7–64).

Geuns, Matthias van, *De eo quod Vitam Constituit in Corpore Animali Disquisitio Physiologica* (Amsterdam: Meijer, 1758).

Girard de Villars, Louis-Marie, 'Quaestio Medica... Carolo Geille De Saint Leger... Praeside ... An ut Sensibilitas, sic Irritabilitas a Nervis?' [M.D. thesis, Paris, 1757], in Giacinto Bartolomeo Fabri (ed.), *Sulla Insensitività ed Irritabilità Halleriana: Opuscoli di vari Autori* (4 vols, Bologna: Corciolani, Colli, 1757–9), ii, 202–11.

——'Si l'Irritabilité Dépend des Nerfs comme la Sensibilité? Extrait d'une Thése Soutenue aux Ecoles de Médecine de Paris... par M. Girard de Villars', *Journal de Médecine*, 6 (1757), 310–15.

——'Mémoire sur la Sensibilité des Parties des Animaux [lue le 20 janvier 1759]', *Mémoires de Mathématique et de Physique, Présentés à l'Académie Royale des Sciences, par divers Savans, & Lus dans les Assemblés*, 4 (1763), 580–96.

Glisson, Francis, *Tractatus de Natura Substantiae Energetica seu de Vitae Naturae eiusque Primis Facultatibus* (London: Flesher, 1672).

Gnauck, Traugott Wilhelm, *De Irritabilitate Partium Corporis Humani Dissertatio... Praeside Carolo Christiano Krause* (M.D. thesis, Leipzig: Langenheim, 1772).

Goldhagen, Johann Friedrich Gottlieb, *Dubitationes de quadam Caussae Motus Muscularis Explicatione* (M.D. thesis, Halle: Beyer, 1765).

Gooch, Benjamin, *A Practical Treatise on Wounds and other Chirurgical Subjects; to which is Prefixed a Short Historical Account of the Rise and Progress of Surgery and Anatomy, Addressed to Young Surgeons* (2 vols, Norwich: Chase 1767).

Gorter, Jan de, *Medicinae Compendium, in Usum Exercitationis Domesticae* (2 vols, Leiden: Van der Aaa, 1737).

——*Exercitationes Medicae Quatuor. I. De Motu Vitali. II. De Somno et Vigilia. III. De Fame. IV. De Siti* (Amsterdam: Ratelband, 1737).

——*Exercitatio Medica Quinta de Actione Viventium Particulari* (Amsterdam: Gerrevink, 1748).

Grima, Michel Angiolo, *Mémoire sur la Sensibilité des Tendons, Prononcé en Italien à l'Académie des Apathistes* (Paris: n.p., 1760; this essay was also printed in the *Journal Etranger*, Août 1760, 169–97).

Grimaud, Jean-Charles-Marguerite-Guillaume de, *Tentamen Physico-Medicum de Irritabilitate... pro Baccalaureatus Gradu Consequendo* (Montpellier. Picot, 1776).

Gruber, Petrus Josephus, *De Excessu Vis Vitalis Vasorum, variisque inde in Machina Animali Pendentibus Phoenomenis* (M.D. thesis, Prague and Dresden: Walther, 1775).

Güldenstedt, Anton Johann, *Theoria Virium Corporis Humani Primitivarum* (M.D. thesis, Frankfurt am Main: Winter, 1767).

Haen, Anton de, *Ratio Medendi in Nosocomio Practico* (15 vols, Vienna: Trattner, 1757–74) [see esp. Vol. 9, 1764, 103–203 ('De Systemate Halleriano'). Vol. 12, 1768, 263–74 ('Finis Quaestionis de Sensibilitate et Irritabilitate') and vol. 14, 1770, 83–123 ('De Inflammatione ac Dolore Membranarum')].

——*Difficultates circa Modernorum Systema de Sensibilitate et Irritabilitate Humani Corporis* (Vienna: Krüchten, 1761).

——*Vindiciae Difficultatum circa Modernorum Systema de Sensibilitate et Irritabilitate Humani Corporis, contra... Alberti v. Haller ad easdem Difficultates Apologiam* (Vienna: Krüchten, 1762).

——*Lettre… à un de ses Amis, au Sujet de la Lettre de Mr. Tyssot à Mr. Hirzel* (Vienna: Krüchten, 1763).

——*Dichiarazione dell'Illustre Signor Alberto Hallero intorno alla Questione della Sensibilità e della Irritabilità, Publicata in Vienna l'Anno 1768: Dal Celebre... Antonio de Haen Tradotta in Lingua Volgare da Giacomo Scovolo* (Vicenza: Modena, 1770).

Haller, Albrecht von, *De Methodico Studio Botanices absque Praeceptore* (Göttingen: Vandenhoeck, 1736).

——(ed.), *Hermanni Boerhaave... Praelectiones Academicae in Proprias Institutiones Rei Medicae* (7 vols, Göttingen: Vandenhoeck, 1739–44).

——*Icones Anatomicae quibus praecipuae aliquae Partes Corporis Humani Delineatae Proponuntur et Arteriarum potissimum Historia Continetur* (7 fascicles, Göttingen: Vandenhoeck, 1743–56).

——[review of the *Praelectiones*], *Bibliothèque Raisonnée*, 33 (1744), 33-50.

——[review of Bonnet's *Traité d'insectologie*], *Bibliothèque Raisonnée*, 36 (1746), 179-92.

Haller, Albrecht von, [review of the correspondence between Leibniz and Bernoulli], *Bibliothèque Raisonnée*, 37 (1746), 178-93.

——*Primae Lineae Physiologiae in Usum Praelectionum Academicarum* (Göttingen: Vandenhoeck, 1747).

——'Mémoire sur une Controverse au Sujet de la Respiration', *Nouvelle Bibliothèque Germanique*, 1748, 412–28.

——[review of Richardson's *Clarissa*], *Bibliothèque Raisonnée*, 42 (1749), 324–36.

——[review of Senac's *Traité de la Structure du Coeur*], *Bibliothèque Raisonnée*, 43 (1749), 323-43.

——'Vorrede [on the use of hypotheses]', in *Allgemeine Historie der Natur nach allen ihren Besondern Theilen Abgehandelt, nebst einer Beschreibung der Naturalienkammer.... Erster Theil* (Hamburg and Leipzig: Grund and Holle, 1750), ix–xxii.

——'The Haller Preface to the German Translation of the Histoire Naturelle (1750)', in John Lyon and Philip R. Sloan (eds), *From Natural History to the History of Nature: Readings from Buffon and his Critics* (Notre Dame, Ind.: University of Notre Dame Press, 1981), 295–310.

——*Opuscula sua Anatomica: De Respiratione, de Monstris aliaque Minora Recensuit, Emendavit, Auxit, aliqua Inedita Novasque Icones Addidit* (Göttingen: Schmidt, 1751).

——*Primae Lineae Physiologiae in Usum Praelectionum Academicarum: Auctae et Emendatae* (Göttingen: Vandenhoeck, 1751).

——[review of Vol. 2 of Buffon's *Histoire Naturelle*], *Bibliothèque Raisonnée*, 46 (1751), 66–88.

——[review of Whytt's *Essay*], *Relationes de Libris Novis*, 3 (1752), 156-90.

——'Oratio Dicta [de Utilitate Societatum Literariarum] cum Die Natali Georgii II. Societas Regia Scientiarum Primum Publice Conveniret', *Commentarii Societatis Regiae Scientiarum Gottingensis*, 1 (Göttingen: Vandenhoeck, 1752), XXXVII–LVI.

——'Experimenta de Cordis Motu a Stimulo Nato: Lecta d. 10. Novembr.', *Commentarii Societatis Regiae Scientiarum Gottingensis*, vol. 1 (Göttingen: Vandenhoeck, 1752), 263–6.

——'De Partibus Corporis Humani Sensilibus et Irritabilibus. Die 22. April. & die 6. Maii 1752', *Commentarii Societatis Regiae Scientiarum Gottingensis*, Vol. 2 (Göttingen: Vandenhoeck, 1753), 114–58.

——*Enumeratio Plantarum Horti Regii et Agri Gottingensis* (Göttingen: Vandenhoeck, 1753)

——*A Dissertation on the Sensible and Irritable Parts of Animals.* Transl. from the Latin [by Donald Monro]; with a Preface by M. Tissot (London: Nourse, 1755).

——*Sull'Insensibilità e Irritabilità di alcune Parti degli Animali: Dissertazioni de' Signori Haller, Zimmerman e Castell.* Trasportate in Lingua Ital. dal P. Gian Vincenzo Petrini... ; Colle Lettere del P. Urbano Tosetti, Sullo Stesso Argomento (Rome: Zempel, 1755).

——'De Respiratione Experimenta Anatomica quibus Aeris inter Pulmonem et Pleuram Absentia Demonstratur, et Musculorum Intercostalium Internorum Officium Adseritur', in *idem, Opuscula Pathologica: Partim Recusa, partim Inedita: Quibus Sectiones Cadaverum Morbosorum Potissimum Continentur: Accedunt Experimenta de Respiratione, Quarta Parte Aucta* (Lausanne: Bousquet, 1755), 163-292.

——(ed.), *Mémoires sur la Nature Sensible et Irritable des Parties du Corps Animal* (4 vols, Lausanne: Bousquet, Arnet, 1756–60).

——*Deux Mémoires sur le Mouvement du Sang et sur les Effets de la Saignée, Fondés sur des Experiences Faites sur des Animaux* (Lausanne: Bousquet, 1756).

——*Elementa Physiologiae Corporis Humani* (8 vols, Lausanne and Bern: Bousquet, Arnet, Grasset, Societas typographica, 1757–66).

——'Mémoire sur Plusieurs Phénomènes Importants de la Respiration: Fondé sur les Expériences', in *idem, Sur la Formation du Coeur dans le Poulet, sur l'Oeil, sur la Structure du Jaune &c.* (2 vols, Lausanne: Bousquet & Comp., 1758), ii, 197–366.

——*Sur la Formation du Coeur dans le Poulet, sur l'Oeil, sur la Structure du Jaune &c.* (2 vols, Lausanne: Bousquet & Comp., 1758).

——*Ad Viri Illustris Antonii de Haen Difficultates Apologia* (n.p., 1761].

——*Vertheidigung gegen die Einwürfe welche Herr Anton von Haen wider die Lehre von der Reitzbarkeit und Empfindlichkeit der Theile des Menschlichen Leibes, Vorgetragen: Aus dem Lateinischen Übersetzt* [with a preface of Johann Kaspar Hirzel] (Zurich: Heidegger, 1761).

Haller, Albrecht von, 'De Partibus Corporis Humani Sentientibus et Irritabilibus: Sermo alter, quo Experimentorum Corollaria Continentur', in *idem, Opera Minora Emendata, Aucta et Renovata* (3 vols, Lausanne: Grasset, 1763–68), i, 405–40.

——'Ad Objectiones contra Experimenta sua Propositas Responsio. Gallice Edita in T. IV. Memoires &c. Lausannae a. 1760 Excusa. Latine Reddita & Aucta', in *idem, Opera Minora Emendata, Aucta et Renovata* (3 vols, Lausanne: Grasset, 1763–68), i, 441–502.

——'Sermonis de Partibus Corporis Humani Sentientibus et Irritabilibus. Pars altera: Missa ad Societ. Reg. Gotting. d. 24 Nov. 1755, qua Experimenta Continentur', in *idem, Opera Minora Emendata, Aucta et Renovata* (3 vols, Lausanne: Grasset, 1763–68), i, 329–404.

——*Opera Minora Emendata, Aucta et Renovata* (3 vols., Lausanne: Grasset, 1763–68).

——*Ad Nuperum Scriptum Roberti Whyttij Apologia; Leopoldi M. Antonij Caldani ad Albertum Allerum [sic] Epistola* (Yverdon: n.p., 1764]).

——*Primae Lineae Physiologiae in Usum Praelectionum Academicarum: Tertio Auctae et Emendatae* (Göttingen: Vandenhoeck, 1765).

——*Commentarius de Formatione Cordis in Pullo [1767].* Edizione Critica a Cura di Maria Teresa Monti (Basel, 2000; Studia Halleriana, VI).

——*Elémens de Physiologie.* Traduction Nouvelle du Latin en François par M. Bordenave (Paris: Guillyn, 1769).

——*Sammlung Kleiner Hallerischer Schriften.* 2., Verbesserte und Vermehrte Aufl. (3 vols, Bern: Emanuel Haller, 1772).

——'Von den Empfindlichen und Reizbaren Theilen des Menschlichen Körpers', in *Sammlung Kleiner Hallerischer Schriften*, 2., Verbesserte und Vermehrte Aufl. (3 vols, Bern: Haller, 1772), ii, 1–103.

——'Faculté Vitale', in *Encyclopédie*, Yverdon edn (q.v.), xviii (1772), 244–50.

——(ed.), *Epistolarum ab Eruditis Viris ad Alb. Hallerum Scriptarum: Pars 1, Latinae* (6 vols, Bern: Societas typographica, 1773–5).

——'De Partibus Corporis Humani Sentientibus et Irritabilibus: Sermo Tertius, Praelectus d. XII. Dec. MDCCLXXII', *Novi Commentarii Societatis Regiae Scientiarum Gottingensis*, t. 3 ad a. 1772, (Göttingen and Gotha: Dietrich, 1773), 1–36.

——'De Partibus Corporis Humani Irritabilibus: Sermo IV, Praelectus d. X. Iul. MDCCLXXIII', *Novi Commentarii Societatis Regiae Scientiarum Gottingensis,* t. 4 ad a. 1773 (Göttingen and Gotha: Dieterich, 1774), 1–24.

——*Bibliotheca Anatomica qua Scripta ad Anatomen et Physiologiam Facientia a Rerum Initiis Recensentur* (2 vols, Zurich: Orell, Gessner, Fuessli et Socc., 1774–77; reprint Hildesheim: Olms, 1969).

——*Briefe über einige Einwürfe Nochlebender Freygeister wider die Offenbarung* (3 vols, Bern: Typographische Gesellschaft, 1775–77).

——'A.v. Hallers "Prospectus d'un Dictionnaire Universel de Médecine" [1776]', ed. Erich Hintzsche, *Gesnerus,* 23 (1966), 48–54.

——'Animal', in *Encyclopédie, Supplément* (q.v.), i (1776), 435–6.

——'Coeur', in *Encyclopédie, Supplément* (q.v.), ii (1776), 490–6.

——'Fibre', in *Encyclopédie, Supplément* (q.v.), iii (1777), 34–6.

—— Foetus', in *Encyclopédie, Supplément* (q.v.), iii (1777), 64–73.

——'Inflammation', in *Encyclopédie, Supplément* (q.v.), iii (1777), 599–601.

——'Irritabilité', in *Encyclopédie, Supplément* (q.v.), iii (1777), 663–5.

——'Ligamens', in *Encyclopédie, Supplément* (q.v.), iii (1777), 744.

——'Méchanisme', in *Encyclopédie, Supplément* (q.v.), iii (1777), 876.

——'Muscle', in *Encyclopédie, Supplément* (q.v.), iii (1777), 977–8.

——'Nerf', in *Encyclopédie, Supplément* (q.v.), iv (1777), 26–31.

——'Oeconomie Animale', in *Encyclopédie, Supplément* (q.v.), iv (1777), 104–5.

——'Physiologie', in *Encyclopédie, Supplément* (q.v.), iv (1777), 344–65.

——'Sensibilité', in *Encyclopédie, Supplément* (q.v.), iv (1777), 776–9.

——'Sommeil', in *Encyclopédie, Supplément* (q.v.), iv (1777), 806–10.

——*Auctarium ad Alberti Halleri Elementa Physiologiae Corporis Humani, Excerptum ex Nova Editione et Adaptatum Veteri* (Leipzig and Francfort: n.p., 1780).

——*First Lines of Physiology.* Transl. from the Correct Latin ed.; Printed under the Inspection of William Cullen (2 vols, Edinburgh and London: Elliot, Robinson, 1786).

——*Grundriß der Physiologie für Vorlesungen.* Nach der 4., Latein. mit den Verbesserungen und Zusätzen des Herrn Hofrath Wrisberg in Göttingen Vemehrten Ausg., von neuem Übers. und mit Anmerkungen Versehen durch Herrn Hofrath Sömmerring in Mainz, mit einigen Anmerkungen Begleitet und Besorgt von P.F. Meckel... (Berlin: Haude and Spener, 1788).

——*Albrecht von Hallers Gedichte*, Ludwig Hirzel (ed.), (Frauenfeld: Huber, 1882).

——*Albrecht von Hallers Briefe an Johannes Gesner (1728–1777)*, Henry E. Sigerist (ed.), (Berlin: Weidmann, 1923; Abhandlungen der Königlichen Gesellschaft der Wissenschaften zu Göttingen. Mathematisch-Physikalische Klasse, N.F., 11/2).

——*Albrecht von Haller – Giambattista Morgagni: Briefwechsel 1745–1768*, Erich Hintzsche (ed.), (Bern and Stuttgart: Huber, 1964).

——*Albrecht von Haller – Ignazio Somis: Briefwechsel 1754–1777*, Erich Hintzsche (ed.), (Bern and Stuttgart: Huber, 1965).

——*Albrecht von Haller – Marc Antonio Caldani: Briefwechsel 1756–1776*, Erich Hintzsche (ed.), (Bern and Stuttgart: Huber, 1966).

——'A. v. Hallers Korrespondenz mit Johann Stephan Bernard', Erich Hintzsche (ed.), *Clio Medica*, 1 (1966), 324–40.

——*Albrecht von Hallers Briefe an Auguste Tissot, 1754–1777*, Erich Hintzsche (ed.), (Bern, Stuttgart, Wien: Huber, 1977).

——*The Correspondence Between Albrecht von Haller and Charles Bonnet*, Otto Sonntag (ed.), (Bern, Stuttgart, Vienna: Huber, 1983; Studia Halleriana, 1).

——*The Correspondence Between Albrecht von Haller and Horace-Bénédict de Saussure*, Otto Sonntag (ed.), (Bern, Stuttgart, Toronto, 1990; Studia Halleriana, 3).

——*John Pringle's Correspondence with Albrecht von Haller*, Otto Sonntag (ed.), (Basel: Schwabe, 1999; Studia Halleriana, 4).

——*Der Nützliche Brief: Die Korrespondenz Zwischen Albrecht von Haller und Christoph Jakob Trew, 1733–1763,* Hubert Steinke (ed.), (Basel: Schwabe, 1999; Studia Halleriana, 5).

Harvey, William, *Exercitationes de Generatione Animalium: quibus Accedunt quaedam De partu: de Membranis atque Humoribus Uteri & de Conceptione* (London: Du-Gardian, 1651).

——*Disputations Touching the Generation of Animals,* transl. with Introduction and Notes by Gweneth Whitteridge (Oxford: Blackwell, 1981).

Heister, Laurenz, *Institutiones Chirurgiae* (2 vols, Amsterdam: Apud Janssonio-Waesbergios, 1750).

——*Chirugie.* Neue viel Vermehrte und Verbesserte Auflage (Nuremberg : Raspe, 1770).

Helwig, Heinrich Christian, *Dissertatio Inauguralis Medica de Irritatione... Praeside Ernesto Antonio Nicolai* (M.D. thesis, Jena: Marggraf, 1760).

Henckel, Joachim Friedrich, *Abhandlung der Chirurgischen Operationen* (8 vols, Berlin Decker and Winter, 1770–6).

Herder, Johann Gottfried, *Ideen zur Philosophie der Geschichte der Menschheit,* Wolfgang Pross (ed.), (2 vols, München: Hanser, 2002).

Heuermann, Georg, *Abhandlungen der Vornehmsten Chirurgischen Operationen* (3 vols, Copenhagen and Leipzig: Pelt, 1754–57).

Hévin, Prudent, *Cours de Pathologie et de Thérapeutique Chirurgicales* (2 vols, Paris: Méquignon, 1793).

Hoin, Jean Jacques Louis, 'Observations et Expériences sur les Plaies du Tendon d'Achille', *Journal de Médecine,* 30 (1769), 56–78.

Housset, Etienne-Jean-Pierre, *Dissertation sur les Parties Sensibles du Corps Animal, Suivie d'un Mémoire sur les Avantages que Procurent les Frictions Mercurielles dans le Traitement de quelques Épilepsies Idiopatiques... Terminé par Deux Observations sur l'Usage du Mercure pour la Guérison du Scorbut & des Dartres* (Lausanne: Grasset, 1770).

——*Mémoires Physiologiques et d'Histoire Naturelle* (2 vols, Auxerre: Fournier, 1787).

Hunter, John, *A Treatise on Blood, Inflammation, and Gun-Shot Wounds* (London: Richardson, 1794).

Hunter, John, *The Works of John Hunter*, J.F. Palmer (ed.), (4 vols, London: Longman *et al.*, 1835).

Hunter, William, 'Dr. Hunter's Thoughts on the Sensibility of the Tendons, Periosteum, Ligaments, Dura and Pia Mater, &c. From his Medical Commentaries, &c.', *The Medical Museum: Or, a Repository of Cases, Experiments, Researches, and Discoveries, Collected at Home and Abroad. Whether in Anatomy, Botany, Medicine, Chemistry, Pharmacy, Surgery, Physiology, &c. By Gentlemen of the Faculty*, 1 (London: Richardson and Clark, 1763), 214–21.

Isenflamm, Jakob Friedrich, *Versuch einiger Praktischen Anmerkungen über die Nerven zur Erläuterung Verschiedener Krankheiten derselben* (Erlangen: Walther, 1774).

Jaucourt, Louis de, 'Maladie', *Encyclopédie* (q.v.), ix (1765), 929–38.

Jausserand, Jean Pierre de, *De Sensibilitate et Irritabilitate Partium Corporis Humani Experimenta* (M.D. thesis, Montpellier: Rochard, 1758).

Kaau-Boerhaave, Abraham, *Impetum Faciens Dictum Hippocrati per Corpus Consentiens Philologice et Physiologice Illustratum Observationibus et Experimentis Passim Firmatum* (Leiden: Luchtmans, 1745).

Kauffmann, Georg Ernst, *De Partium Corporis Humani Irritabilitate et Sensibilitate Commentatio sub Praesidio... Christiani Dogof. Gruneri* (M.D. thesis, Jena: Fickelscher, 1777).

Kergel, Johannes Levi Benjamin, *De Partibus Irritabilibus Humani Corporis Dissertatio altera* (M.D. thesis, Leipzig: Langenheim, 1777).

Klinkosch, Joseph Thaddaeus, 'Observatio de Sensibilitate Tendinis', in *idem* (ed.), *Dissertationes Medicae Selectiores Pragenses*, Vol. 1 (Prague and Dresden: Walther, 1775), 335–38.

Köhler, Gabriel Valentin, *De Destructa Partium Irritabilitate per Incongruum Adstringentium Usum... Praeside... Andrea Elia Büchnero...* (M.D. thesis, Halle: Hendel, 1763).

Kölliker, Albert, 'Physiologische Untersuchungen über die Wirkung einiger Gifte', *Archiv für Pathologische Anatomie und Physiologie und für Klinische Medicin*, 10 (1856), 3–77, 235–96.

——'Note sur l'Action du Curare sur le Système Nerveux', *Compte Rendu des Séances de l'Académie des Sciences*, 43 (1856), 791–2.

Krause, Karl Christian, *Prüfung der Preisschrift des Herrn Le Cat von der Muskelbewegung* (Leipzig: Langenheim, 1755).

——*Abhandlung des Herrn von Haller von den Empfindlichen und Reizbaren Theilen des Menschlichen Leibes.* Verdeutscht und Geprüft von Carl Christian Krausen (Leipzig: Jacobi, 1756).

Labroquere, François, 'Dissertatio Medica de Motum Vitalium Causa... Praeside D.D. Francisco De Sauvages' [Montpellier, 1740], in *Disputationum Anatomicarum Selectarum*, Collegit, Edidit, Praefatus est Albertus Haller, Vol. 4 (Göttingen: Vandenhoeck, 1749), 481–515.

La Faye, George de, *Principes de Chirurgie* (5th edn, Paris: Cavelier, 1761).

Laghi, Tommaso, 'De Insensibilitate, atque Irritabilitate Halleriana', *De Bononiensi Scientiarum et Artium Instituto atque Academia Commentarii*, vol. 4 (Bologna: Vulpe, 1757), 208–17.

——*Jacobo Bartholomaeo Beccario... De Sensivitate, atque Irritabilitate Halleriana. Sermo alter. Accedit Epitome Experimentorum quibus Praecedentis Sermonis Propositiones Nituntur* (Bologna: Vulpe, 1757).

Lamberti, Giovanni Michele, *Lettera... al Illustrissimo Signor D. Giambatista Bianchi... Unite a quella diverse Osservazioni sopra la Sensibilità del Pericranio, e Tendini Nelli Uomini* (Alessandria: Vimercati, 1756).

La Mettrie, Julien Offray de, *Oeuvres Philosophiques* (London: Nourse, 1751).

——*La Mettrie's L'Homme Machine: A Study in the Origins of an Idea.* Critical Edition with an Introductory Monograph and Notes by Aram Vartanian (Princeton: Princeton University Press, 1960).

——*Le Traité de l'Ame de la Mettrie. Edition Critique du Texte avec une Introduction et un Commentaire Historiques*, Th. Verbeek (ed.), (2 vols, Utrecht: OMI-Grafisch Bedrijf, 1988).

La Motte, Jacobus Gourlez de, *Quaestio Medica... Discutienda in Scholis Medicorum Die Decimo-Tertio Januarii MDCCLII. Guillelmo De Magny Praeside. An a Vasorum Aucta aut Imminuta Irritabilitate omnis Morbus?* (M.D. thesis, Paris: Quillau, 1752).

La Roche, François de, *Encyclopédie Méthodique: Chirurgie* (3 vols, Paris. Panckoucke, 1790–92).

[Le Cat, Claude-Nicolas] *Dissertation qui a Remporté le Prix Proposé par l'Académie Royale des Sciences et Belles-Lettres de Prusse, sur le Principe de l'Action des Muscles avec les Pieces qui ont Concouru* (Berlin: Haude und Spener, 1753).

——'Dissertation sur la Sensibilité de la Dure-mere, de la Pie-mere, des Membranes, des Ligamens, des Tendons, &c. sur l'Insensibilité du Cerveau & sur l'Irritabilité Hallerienne: Extrait du Plan d'une Nouvelle Édition du Mémoir qui fait la Suite de Celui sur le Mouvement Musculaire, Couronné par l'Académie de Berlin en 1753', *Journal Encyclopédique*, 1757, IV/iii, 129–41; V/i, 128–40; V/ii, 131–43; V/iii, 129–41; VI/i, 132–41.

——*Traité de l'Existance, de la Nature et des Propriétés du Fluide des Nerfs, et Principalement de son Action dans le Mouvement Musculaire, Ouvrage Couronné en 1753 par l'Académie de Berlin; Suivie des Dissertations sur la Sensibilité des Meninges, des Tendons &c., l'Insensibilité du Cerveau, la Structure des Nerfs, l'Irritabilité Hallérienne, &c.* (Berlin n.p., 1765).

Leppentin, Christoph Nicolaus, *De Irritabilitate Ultimo Termino Cognitionis Motus Animalis* (M.D. thesis, Halle: Hendel, 1771).

Lieutaud, Joseph, *Précis de la Médecine Pratique.* Nouvelle Édition (2 vols, Paris: Barrois, 1781).

Lorry, Anne Charles, 'Nouvelles Expériences sur l'Irritabilité', *Recueil Périodique d'Observations de Médecine, Chirurgie, Pharmacie &c.*, Vol. 5 (Paris: Vincent, 1756), 322–40, 403–421 and Vol. 6 (1757), 7–19.

Lotteri, Carlo Michele, 'Dissertazione... Recitata nel Teatro Anatomico della Detta Regia Università [di Torino] per Principio del Pubblico Corso Cerusico li 25. Febbrajo 1757: Sulla Sensibilità, ed Irritabilità delle Parti de' Bruti, e degli Uomini', in Giacinto Bartolomeo Fabri (ed.), *Sulla Insensitività ed Irritabilità Halleriana: Opuscoli di Vari Autorii* (4 vols, Bologna: Corciolani, Colli, 1757–59), 278–306.

Louis, Antoine, *Dictionnaire de Chirurgie, Communiqué à l'Encyclopédie,* Extrait et Rédigé par M.P.F. [i.e. François Paul] (2 vols, Paris: Saillant, 1772).

——'Éloge de Haller, Lu dans la Séance Publique du 30 Avril 1778', *Éloges Lus dans les Séances Publiques de L'Académie Royale de Chirurgie de 1750 a 1792 par A. Louis,* Recueillis et Publiés pour la Première Fois par E.-Fréd. Dubois (Paris: Barillière, 1859), 265–81.

Ludwig, Christian Gottlieb, *Institutiones Pathologiae Praelectionibus Academicis Accomodatae* (Leipzig: Gleditsch, 1754).

Lups, Johannes, *Dissertatio Physiologico-Medica Inauguralis de Irritabilitate* (M.D. thesis, Leiden: Luzac, 1748).

Magendie François, *Elementary Compendium of Physiology; for the Use of Students.* Transl. from the French by E. Milligan (4th edn, Edinburgh: Carfrae, 1831).

Magenise, Daniel, *The Doctrine of Inflammations Founded upon Rreason and Experience; and Intirely Cleared from the Contradictory Systems of Boerhaave, Van Swieten, and Others* (London: Owen, 1768).

Man, Maximilian Jacob de, *Specimen Medicum Inaugurale de Natura Hominis quatenus Sanitatis Praesentis est Custos, Absentis Vindex* (M.D. thesis, Leiden: Corts 1754).

Manitius, Wolfgang, *Dissertatio Medica Inauguralis, de Idiosincrasia, ex Diversa Solidorum c. h. Irritabilitate Optime Dijudicanda* (M.D. thesis, Leiden: Wishoff, 1749).

Martin, Claudius, *Dissertatio Medica circa Sensibilitatem et Irritabilitatem Partium Solidarum Speciatim Textus Cellulosi* (M.D. thesis, Regensburg, Zunkel, 1769).

Martini, Ferdinand, *Versuche und Erfahrungen über die Empfindlichkeit der Sehnen* (Copenhagen: Rothen and Profft, 1770).

Medicus, Friedrich Casimir, *Von der Lebenskraft* (Mannheim: Hof- und Akademische Buchdruckerei, 1774).

Ménuret de Chambaud, Jean-Jacques, 'Inflammation, Inflammatoires Maladies', *Encyclopédie* (q.v.), viii (1765), 708–27.

——'Observateur', in *Encyclopédie* (q.v.), xi (1765), 310–13.

——'Observation', in *Encyclopédie* (q.v.), xi (1765), 313–23.

——'Paraphrénésie', in *Encyclopédie* (q.v.), xi (1765), 921–2.

——'Pouls', in *Encyclopédie* (q.v.), xiii (1765), 205–40.

——'Sciatique', in *Encyclopédie* (q.v.), xiv (1765), 781–4.

Metzger, Johann Daniel, *Ueber Irritabilität und Sensibilität als Lebensprincipien in der Organisierten Natur* (Königsberg: Hartung, 1794).

Monro, Alexander, 'An Attempt to Determine by Experiments, How Far Some of the Most Powerful Medicines, viz. Opium, Ardent Spirits, and Essential Oils, Affect Animals by Acting on those Nerves to which they are Primarily Applied, and thereby Bringing the Rest of the Nervous System into Sufferance, by what Is Called Sympathy of Nerves', *Essays and Observations Physical and Literary: Read Before a Society in Edinburgh, and Published by them*, Vol. 3 (Edinburgh: Hamilton and Balfour, 1771), 282–365.

——*Observations on the Structure and Functions of the Nervous System* (Edinburgh: Creech, 1783).

Morin, François V., *An a Nervis Irritabilitas Fibrae Muscularis ?... Praeside Henri-François-Anne Roussel* (M.D. thesis, Caen, 1778).

Mühlmann, Ferdinand Wilhelm, *Kurze Untersuchung, ob durch die Verletzung der Sehnen und des Beinhäutchens Schmerz und Brand Entstehen Könne* (Königsberg: n.p., 1754).

Müller, Johann Rudolph, *De Irritabilitate Iridis hincque Pendente Motu Pupillae... Sub Praesidio Io. Rudolphi Zuingeri* (M.D. thesis, Basel: Decker, 1760).

Müller, Johannes, *Handbuch der Physiologie des Menschen für Vorlesungen* (3rd edn, 2 vols, Coblenz: Hölscher, 1838–40).

Musschenbroek, Pieter van, *Oratio de Methodo Instituendi Experimenta Physica. Habita Ultrajecti XXVII. Martii MDCCXXX* (Leiden: Verbeek, 1731).

Needham, John Tuberville, *Nouvelles Observations Microscopiques, avec des Découvertes Intéressantes sur la Composition et la Décomposition des Corps Organisés* (Paris: Ganeau, 1750).

Nicolai, Ernst Anton, *Programma de Sensatione ac Sensibilitate quo Praelectiones suas Futuro Semestri Habendas Indicat* (Jena: Fickelscher, 1758).

Niderhueber, Ignaz, *Irritabilitas Muscularis ad Simplices Naturae Leges Revocata* (M.D. thesis, Ingolstadt: Lutzenberger, 1777).

Noortwyck, Willem van, 'Defense de Mr. Boerhaave', *Bibliothèque Britannique*, 23 (1746), 336–67 and 24 (1746), 187–216.

Odendahl, Aegidius, *Inflammationis Ratio Dissertatione Inaugurali Praeside G. M. Gattenhof... Proposita* (M.D. thesis, Heidelberg: Haener, 1773).

Pagani, Orazio Maria and Camillo Bonioli, *Delle Parti Insensibili, ed Irritabili degli Animali Discorso Teorico-pratico* (Venice: Occhi, 1757).

Pakendorf, Johann August, *De Legibus Irritabilitatis Generalioribus... Praeside... Andrea Elia Büchnero...* (M.D. thesis, Halle: Hendel, 1765).

Pallas, Simon, *Anleitung zur Practischen Chirurgie, und der Lehre von den Knochenkrankheiten* (2nd edn, Berlin Lange, 1777).

Perenotti, P.A., 'Lettre, à l'Auteur du Mercure [sur l'Insensbilité des Tendons]', *Mercure de France*, Aoust 1760, 144–50.

[Petrioli, Gaetano], 'Riflessioni Anatomiche di un Dottor Fisico intorno alla Moderna Dissertazione del Signor Alberto Haller, dal quale con il Rapporto de' Quadrupedi Aperti Vivi, si Vogliono Sostenere per Insensibili Molte Parti del Corpo Umano, contro l'Esperienze Anatomiche, e contro le Autorità di Gravissimi Scrittori di Medicina, e Cirurgia', in Giacinto Bartolomeo Fabri (ed.), *Sulla Insensitività ed Irritabilità Halleriana: Opuscoli di vari Autorii* (4 vols, Bologna: Corciolani, Colli, 1757–9), ii, 249–77.

Pfaff, Christoph Henrich, *Über Thierische Elektricität und Reizbarkeit: Ein Beytrag zu den Neuesten Entdeckungen über diese Gegenstände* (Leipzig: Crusius, 1795).

Pitcairne, Archibald, *The Works... wherein are Discovered the True Foundation and Principles of the Art of Physic* (London: Curll *et al.*, 1715).

——*The Philosophical and Mathematical Elements of Physick: In Two Books* (London: Bell and Osborn, 1718).

Planque, François, *Chirurgie Complète, Suivant le Système des Modernes* (2 vols, 2nd edn, Paris: D'Houry 1757; 3rd edn, *ibid.* 1770).

Platner, Ernst, *Zusätze zu seines Vaters Einleitung in die Chirurgie* (Leipzig: Dyck, 1776).

——'Über einige Schwierigkeiten des Hallerischen Systems', in *Anton von Haens... Heilungsmethode in dem Kaiserlichen Krankenhause zu Wien.* Aus dem Lateinischen, III. Band. Mit einigen Aufsätzen Begleitet von Herrn D. Ernst Platner (Leipzig: Weygand, 1781), 373–87.

Platner, Johann Zacharias, *Gründliche Einleitung in die Chirurgie* (2 vols, Leipzig: Fritsch, 1757; 3rd edn, *ibid.*, 1770).

Portal, Antoine, *Précis de Chirurgie Pratique* (2 vols, Paris: Vincent, 1768).

——*Histoire de l'Anatomie et de la Chirurgie* (7 vols, Paris: Didot, 1770).

Porterfield, William, 'Essay Concerning the Motions of our Eyes', *Medical Essays and Observations*, 3 (Edinburgh: Hamilton and Balfour, 1735), 160–261, iv (1737), 124–294.

——*A Treatise on Eye, the Manner and Phaenomena of Vision* (2 vols, Edinburgh: Miller, 1759).

Pozzi, Cesare Giuseppe, *Epistola Cl. Viro Tomae Laghi* [Florence, 30 Septemer 1755], in Giacinto Bartolomeo Fabri (ed.), *Sulla Insensitività ed Irritabilità Halleriana: Opuscoli di Vari Autorii* (4 vols, Bologna: Corciolani, Colli, 1757–59), i, 230–6 (1st edn, Florence: Giovannelli, 1755).

Prochaska, Georg, *Adnotationum Aacademicarum Fasciculus Tertius: De Functionibus Systematis Nervosi Commentatio* (Prague: Gerle, 1784; engl. transl. in Karel B. Absolon (ed.), *From Georg Prochaska to Jan Evangelista Purkinje* [Rockville (Md.), 1990)]).

Pütter, Johann Stephan, *Versuch einer Academischen Gelehrten-Geschichte von der Georg-Augustus-Universität zu Göttingen* (Göttingen: Vandenhoeck, 1765).

Quesnay, François, 'Préface', *Mémoires de l'Académie Royale de Chirurgie*, 1 (Paris: Osmont, 1743), ix–xxxi.

——*Essai Physique sur l'Oeconomie Animale* (2nd edn, 3 vols, Paris: Cavelier, 1747).

Radniczky, Ignaz, *Specimen Inaugurale Medicum Sistens Experimenta quaedam quibus Constitit, eas Partes Esse Sensu Praeditas, quibus Hallerus cum aliis quibusdam omnem Sentiendi Facultatem cum Irritabilitate Denegat* (M.D. thesis, Prague: Pruscha, 1756).

Ravaton, Hugues, *Chirurgie d'Armée* (Paris: Barrois, 1768).

Rehfeld, Wilhelm Samuel, *Dissertatio Inauguralis Medica Quaestionem Perquirens: 'An Vis Irritabilis Fibrarum Muscularium Innata ipsis Inhaereat, an aliunde ad eas Accedat?'* (M.D. thesis, Greifswald: Röse, 1771).

Reil, Johann Christian Reil, 'Über die Lebenskraft', *Archiv für die Physiologie*, 1 (Halle: Curtsche Buchhandlung, 1796), 8–162 (reprint Leipzig: Barth, 1910)

Reimarus, Johann Albrecht Friedrich, *Lebensbeschreibung von ihm selbst Aufgesetzt: Nebst dem Entwurf einer Teleologie zu seinen Vorlesungen Bestimmt* (Hamburg: Campe, 1814).

Richter, August Gottlob, *Anfangsgründe der Wundarzneykunst* (2nd edn, 6 vols, Frankenthal: Gegel, 1788–1800).

Sabatier, Raphael Bienvenu, *De la Médecine Opératoire* (3 vols, Paris: Didot, 1796).

Sanseverini, Domenico, 'De Fibrarum Sensibilitate atque Irritabilitate', in Giacinto Bartolomeo Fabri (ed.), *Sulla Insensitività ed Irritabilità Halleriana: Opuscoli di vari Autorii* (4 vols, Bologna: Corciolani, Colli, 1757–59), ii, 67–78.

Sauvages, François Boissier de, *Pathologia Methodica, seu de Cognoscendis Morbis* (Amsterdam: De Tournes, 1752).

——'A Messieurs les Journalistes de la Bibliothèque Raisonnée', *Recueil Périodique d'Observations de Médecine, Chirurgie, Pharmacie &c.*, Vol. 3 (Paris: Vincent, 1755), 83–116.

——'Von der Reitzbarkeit der Theile des Menschlichen Körpers: Auszug eines Schreibens: Aus dem Lateinischen', *Fränkische Sammlungen von Anmerkungen aus der Naturlehre, Arzneygelahrtheit, Oekonomie und den damit Verwandten Wissenschaften,* 1 (Nuremberg: Monath, 1756), 146–50.

——'Sendschreiben... von der Herrschaft der Seele über den Körper, an Herrn HofRath Delius', *Fränkische Sammlungen von Anmerkungen aus der Naturlehre, Arzneygelahrtheit, Oekonomie und den damit Verwandten Wissenschaften,* 5 (1760), 474–81.

——*Nosologia Methodica Sistens Morborum Classes iuxta Sydenhami Mentem & Botanicorum Ordinem... Editio Ultima* (2 vols, Amsterdam: de Tournes, 1768)

——*Nosologie Méthodique, ou Distribution des Maladies en Classes, en Genres et en Especes, Traduite sur la Derniere Édition Latine, par M. Gouvion* (10 vols, Lyons: Bruyset, 1772).

Scheid, Frederick Albert, *Specimen Inaugurale de Irritabilitate a Materialismo Vindicata* (M.D. thesis, Leiden: Hoogeveen and Heyligert, 1772).

Schmucker, Johann Leberecht, *Chirurgische Wahrnehmungen* (2 vols, Berlin and Stettin Nicolai, 1774).

Schneider, Lebrecht Ehregott, *Chirurgische Geschichte mit Theoretischen und Practischen Anmerkungen* (7 vols, Chemnitz: Stößel, 1762–75).

Schultz, Karl Wilhelm, *De Differentia Sensationis et Irritationis... Praeside ... Andrea Elia Büchnero...* (M.D. thesis, Halle: Hilliger, 1755).

Senebier, Jean, *L'Art d'Observer* (Geneva: Chirol, 1775).

Sèze, Paul Victor de, *Recherches Physiologiques et Philosophiques sur la Sensibilité ou la Vie Animale* (Paris: Prault, 1786).

Sichi, Lucca, *De Irritabilitate et Sensibilitate Partium Humani Corporis Propositiones Medico-chirurgicae* (Pisa: Carotti, 1764).

Simon, Jean-François, *Cours de Pathologie et de Thérapeutique Chirurgicales...* Revû, Mis en Ordre... et Augmenté par M. [Prudent] Hévin (Paris: Méquignon, 1780).

Smith, Thomas, *Tentamen Physiologicum Inaugurale de Actione Musculari: Quod ex Auctoritate Gulielmi Robertson pro Gradu Doctoris Eruditorum Examini Subjicit* (Edinburgh: Balfour, Auld and Smellie, 1767).

Spallanzani, Lazzaro, *Dissertazioni di Fisica Animale, e Vegetabile* (Modena: Soc. Tipografica, 1780).

——*I Giornali delle Sperienze e Osservazioni*, Carlo Castellani (ed.), (6 vols, Florence: Giunti, 1994).

——*Edizione Nazionale delle Opere di Lazzaro Spallanzani: Parte Prima. Carteggi*, Pericle di Pietro (ed.), (12 vols, Modena: Mucchi, 1984–90).

Soemmering, Samuel Thomas, *Rezensionen für die Göttingischen Gelehrten Anzeigen*, Ulrike Enke (ed.), (2 vols, Stuttgart: Fischer, 1995).

Sprengel, Kurt, *Versuch einer Pragmatischen Geschichte der Arzneikunde* (5 vols, Halle: Gebauer, 1792–1803).

Stoll, Maximilian, *Aphorismi de Cognoscendis et Curandis Febribus* (Vienna: Kurzbek, 1786).

[Störck, Anton] *Instituta Facultatis Medicae Vindobonensis* (Vienna: Trattner, 1775).

Struve, Johann Wilhelm, *De Sensilibus Partibus Humani Corporis Praeside Carolo Christiano Krause...* (M.D. thesis, Leipzig: Langenheim, 1765).

Sue, Jean-Joseph, *Elémens de Chirurgie* (Paris: d'Houry, 1755).

Sue, Pierre, *Elémens de Chirurgie, en Latin et en Français, avec des Notes* (Paris: Vincent, 1774).

——*Dictionnaire Portatif de Chirurgie* (Paris: Vincent, 1771; 2nd edn, *ibid.*: Barbou, 1777).

Swainston, Allen, *Dissertatio Inauguralis Medica de Purpura... Praeside... Alberto de Haller...* (M.D. thesis, Göttingen: Vandenhoeck, 1752).

Swammerdam, Jan, *Bybel der Natuure..., Hier by komt een Voorreeden, waar i het Leven van de Auteur Beschreben is door Herman Boerhaave... De Latynsche Overzetting heft Bezorgt Hieronimus David Gaubius* (2 vols, Leiden: Severinus, 1737–38).

Swieten, Gerard van, *Commentaria in Hermanni Boerhaave Aphorismos de Cognoscendis et Curandis Morbis* (5 vols; Leiden: Verbeek, 1742–72).

Taube, Wilhelm Ludwig, *Dissertationem Inauguralem de Vera Nervi Intercostalis ... Praeside D. Alberto Haller... Publice Defendet...* (M.D. thesis, Göttingen: Vandenhoeck, 1743).

Teckel, John, 'An Observation of the Insensibility of Tendons by Mr. John Teckel, Surgeon; with an Introduction by Dr. Hunter, and by him Communicated to the Society. Read October 1, 1770', *Medical Observations and Inquiries by a Society of Physicians in London*, Vol. 4 (London: Johnston, 1771), 343–6.

Thaer, Albrecht Daniel, *De Actione Systematis Nervosi in Febribus* (M.D. thesis, Göttingen: Dieterich, 1774).

Theden, Johann Christian Anton, *Unterrricht für die Unterwundärzte bey Armeen, besonders bey dem Königl. Preussischen Artilleriecorps* (2 vols., Berlin: Nicolai, 1774).

Tissot, Samuel-Auguste-André-David, 'Discours Préliminaire du Traducteur', in Albrecht von Haller, *Dissertation sur les Parties Irritables et Sensibles des Animaux...* trad. du Latin par M. Tissot (Lausanne: Bousquet, 1755), iii–xlix.

——*Lettre a Monsieur Hirzel... sur quelques Critiques de M. de Haen* (Lausanne: Grasset, 1762).

——*Traité des Nerfs et de Leurs Maladies* (4 vols, Paris and Lausanne: Didot *et al.*, 1778–80).

——*Vie de Zimmermann* (Lausanne: Fischer & Vincent, 1797).

Tosetti, Urbano, 'Sull'Insensibilità di alcune Parti degli Animali: Lettera Prima [-terza] al Signor Dottore Giuseppe Valdambrini', in *Sull' Insensibilità e Irritabilità di alcune Parti degli Animali: Dissertazioni de' Signori Haller, Zimmerman e Castell*; Trasportate in Lingua Ital. dal P. Gian Vincenzo Petrini... ; Colle Lettere del P. Urbano Tosetti, Sullo stesso Argomento (Rome: Zempel, 1755), 181–220.

——'Sull'Insensibilità di alcune Parti degli Animali, Lettera Quarta... al Signor Dottore Giuseppe Valdembrini', in Giacinto Bartolomeo Fabri (ed.), *Sulla Insensitività ed Irritabilità Halleriana: Opuscoli di vari Autorii* (4 vols, Bologna: Corciolani, Colli, 1757–9), i, 237–67.

Trembley, Abraham, *Mémoires pour Servir à l'Histoire d'un Genre de Polypes d'Eau Douce, à Bras en Forme de Cornes* (2 vols, Paris: Durand, 1744).

Trzebiczky, Franz Xaver Kaspar, *De Irritabilitate & Sensibilitate Partium Corporis Animalis* (M.D. thesis, Prague: Hochenberger, 1772).

Unzer, Johann August, '[On the Powers of the Soul]', *Der Arzt* (Hamburg: Grund, 1759), 273–88.

——*Erste Gründe einer Physiologie der Eigentlichen Thierischen Natur Thierischer Körper* (Leipzig: Weidmann, 1771).

——*Physiologische Untersuchungen: Auf Veranlassung der Göttingischen, Frankfurter, Leipziger und Hallischen Recensionen seiner Physiologie der Thierischen Natur* (Leipzig: Weidmann, 1773).

Vandelli, Domenico, *Epistola de Sensibilitate Pericranii, Periostii, Medullae, Durae Meningis, Corneae, et Sendinum* (Padua: Conzatti, 1756).

——*Epistola Secunda & Tertia de Sensitivitate Halleriana* (Padua: Conzatti, 1758).

——*Apologia contra... Hallerum* (Padua: Conzatti, 1760).

Venel, Gabriel-François, 'Irritabilité', in *Encyclopédie* (q.v.), viii (1765), 909.

Verschuir, Walther Forsten, *De Arteriarum et Venarum Vi Irritabili: Ejusque in Vasis Excessu: et inde Oriunda Sanguinis Directione Abnormi* (M.D. thesis, Groningen: Bolt, 1766).

——*Oratio Inauguralis de Recentiorum Medicorum, inprimis Belgarum, Meritis, in Phaenomenis et Effectibus Principii, quod Vitam Animalem Constituit, Indagandis, et ad Statum Corporis Humani Sanum et Morbosum Applicandis* (Groningen: Spandaw, 1781).

Vicq d'Azyr, Félix, *Œuvres de Vicq d'Azyr,* J.-L. Moreau de la Sarthe (ed.), (6 vols, Paris: Duprat-Duverger, 1805).

Vieillard, Pierre-Marie, *Quaestio Medica Quodlibetariis Disputationibus Mane Discutienda in Scholis Medicorum... Joanne-Antonio Elia De La Poterie Praeside. An Corpori Viventi Peculiare Sit Motus Principium, Innata Fibrae Irritabilitas* (Paris: Quillau, 1766).

Virchow, Rudolf, 'Reizung und Reizbarkeit', *Archiv für Pathologische Anatomie und Physiologie und für Klinische Medicin,* 14 (1858), 1–63.

Walstorff, Johann Dietrich, *Dissertatio Inauguralis Medica Sistens Experimenta circa Motum Cerebri, Cerebelli, Durae Matris et Venarum in Vivis Animalibus Instituta* (M.D. thesis, Göttingen: Schulz, 1753).

Weber, August Gottlob, *Commentatio de Initiis ac Progressibus Irritabilitatis cum Historia Sensibilitatis atque Irritabilitatis Partium Morbosae* (Halle: Litteris Orphanotrophei, 1783).

Weise, Johannes, *De Irritabilitate Morborum Genitrice... Praeside Ernesto Godofredo Baldinger* (M.D. thesis, Jena: Heller, 1772).

Whytt, Robert, *An Essay on the Vital and other Involuntary Motions of Animals* (Edinburgh: Hamilton, Balfour and Neill, 1751).

——*Physiological Essays, Containing, I. An Inquiry into the Causes which Promote the Circulation of the Fluids in the very Small Vessels of Animals. II. Observations on Sensibility and Irritability of the Parts of Men and other Animals; occasioned by M. de Haller's Late Treatise on these Subjects* (Edinburgh: Hamilton, Balfour and Neill, 1755).

——*Physiological Essays, Containing, I. An Inquiry into the Causes which Promote the Circulation of the Fluids in the very Small Vessels of Animals. II. Observations on Sensibility and Irritability of the Parts of Men and other Animals; occasioned by M. de Haller's Late Treatise on these Subjects. The Second Edition, Corrected and Enlarged. With an Appendix, Containing an Answer to M. de Haller's Remarks in the 4th Volume of the Memoires sur les Parties Sensibles et Irritables* (Edinburgh: Hamilton, Balfour, and Neill, 1761).

——*The Works of Robert Whytt* (Edinburgh: Balfour, 1768).

Winter Frederick, *Dissertatio Medica Inauguralis de Motu Musculorum* (M.D. thesis, Leiden: Luzac, 1736).

——*Oratio Inauguralis de Certitudine in Medicina Practica Dicta A.D. XXVI. Aprilis MDCCXLVI* (Franeker: Coulon, 1746).

Wolff, Caspar Friedrich, *Theoria Generationis* (M.D. thesis, Halle: Hendel, 1759; reprint Hildesheim, 1966).

——*Theorie der Generation: In Zwo Abhandlungen Erklärt und Bewiesen* (Berlin: Birnstiel, 1764; reprint Hildesheim: Olms, 1966).

Zanotti, Francesco Maria, 'De quibusdam Animalium Partibus, an Sensu Sint Compotes, et unde Irritabiles', *De Bononiensi Scientiarum et Artium Instituto atque Academia Commentarii*, 4 (Bologna: Vulpo, 1757), 48–57.

Zimmermann, Johann Georg, *Dissertatio Physiologica de Irritabilitate* (M.D. thesis Göttingen: Vandenhoeck, 1751).

——*Das Leben des Herrn von Haller* (Zurich: Heidegger, 1755).

——*Von der Erfahrung in der Arzneykunst* (2 vols, Zurich: Heidegger, 1763–64).

——*A Treatise on Experience in Physic* (2 vols, London: Wilkie, 1778).

——'J.G Zimmermanns Briefe an Haller nach dem Manuskript der Stadtbibliothek', Rudolf Ischer (ed.), *Neues Berner Taschenbuch auf das Jahr 1904–1912* (1903–1911).

——*Johann Georg Zimmermann: Mit Skalpell und Federkiel – ein Lesebuch* , Andreas Langenbacher (ed.), (Bern, Stuttgart, Vienna: Haupt, 1995; Schweizer Texte, N.F., 5).

Zinn, Johann Gottfried, *Dissertatio Inauguralis Medica Sistens Experimenta quaedam circa Corpus Callosum, Cerebellum, Duram Meningem, in Vivis Animalibus Instituta... Praeside... D. Alberto de Haller* (M.D. thesis, Göttingen: Vandenhoeck, 1749).

Secondary Sources

Abe, Horst Rudolf and Kaiser, Wolfram, 'Beiträge zum Leben und Werk von Andreas Elias Büchner (1701–1769)', *Beiträge zur Geschichte der Universität Erfurt (1392–1816). Zur Erfurter Medizin- und Hochschulgeschichte*, Heft 18 (1975–78), 91–138.

Ackerknecht, Erwin H., *Medicine at the Paris Hospital 1794–1848* (Baltimore: Johns Hopkins University Press, 1967).

Albury, William Randall, 'Experiment and Explanation in the Physiology of Bichat and Magendie', *Studies in History of Biology*, 1 (1977), 47-131.

Alexander, Peter, 'Robert Boyle', in Andrew Pyle (ed.), *The Dictionary of Seventeenth-Century British Philosophers* (2 vols, Bristol: Thoemmes Press, 2000), i, 109–16.

Azouvi, François, 'Le Vitalisme en Allemagne, en France et en Grande-Bretagne au XVIIIe siècle', *La lettre de la Maison française d'Oxford*, 14 (2001), 9–19.

Baasner, Frank, *Der Begriff 'Sensibilité' im 18. Jahrhundert: Aufstieg und Niedergang eines Ideals* (Heidelberg: Winter, 1988; Studia Romanica, 69).

Ballstadt, Kurt, *The Natural Philosophical Thought of Denis Diderot* (Oxford Univ. D.Phil. thesis, 2002).

Barsanti, Giulio, 'Les Phénomènes "Étranges" et "Paradoxaux" aux Origines de la Première Révolution Biologique (1740–1810)', in Guido Cimino and François Duchesneau (eds), *Vitalisms from Haller to the Cell Theory* (Florence: Olschki, 1997), 67–82.

Bastholm, Evynd, *The History of Muscle Physiology from the Natural Philosophers to Albrecht von Haller* (Copenhagen: Munksgaard, 1950; Acta Historica Scientiarum Naturalium et Medicinalium, 7).

Bazerman, Charles, *Shaping Written Knowledge: The Genre and Activity of the Experimental Article in Science* (Madison: University of Wisconsin Press, 1988).

Belloni, Luigi, 'L'Opera di Giambattista Morgagni – dalla Strutturazione Meccanica dell'Organismo Vivente all'Anatomia Patologica', *Per la Storia della Medicina* (Bologna: Forni, 1980), 239–246.

Berg, Alexander, 'Die Lehre von der Faser als Form und Funktionselement des Organismus', *Virchows Archiv für Pathologische Anatomie und Physiologie*, 309 (1942), 333–460.

Bergmann, Ernst, *Die Satiren des Herrn Maschine: Ein Beitrag zur Philosophie- und Kulturgeschichte des 18. Jahrhunderts* (Leipzig: Wiegandt, 1913).

Bernardi, Walter, *Le Metafisiche dell'Embrione: Scienze della Vita e Filosofia da Malpighi a Spallanzani (1672–1793)* (Florence: Olschki, 1986).

Bernardi, Walter, *I Fluidi della Vita: Alle Origni della Controversia sull'Elettricità Animale* (Florence: Olschki, 1992).

Beukers, H., 'Boerhaavianism in the Netherlands', *Journal of the Japan–Netherlands Institute*, 1 (1989), 116–29.

Bodemann, Eduard (ed.), *Von und über Albrecht von Haller: Ungedruckte Briefe und Gedichte Hallers sowie Ungedruckte Briefe und Notizen über denselben* (Hanover: Meyer, 1885).

Boschung, Urs, 'Albrecht von Haller als Arzt: zur Geschichte des Elixir Acidum Halleri', *Gesnerus*, 34 (1977), 267–93.

——'Iatromechanik und Chirurgie bei Lorenz Heister', *Gesnerus*, 40 (1983), 31–41.

——'Heimweh, die "Schweizer Krankheit"': vor 300 Jahren erstmals Beschrieben', *Inselbote: Hauszeitschrift für das Personal des Inselspitals Bern*, 2 (June 1988), 22–28.

——'Une Manifestation de la Sagesse Divine: Le Foetus Bicéphale Préparé par Albert Haller en 1735', in Christoph Mörgeli (ed.), *Les Objets Médicaux et Leurs Textes: Actes du 7ième Colloque des Conservateurs des Musées d'Histoire des Sciences Médicales, 8.–10. Septembre 1994* (Lyon: Fondation Mérieux, 1996), 26–38.

——'Neurophysiologische Grundlagenforschung: "Irritabilität" und "Sensibilität" bei Albrecht von Haller', in Heinz Schott (ed.), *Meilensteine der Medizin* (Dortmund: Harenberg, 1996), 242–9.

——' "Multa pro nostra Innocentia": L'Implication de Tissot dans la Querelle Haller–de Haen', in Vincent Barras and Micheline Louis-Courvoisier (eds), *La Médecine des Lumières: Tout autour de Tissot* (Chêne-Bourg: Georg, 2001; Bibliothèque d'Histoire des Sciences, 3), 113–47.

——*et al.* (eds), *Repertorium zu Albrecht von Haller's Korrespondenz, 1724–1777* (2 vols., Basel: Schwabe, 2002).

Bosse, Heinrich, 'Die Gelehrte Republik', in Hans-Wolf Jäger (ed.), *Öffentlichkeit im 18. Jahrhundert* (Göttingen: Wallstein, 1997), 51–76.

Bots, Hans (ed.), *La Diffusion et la Lecture des Journaux de Langue Française sous l'Ancien Régime* (Amsterdam and Maarssen: APA-Holland University Press, 1988).

Bresadola, Marco, *Medicina e Filosofia Naturale in Luigi Galvani (1737–1798)* (Florence Univ. D.Phil. thesis, 1998).

Brissenden, Robert Francis, *Virtue in Distress: Studies in the Novel of Sentiment from Richardson to Sade* (London: Macmillan, 1974).

Brockliss, Laurence W.B., 'Before the Clinic: French Medical Teaching in the Eighteenth Century', in Caroline Hannaway and Ann La Berge (eds), *Constructing Paris Medicine* (Amsterdam: Rodopi, 1998; Clio Medica, 50), 71–115.

——*Calvet's Web: Enlightenment and the Republic of Letters in Eighteenth-Century France* (Oxford: Oxford University Press, 2002).

——and Colin Jones, *The Medical World of Early Modern France* (Oxford: Oxford University Press, 1997).

Broman, Thomas, 'Rethinking Professionalization: Theory, Practice, and Professional Ideology in Eighteenth-Century German Medicine', *The Journal of Modern History*, 67 (1995), 835–72.

——*The Transformation of German Academic Medicine, 1750–1820* (Cambridge: Cambridge University Press, 1996).

——'The Habermasian Public Sphere and "Science in the Enlightenment"', *History of Science*, 36 (1998), 123–149.

Brown, Theodore Morris, *The Mechanical Philosophy and the 'Animal Oeconomy': A Study in the Development of English Physiology in the Seventeenth and Early Eighteenth Century* (Princeton Univ. D.Phil. thesis, 1968).

——'The College of Physicians and the Acceptance of Iatromechanism in England, 1665–1695', *Bulletin of the History of Medicine*, 44 (1970), 12–30.

——'From Mechanism to Vitalism in Eighteenth-Century English Physiology', *Journal of the History of Biology*, 7 (1974), 179–216.

Buess, Heinrich, 'Zur Entwicklung der Irritabilitätslehre', *Festschrift für Jacques Brodbeck-Sandreuter zu Zeinem 60. Geburtstag* (Basel: [n.p.], 1942), 299–333.

——'Experientia Majorum: ein Physiologenstreit vor 200 Jahren', *Experientia*, 2/4 (1946), 150–1.

Buess, Heinrich, 'Die Anfänge der Pathologischen Physiologie auf dem Gebiet der Kreislaufforschung nach Albrecht Hallers "Elementa Physiologiae" (1756/1760)', *Gesnerus*, 11 (1954), 121–51.

Buscaglia, Marino, 'Pour une Histoire Spécifique de la Méthode en Biologie', *Archives des Sciences et Compte Rendu des Séances de la Société de Physique et d'Histoire Naturelle de Genève*, 47 (1994), 137–54.

——'The History of the Experimental Method in the Life Sciences as an Illustration of the Versatility in Interpretation', in J. Montangero *et al.* (eds), *Conceptions of Time over Change* (Cahiers de la Fondation Archives Jean Piaget, 13, 1994), 45–64.

——'Bonnet dans l'Histoire de la Méthode Expérimentale', in *idem, et al.* (eds), *Charles Bonnet: Savant et Philosphe (1720–1793)* (Geneva: Editions Passé Présent, 1994), 283–313.

——'La Pratique, la Figure et les Mots dans les Mémoires d'Abraham Trembley sur les Polypes (1744) comme Exemple de Communication Scientifique', in Massimo Galuzzi *et al.* (eds), *Le Forme della Comunicazione Scientifica* (Milan: Angeli, 1998), 313–46.

Bynum, W.F., 'Cullen and the Study of Fevers in Britain, 1760–1820', in *idem* and V. Nutton (eds), *Theories of Fever from Antiquity to the Enlightenement* (London: Wellcome Institute for the History of Medicine, 1981; Medical History, Suppl. 1), 135–47.

——'Cullen and the Nervous System', in A. Doig *et al.* (eds), *William Cullen and the Eighteenth Century Medical World* (Edinburgh: Edinburgh University Press, 1993), 152–62.

——and Roy Porter (eds), *Medical Fringe & Medical Orthodoxy 1750–1850* (London: Croom Helm, 1987).

Canguilhem, Georges, *La Formation du Concept de Réflexe aux XVIIe et XVIIIe Siècles* (2nd edn, Paris: Vrin, 1977; 1st edn, 1955).

Cavazza, Marta, 'La Recezione della Teoria Halleriana dell'Irritabilità nell'Accademia delle Scienze di Bologna', *Nuncius*, 12 (1997), 359–77.

Clark, William, Golinski, Jan and Schaffer, Simon (eds), *The Sciences in Enlightened Europe* (Chicago and London: University of Chicago Press, 1999).

Clarke, E., 'The Doctrine of the Hollow Nerve in the Seventeenth and Eighteenth Century', in L.G. Stevenson and R.P. Multhauf (eds),

Medicine, Science, and Culture: Historical Essays in Honor of Owsei Temkin (Baltimore: Johns Hopkins University Press, 1968), 123–41.

Cohen, E., 'Katalog der Wiedergefundenen Manuskripte und Briefwechsel von Herman Boerhave', *Verhandelingen der Koninklijke Nederlandes Akademie van Wetenschappen, Afdeling Natuurkunde*, 2nde Sectie, xl/2 (1941), 1–45.

Coleman, William and Holmes, Frederic L. (eds), *The Investigative Enterprise: Experimental Physiology in Nineteenth-Century Medicine* (Berkeley: University of California Press, 1988).

Cook, Harold J., 'The New Philosophy in the Low Countries', in Roy Porter and Mikulás Teich (eds), *The Scientific Revolution in National Context* (Cambridge: Cambridge University Press, 1992), 115–49.

——'Boerhaave and the Flight from Reason in Medicine', *Bulletin of the History of Medicine*, 74 (2000), 221–40.

Cross, Stephen J., 'John Hunter, the Animal Oeconomy and Late Eighteenth Century Physiological Discourse', *Studies in History of Biology*, 5 (1981), 1–110.

Cunningham, Andrew, 'Medicine to Calm the Mind. Boerhaave's Medical System, and Why it Was Adopted in Edinburgh', in *idem* and Roger French (eds), *The Medical Enlightenment of the Eighteenth Century* (Cambridge: Cambridge University Press, 1990), 40–66.

——'The Pen and the Sword: Recovering the Disciplinary Identity of Physiology and Anatomy before 1800', *Studies in History and Philosophy of the Biological and Biomedical Sciences*, 33 (2002), 631–65; 34 (2003), 51–76.

——and Perry Willliams (eds), *The Laboratory Revolution in Medicine* (Cambridge: Cambridge University Press, 1992).

Curran, William S., 'Dr. Brocklesby of London (1722–1797). An 18th-Century Physician and Reformer', *Journal of the History of Medicine*, 17 (1962), 509–21.

De Angelis, Simone, *Von Newton zu Haller: Studien zum Naturbegriff zwischen Empirismus und Deduktiver Methode in der Schweizer Frühaufklärung* (Tübingen: Niemeyer, 2003; Frühe Neuzeit, 74).

Dini, Alessandro, *Vita e Organismo: Le Origini della Fisiologia Sperimentale in Italia* (Florence: Olschki, 1991).

Dini, Alessandro, 'Spallanzani e la Teoria dell'Irritabilità di Haller', in Walter Bernardi and Marta Stefani (eds), *La Sfida della Modernità: Atti del Convegno Internazionale di Studi del Bicentenario della Morte di Lazzaro Spallanzani* (Florence: Olschki, 2000; Biblioteca di Storia della Scienza, 43), 197–208.

Donoghue, Frank, 'Colonizing Readers: Review Criticism and the Formation of a Reading Public', in Ann Bermingham and John Brewer (eds), *The Consumption of Culture 1600–1800: Image, Object, Text* (London and New York: Routledge, 1995), 54–74.

Driesch, Hans, *Der Vitalismus als Geschichte und als Lehre* (Leipzig: Barth, 1905).

Duchesneau, François, *La Physiologie des Lumières: Empirisme, Modèles, Théories* (The Hague: Nijhoff, 1982; Archives Internationales d'Histoire des Idées, 95).

——'Vitalism in Late 18th-Century Physiology: The Cases of Barthez, Blumenbach, and John Hunter', in W. F. Byum and Roy Porter (eds), *William Hunter an the Eighteenth Century Medical World* (Cambridge: Cambridge University Press, 1985), 259–95.

——'Territoires et Frontières du Vitalisme (1750–1850)', in Guido Cimino and François Duchesneau (eds), *Vitalisms: From Haller to the Cell Theory* (Florence: Olschki, 1997), 297–349.

Dull, Monique Willemine, *Irritability and Fiction, 1814–1837* (Virginia Univ. D.Phil. thesis, 1997).

Engelhardt, H. Tristram and Arthur L. Caplan (eds), *Scientific Controversies. Case Studies in the Resolution and Closure of Disputes in Science and Technology* (Cambridge: Cambridge University Press, 1987).

Ferrone, Vincenzo, *The Intellectual Roots of the Italian Enlightenment: Newtonian Science, Religion, and Politics in the Early Eighteenth Century*, Engl. transl. (Atlantic Highlands, N.J: Humanities Press, 1995).

Figlio, Karl M., 'Theories of Perception and the Physiology of Mind in Late Eigtheenth Century', *History of Science*, 12 (1975), 177–212.

Fontius, Martin, 'Tendenzen der Literaturkritik in Frankreich und Deutschland im 18. Jahrhundert', in Siegfried Jüttner and Jochen Schlobach (eds), *Europäische Aufklärung(en): Einheit und Nationale Vielfalt* (Hamburg: Meiner, 1992), 127–140.

Forster, Antonia, '"The Self-Impannelled Jury": The Reception of the Review Journals, 1749–1760', *Studies in Newspaper and Periodical History* (1993), 27–51.

Foucault, Michel, *The Birth of the Clinic: An Archaeology of Medical Perception.* Transl. from the French by A.M. Sheridan Smith (London: Tavistock, 1973).

Frank, Robert G., *Harvey and the Oxford Physiologists* (Berkeley: University of California Press, 1980).

Franklin, Allan, *The Neglect of Experiment* (Cambridge: Cambridge University Press, 1986).

French, Roger, *Robert Whytt, the Soul, and Medicine* (London: The Wellcome Institute of the History of Medicine, 1969).

——'Sauvages, Whytt and the Motion of the Heart: Aspects of Eighteenth Century Animism', *Clio Medica*, 7 (1972), 35–54.

——'Ether and Physiology', in G.N. Cantor and M.S. Hodge (eds), *Conceptions of Ether: Studies in the History of Ether Theories 1740–1900* (Cambridge: Cambridge University Press, 1981), 110–34.

——'Sickness and the Soul: Stahl, Hoffmann and Sauvages on Pathology', in Andrew Cunningham and Roger French (eds), *The Medical Enlightenment of the Eighteenth Century* (Cambridge: Cambridge University Press, 1990), 88–110.

——*William Harvey's Natural Philosophy* (Cambridge: Cambridge University Press, 1994).

Freschi, Francesco, *Storia della Medicina in Aggiunta e Continuazione a Quella di Curzio Sprengel, Vol. VII, Parte prima* (Florence: Speranza, 1845).

Fuchs, Thomas, *The Mechanization of the Heart: Harvey and Descartes.* Transl. from the German by Marjorie Grene (Rochester: University of Rochester Press, 2001; Rochester Studies in Medical History, 1).

Galison, Peter, *How Experiments End* (Chicago: The University of Chicago Press, 1987).

Garrison, Fielding H., *An Introduction to the History of Medicine* (4th edn, Philadelphia and London: Saunders, 1929).

Gelfand, Toby, 'Empiricism and Eighteenth-Century French Surgery', *Bulletin of the History of Medicine*, 44 (1970), 40–53.

Gelfand, Toby, *Professionalizing Modern Medicine: Paris Surgeons and Medical Science and Institutions in the 18th Century* (London: Greenwood, 1980).

Geyer-Kordesch, Johanna, *Pietismus, Medizin und Aufklärung in Preußen im 18. Jahrhundert: Das Leben und Werk Georg Ernst Stahls* (Tübingen: Niemeyer, 2000).

Gierl, Martin, 'Kompilation und die Produktion von Wissen im 18. Jahrhundert', in Helmut Zedelmaier and Martin Mulsow (eds), *Die Praktiken der Gelehrsamkeit in der Frühen Neuzeit* (Tübingen: Niemeyer, 2001), 63–94.

Goldgar, Anne, *Impolite Learning: Conduct and Community in the Republic of Letters, 1680–1750* (New Haven and London: Yale University Press, 1995).

Gooding, David, *Experiment and the Making of Meaning: Human Agency in Scientific Observation and Experiment* (Dordrecht: Kluwer, 1990).

——Pinch, Trevor and Schaffer, Simon (eds), *The Uses of Experiment. Studies in the Natural Sciences* (Cambridge: Cambridge University Press, 1989).

Grmek, Mirko, 'La Notion de la Fibre Vivante chez les Médecins de l'École Iatrophysique', *Clio Medica*, 5 (1970), 297–318.

——'La Théorie et la Pratique dans l'Expérimentation Biologique au Temps de Spallanzani', in G. Montalenti and P. Rossi (eds), *Lazzaro Spallanzani e la Biologia del Settecento: Teorie, Esperimenti, Insituzioni Scientifiche* (Florence: Olschki, 1980), 321–44.

——'L'Apport des Correspondants Genevois de Spallanzani à la Méthodologie Expériementale', *Gesnerus*, 48 (1991), 287–306.

——'Il 'de Fibra Motrice e Morbosa' di Giorgio Baglivi', *Medicina nei Secoli*, 12 (2000), 19–27.

Guerrini, Anita, 'Archibald Pitcairne and Newtonian Medicine', *Medical History*, 31 (1987), 70–83.

——'Isaac Newton, George Cheyne and the *Principia Medicinae*', in Roger French and Andrew Wear (eds), *The Medical Revolution of the Seventeenth Century* (Cambridge: Cambridge University Press, 1989), 222–45.

Guthke, Karl S., *Haller und die Literatur* (Göttingen: Vandenhoeck & Ruprecht, 1962).

——'Haller und die *Bibliothèque Raisonnée*', *Jahrbuch des Freien Deutschen Hochstifts* (1973), 1–13.

Hacking, Ian, *Representing and Intervening. Introductory Topics in the Philosophy of Natural Science* (Cambridge: Cambridge University Press, 1983).

Hackmann, W.D., 'The Growth of Science in the Netherlands in the Seventeenth and Early Eighteenth Centuries', in Maurice Crosland (ed.), *The Emergence of Science in Western Europe* (London: Macmillan, 1975), 89–109.

Haigh, Elizabeth, *Xavier Bichat and the Medical Theory of the Eighteenth Century* (London: Wellcome Institute for the History of Medicine, 1984; Medical History, Suppl., 4).

Hall, Thomas S., 'On Biological Analogs of Newtonian Paradigms', *Philosophy of Science*, 35 (1968), 6–27.

——*Ideas of Life and Matter: Studies in the History of General Physiology, 600 B.C. – 1900 A.D.* (2 vols, Chicago: University of Chicago Press, 1969).

Hardesty Doig, Kathleen, 'The Encyclopédie of Yverdon', in Frank A. Kafker (ed.), *Notable Enyclopedias of the Late Eighteenth Century: Eleven Successors of the Encyclopédie* (Oxford: Voltaire Foundation, 1994; Studies on Voltaire and the Eighteenth Century, 315), 85–116.

Heimann, P. and McGuire, J.E., 'Newtonian Forces and Lockean Powers: Concepts of Matter in the Eigtheenth-Century Thought', *Historical Studies in the Physical Sciences*, 1971, 233–306.

Henkelmann, Thomas, *Zur Geschichte des Pathophysiologischen Denkens: John Brown (1735–1788) und sein System der Medizin* (Berlin: Springer, 1981).

Hintzsche, Erich, 'Einige Kritische Bemerkungen zur Bio- und Ergographie Albrecht von Hallers', *Gesnerus*, 16 (1959), 1–15.

——'Albrecht von Hallers Tätigkeit als Enzyklopädist', *Clio Medica*, 1 (1966), 235–54.

——'Neue Funde zum Thema 'L'Homme Machine' und Albrecht Haller', *Gesnerus* 25 (1968), 135–66.

Hintzsche, Erich, 'Haller, Albrecht von', in *Dictionary of Scientific Biography*, Vol. 6 (New York: Scribner, 1972), 61–67.

Holmes, Frederic Lawrence, *Lavoisier and the Chemistry of Life: An Exploration of Scientific Creativity* (Madison: University of Wisconsin Press, 1985).

Horst, Christoph auf der, 'Vorstellungen, Ideen, Begriffe: Intellectual History in der Medizingeschichtsschreibung am Beispiel des Naturbegriffs', in Norbert Paul and Thomas Schlich (eds), *Medizingeschichte: Aufgaben, Probleme, Perspektiven* (Frankfurt am Main and New York: Campus, 1998), 186–215.

Hufbauer, Karl, *The Formation of the German Chemical Community (1720–1795)* (Berkeley: University of California Press, 1982).

Huisman, Frank, 'Medicine and Health Care in the Netherlands, 1500–1800', in Klaas van Berkel, Albert van Helden and Lodewijk Palm (eds), *A History of Science in the Netherlands. Survey: Themes and Reference* (Leiden: Brill, 1999), 239–78.

Jacyna, Stephen, 'Physiological Principles in the Surgical Writings of John Hunter', in Christopher Lawrence (ed.), *Medical Theory, Surgical Practice: Studies in the History of Surgery* (London: Routledge, 1992), 135–52.

Jantzen, Jörg, 'Physiologische Theorien', in *idem*, Manfred Durner and Francesco Moiso (eds), *Wissenschaftshistorischer Bericht zu Schellings Naturphilosophischen Schriften 1797–1800*, (Stuttgart: Frommann-Holzboog, 1994; Friedrich Wilhelm Schelling. Ergänzungsband zu Werke Band 5 bis 9), 375–668.

Jauch, Ursula Pia, *Jenseits der Maschine: Philosophie, Ironie und Ästhetik bei Julien Offray de La Mettrie (1709–1751)* (München, Wien: Hanser, 1998).

Jewson, N.D., 'Medical Knowledge and the Patronage System in Eighteenth-Century England', *Sociology*, 8 (1974), 369–85.

Jordanova, Ludmilla, 'The Social Construction of Medical Knowledge', *Social History of Medicine*, 8 (1995), 361–81.

Kafker, Frank A. and Kafker, Serena (eds), *The Encyclopedists as Individuals: A Biographical Dictionary of the Authors of the Encyclopédie* (Oxford: Voltaire Foundation, 1988; Studies on Voltaire and the Eighteenth Century, 257).

Kaiser, Wolfram, 'Heinrich Friedrich Delius (1720–1791)', *Harz-Zeitschrift* 31 (1979), 65–82.

King, Lester S., *The Medical World of the Eighteenth Century* (Chicago: University of Chicago Press, 1958).

——'Rationalism in Early Eighteenth Century Medicine', *Journal of the History of Medicine*, 18 (1963), 257–71.

——'Introduction', in Albrecht von Haller, *First Lines of Physiology* (New York and London: Johnson Reprint, 1966), ix–lxxii.

——*The Philosophy of Medicine. The Early Eighteenth Century* (Cambridge: Mass.: Harvard University Press, 1978).

Knabe, Peter-Eckhard, *Die Rezeption der Französischen Aufklärung in den 'Göttingischen Gelehrten Anzeigen' (1739–1779)* (Frankfurt am Main: Klostermann, 1978; Analecta Romanica, 42).

Knoeff, Rina, *Herman Boerhaave (1668-1738): Calvinist Chemist and Physician* (Amsterdam: Koninklijke Nederlandse Akademie van Wetenschappen, 2002; History of Science and Scholarship in the Netherlands, 3).

Knorr-Cetina, Karin, *The Manufacture of Knowledge: An Essay on the Constructivist and Contextual Nature of Science* (Oxford: Pergamon Press, 1981).

Knott, Sarah, *A Cultural History of Sensibility in the Era of the American Revolution* (Oxford Univ. D.Phil. thesis, 1999).

Korst, J.K. van der, '"Anatomia Actuosa et Apta". De Mechanicistische 'Proto'-fysiologie van B.S. Albinus', *Gewina*, 16 (1993), 308–23.

Kuhles, Doris, *Deutsche Literarische Zeitschriften von der Aufklärung bis zur Romantik* (2 vols., Munich: Saur, 1994).

Kuhn, Thomas S., *The Structure of Scientific Revolutions* (Chicago: University of Chicago Press, 1962; 3rd edn. 1996).

——'Mathematical versus Experimental Traditions in the Development of Physical Science', in *idem*, *The Essential Tension. Selected Studies in Scientific Tradition and Change* (Chicago: University of Chicago Press, 1977), 31–65.

Lakatos, Imre, 'Falsification and the Methodology of Scientific Research Programmes', in *idem* and Alan Musgrave (eds), *Criticism and the Growth of Knowledge* (Cambridge: Cambridge University Press, 1970), 91–196.

Larson, James L., 'Vital Forces: Regulative Principles or Constitutive Agents? A Strategy in German Physiology, 1786–1802', *Isis*, 70 (1979), 235–49.

Latour, Bruno, *Science in Action: How to Follow Scientists and Engineers through Society* (Cambridge, Mass.: Harvard University Press, 1987).

——and Steve Woolgar, *Laboratory Life: The Construction of Scientific Facts* (Beverly Hills: Sage, 1979).

Lawrence, Christopher J., 'The Nervous System and Society in the Scottish Enlightenment', in Barry Barnes and Steven Shapin (eds), *Natural Order: Historical Studies of Scientific Culture* (Beverly Hills: Sage, 1979), 19–40.

——'Ornate Physicians and Learned Artisans: Edinburgh Medical Men, 1727–1776', in W.F. Bynum and Roy Porter (eds), *William Hunter and the Eighteenth-Century Medical World* (Cambridge: Cambridge University Press, 1985), 153–76.

Lenoir, Timothy, *The Strategy of Life: Teleology and Mechanics in Nineteenth Century German Biology* (Dordrecht and London: Reidel, 1982).

Lesch, John E., *Science and Medicine in France: The Emergence of Experimental Physiology, 1790–1855* (Cambridge, Mass.: Harvard University Press, 1984).

Lesky, Erna, 'Albrecht von Haller, Gerard van Swieten und Boerhaavens Erbe', *Gesnerus* 15 (1958), 120–40.

——'Albrecht von Haller und Anton de Haen im Streit um die Lehre von der Sensibilität', *Gesnerus*, 16 (1959), 16–46.

——'Vom Hippokratismus Boerhaaves und De Haens', *Boerhaave and his Time* (Leiden: Brill, 1970), 123–43.

——'Neue Dokumente zum Streit Haller – van Swieten', *Clio Medica*, 7 (1972), 120–7.

Licoppe, Christian, *La Formation de la Pratique Scientifique: Le Discours de l'Expérience en France et en Angleterre (1630–1820)* (Paris: Editions la découverte, 1996).

Lindeboom, Gerrit A., *Herman Boerhaave: The Man and his Work* (London: Methuen, 1968).

——'Boerhaave's Concept of the Basic Structure of the Body', *Clio Medica*, 5 (1970), 203–8.

Lindeboom, Gerrit A.,'Dog and Frog: Physiological Experiments at Leiden During the Seventeenth Century', in T.H. Lunsing Scheeurler and G.H.M. Posthumus Meyjes (eds), *Leiden University in the Seventeenth Century: An Exchange of Learning* (Leiden: Brill, 1975), 278–93.

Lohff, Brigitte, *Die Suche nach der Wissenschaftlichkeit der Physiologie in der Zeit der Romantik* (Stuttgart and New York: Fischer, 1990; Medizin in Geschichte und Kultur 17).

——'Johannes Müller und das Physiologische Experiment', in M. Hagner and B. Wahrig-Schmidt (eds), *Johannes Müller und die Philosophie* (Berlin: Akademie-Verlag, 1993), 105–23.

——'The Concept of Vital Force as a Research-Program', in Guido Cimino and François Duchesneau (eds), *Vitalisms from Haller to the Cell Theory* (Florence: Olschki, 1997), 127–42.

Lordat, Jacques, *Exposition de la Doctrine Médicale de P.-J. Barthez, et Mémoires sur la Vie de ce Médecin* (Paris: Gabon, 1818).

Lovejoy, Arthur O., *The Great Chain of Being: The Study of the History of an Idea* (Cambridge, Mass.: Harvard University Press, 1936).

Lund, Wilhelm, *Physiologische Resultate der Vivisectionen Neuerer Zeit: Eine von der Kopenhagener Universität Gekrönte Preisschrift: Aus dem Dänischen Übersetzt* (Copenhagen: Brummer, 1825).

Luyendijk-Elshout, Antonie M., 'Oeconomia Animalis, Pores and Particles: The Rise and Fall of the Mechanical Philosophical School of Theodoor Craanen (1621/1690)', in T.H. Lunsing Scheeurler and G.H.M. Posthumus Meyjes (eds), *Leiden University in the Seventeenth Century: An Exchange of Learning* (Leiden: Brill, 1975), 294–307.

——'Mechanisme contre Vitalisme: De School van Herman Boerhaaven en de Beginselen van het Leven', *Gewina*, 5 (1982), 16–26.

Maehle, Andreas-Holger, *Kritik und Verteidigung des Tierversuchs: Die Anfänge der Diskussion im 17. und 18. Jahrhundert* (Stuttgart: Steiner, 1992).

——*Drugs on Trial: Experimental Pharmacology and Therapeutic Innovation in the Eighteenth Century* (Amsterdam: Rodopi, 1999; Clio Medica, 53).

Maehle, Andreas-Holger, and Ulrich Tröhler, 'Animal Experimentation from Antiquity to the End of the Eighteenth Century: Attitudes and Arguments', in Nicolaas A. Rupke (ed.), *Vivisection in Historical Perspective* (London and New York: Routledge, 1987), 14–47.

Maerker, Dietrich von, 'Die Zahlen der Studierenden an der Georg-August-Universität in Göttingen 1734/37 bis 1984', *Göttinger Jahrbuch*, 1979, 141–58.

Mallepree, Klaus, *Die Physiologischen Reizmethoden Albrecht von Hallers* (Münster Univ. M.D. thesis, 1966).

Mann, Gunter, 'Ernst Gottfried Baldinger und sein Magazin für Aerzte', *Sudhoffs Archiv*, 42 (1958), 312–18.

Martin, Julian, 'Sauvages's Nosology: Medical Enlightenment in Montpellier', in Andrew Cunningham and Roger French (eds), *The Medical Enlightenment of the Eighteenth Century* (Cambridge: Cambridge University Press, 1990), 111–37.

Mattauch, Hans, *Die Literarische Kritik der Frühen Französischen Zeitschriften (1665–1748)* (München: Hueber, 1968).

Mazzolini, Renato G., 'Sugli Studi Embriologici di Albrecht von Haller negli anni 1755–1758', *Annali dell'Istituto Storico Italo-germanico in Trento = Jahrbuch des Italienisch-deutschen Historischen Instituts in Trient*, 1977, 183–242.

——'Le Dissertazioni degli Allievi di Albrecht von Haller a Göttingen (1736–53): Una Indagine Bio-bibliografica', *Nuncius*, 2 (1987), 125–94.

——and Shirley A. Roe, *Science against the Unbelievers: The Correspondence of Bonnet and Needham. 1760–1780* (Oxford: Voltaire Foundation, 1986; Studies on Voltaire and the Eighteenth Century, 243).

McClellan, James E., 'The Scientific Press in Transition: Rozier's Journal and the Scientific Societies in the 1770s', *Annals of Science*, 36 (1979), 425–49.

——*Science Reorganized: Scientific Societies in the Eighteenth Century* (New York: Columbia University Press, 1985).

McLaughlin, Peter, 'Blumenbach und der Bildungstrieb: Zum Verhältnis von Epigenetischer Embryologie und Typologischem Artbegriff', *Medizinhistorisches Journal*, 17 (1982), 357–72.

Meyer, Georg Hermann, 'Glisson's Irritabilitäts- und Sensibilitätslehre', *Archiv für die Gesamte Medizin*, 5 (1843), 1–17.

Mocek, Reinhard, 'Caspar Friedrich Wolffs Epigenesis-Konzept: Ein Problem im Wandel der Zeit', *Biologisches Zentralblatt*, 114 (1995), 179–90.

Möller, Hans-Jürgen, *Die Begriffe 'Reizbarkeit' und 'Reiz': Konstanz und Wandel ihre Bedeutungsgehaltes sowie Problematik ihrer Exakten Definition* (Stuttgart: Fischer, 1975; Medizin in Geschichte und Kultur, 11).

Mondella, Felice, 'L'Irritabilità ed il Programma di Ricerca nella Fisiologia di C. Bernard', in M. Di Giandomenico (ed.), *Claude Bernard: Scienza, Filosofia, Letteratura* (Verona: Bertani, 1982), 177–88.

Monti, Maria Teresa (ed.), *Catalogo del Fondo Haller della Biblioteca Nazionale Braidense di Milano* (13 vols., Milan: Angeli, 1983–94).

——'Inediti di Pietro Moscati: Un Episodio del Dibattito Italiano sull'Insensibilità della Dura Madre', *Nuncius*, 2 (1987), 157–70.

——'Théologie Physique et Mécanisme dans la Physiologie de Haller', in Änne Bäumer and Manfred Büttner (eds), *Science and Religion: Proceedings of the Symposium of the XVIIIth International Congress of History of Science at Hamburg–Munich, 1.–9. August 1989* (Bochum: Brockmeyer, 1989), 68–79.

——*Congettura ed Esperienza nella Fisiologia di Haller: La Riforma dell' Anatomia Animata e il Sistema della Generazione* (Florence: Olschki, 1990).

——'Haller et les Monstres: Objets de Recherche Positive ou Preuves Éclatantes de la Liberté Infinie de Dieu', in Eric Fierens *et al.* (eds), *Actes du XXXIIe Congrès International d'Histoire de la Médecine, Anvers, 3–7 septembre 1990* (Bruxelles: Societas Belgica Historiae Medicinae, 1991), 181–87.

——'Les Dynamismes du Corps et les Forces du Vivant dans la Physiologie de Haller', in Guido Cimino and François Duchesneau (eds), *Vitalisms: From Haller to the Cell Theory* (Florence: Olschki, 1997), 41–66.

——'Forma della Scrittura e Forme della Comunicazione: Dai Mémoires di Haller al Prodromo di Spallanzani', in Massimo Galuzzi *et al.* (eds), *Le Forme della Comunicazione Scientifica* (Milan: Angeli, 1998), 277–311.

Moravia, Sergio, *Il Pensiero degli Idéologues: Scienza e Filosofia in Francia (1780–1815)* (Florence: La Nuova Italia, 1974).

——'From Homme Machine to Homme Sensible: Changing Eighteenth-Century Models of Man's Image', *Journal of the History of Ideas*, 39 (1978), 45–60.

Neuburger, Max, *The Historical Development of Experimental Brain and Spinal Cord Physiology before Flourens.* Translated and Edited, with Additional Material, by Edwin Clarke (Baltimore: Johns Hopkins University Press, 1981).

Nohr, René, Papacek, Ellen and Vetter, Andreas, '"Das Richtige Urteil über den Zustand der Vaterländischen Literatur?": Zum Anteil des Rezensionswesens an der Französisch-deutschen Kulturvermittlung im Zeitalter der Aufklärung', in Hans-Jürgen Lüsebrink *et al.* (eds), *Kulturtransfer als Epochenumbruch Frankreich – Deutschland: 1770 bis 1815* (Leipzig: Leipziger Universitätsverlag, 1997), 499–535.

O'Keefe, Cyril B., *Contemporary Reactions to the Enlightenment (1728–1762): A Study of Three Critical Journals, the Jesuit Journal de Trévoux, the Jansensist Nouvelles Ecclésiastiques, and the Secular Journal des Savants* (Geneva: Slatkine, 1974).

Ongaro, Giuseppe, 'Giambattista Morgagni e la Questione dell'Irritabilità', in Vincenzo Cappelletti and Federico Di Trocchio (eds), *De Sedibus et Causis: Morgagni nel Centenario* (Rome: Istituto della Enciclopedia Italiana, 1986), 27–45.

——'L'Irritabilità Halleriana in Italia', in Giorgio Zanchin and Loris Premuda (eds), *Lo Sviluppo Storico della Neurologia Italiana: Lo Studio delle Fonti* (Padua: La Garangola, 1990), 113–25.

Pagel, Walter, 'Harvey and Glisson on Irritability: With a Note on van Helmont', *Bulletin for the History of Medicine*, 38 (1967), 497–514.

Pera, Marcello, *The Ambiguous Frog: The Galvani–Volta Controversy on Animal Electricity*, Eng. trans. (Princeton: Princeton University Press, 1992).

Pogány-Wnendt, Peter, *Das Mechanistische Denken in der Modernen Medizin im Spiegel ihrer Geschichtlichen Entwicklung: Hieronimus David Gaub (1705–1780)* (Frankfurt am Main: Lang, 1991).

Porter, Roy, 'Laymen, Doctors and Medical Knowledge in the 18th Century: The Evidence of the Gentleman's Magazine', *Medical History*, 29 (1985), 138–68.

——'Medical Science', in *idem* (ed.), *Cambridge Illustrated History of Medicine* (Cambridge: Cambridge University Press, 1996), 154–201.

——*The Greatest Benefit to Mankind: A Medical History of Humanity from Antiquity to the Present* (London: Fontana, 1997).

Premuda, Loris, 'Beobachtungen und Kritische Bemerkungen über die Methodologische Grundlage von Herman Boerhaave', *Boerhaave and his Time* (Leiden: Brill, 1970), 40–59.

Pross, Wolfgang, '"Ein Reich Unsichtbarer Kräfte": Was Kritisiert Kant an Herder?', *Scientia Poetica: Jahrbuch für Geschichte der Literatur und der Wissenschaften*, 1 (1997), 62–119.

Punt, H., *Bernard Siegfrid Albinus (1697–1770) on 'Human Nature': Anatomical and Physiological Ideas in Eighteenth Century Leiden* (Amsterdam: Israël, 1983).

Rather, Lelland Joseph, *Mind and Body in Eighteenth-Century Medicine: A Study Based on Jerome Gaub's De Regime Mentis* (London: The Wellcome Historical Medical Library, 1965).

——'Some Relations Between Eigtheenth-Century Fiber Theory and Nineteenth-Century Cell Theory', *Clio Medica*, 4 (1969), 191–202.

Reiber, Matthias, *Anatomie eines Bestsellers: Johann August Unzers Wochenschrift 'Der Arzt'* (1759–1764) (Göttingen: Wallstein, 1999).

Rey, Roselyne, 'La Théorie de la Sécrétion chez Bordeu, Modèle de la Physiologie et de la Pathologie des Vitalistes', *Dix–huitième Siècle*, 23 (1991), 45–58.

——'L'École de Santé de Paris sous la Révolution', *Histoire de l'Éducation*, 57 (1993), 23–57.

——'Vitalism, Disease and Society', in Roy Porter (ed.), *Medicine in the Enlightenment* (Amsterdam: Rodopi, 1995; Clio Medica, 29), 274–88.

——*Naissance et Développement du Vitalisme en France de la Deuxième Moitié du 18e Siècle à la Fin du Premier Empire* (Oxford: Voltaire Foundation, 2000; Studies on Voltaire and the Eighteenth Century, 381).

Rheinberger, Hans-Jörg and Michael Hagner (eds), *Die Experimentalisierung des Lebens: Experimentalsysteme in den Biologischen Wissenschaften, 1850/1950* (Berlin: Akademie-Verlag, 1993).

Richard, Robert J., 'Kant and Blumenbach on the 'Bildungstrieb': A Historical Misunderstanding', *Studies in History and Philosophy of Biological and Biomedical sciences*, 31 (2000), 11–32.

Risse, Guenter B., *The History of John Brown's Medical System in Germany during the Years 1790–1806* (Chicago Univ. D.Phil. thesis, 1971).

Roe, Shirley A., *Matter, Life and Generation: Eighteenth-Century Embryology and the Haller–Wolff Debate* (Cambridge: Cambridge University Press, 1981).

——'Anatomia Animata: The Newtonian Physiology of Albrecht von Haller', in Everett Mendelsohn (ed.), *Transformation and Tradition in the Sciences: Essays in Honor of I. Bernard Cohen* (Cambridge: Cambridge University Press, 1984), 273–300.

——(ed.), *The Natural Philosophy of Albrecht von Haller* (New York: Arno Press, 1981).

Roethe, Gustav, 'Göttingische Zeitungen von gelehrten Sachen', in *Festschrift zur Feier des Hundertfünfzigjährigen Bestehens der Königlichen Gesellschaft der Wissenschaften zu Göttingen: Beiträge zur Gelehrtengeschichte Göttingens* (Berlin: Weidmann, 1901), 569–688.

Roger, Jacques, *Les Sciences de la Vie dans la Pensée Française au XVIIIème Siècle: La Génération des Animaux de Descartes à l'Encyclopédie* (Paris: Colin, 1963).

Rosner, Lisa, 'Eighteenth-Century Medical Education and the Didactic Model of Experiment', in Peter Dear (ed.), *The Literary Structure of Scientific Argument: Historical Studies* (Philadelphia: University of Pennsylvania Press, 1991), 182–94.

Rössler, Emil Franz (ed.), *Die Gründung der Universität Göttingen: Entwürfe, Berichte und Briefe der Zeitgenossen* (Göttingen: Vandenhoeck & Ruprecht, 1855; reprinted Aalen: Scientia-Verlag, 1987).

Rothschuh, Karl Eduard, 'Vom Spiritus Animalis zum Nervenaktionsstrom', *Ciba–Zeitschrift*, 8 (1958), 2950–80.

——'Geschichtliches zur Lehre der Automatie, Unterhaltung und Regelung der Herztätigkeit', *Gesnerus*, 27 (1970), 1–19.

——*Konzepte der Medizin in Vergangenheit und Gegenwart* (Stuttgart: Hippokrates-Verlag, 1978).

Rousseau, George S., 'Nerves, Spirits and Fibres: Towards Defining the Origins of Sensibility', *Studies in the Eighteenth Century*, 3 (1976), 137–57.

——'Discourses of the Nerve', in Frederick Amrine (ed.), *Literature and Science as Modes of Expression* (Dordrecht: Kluwer, 1989), 29–60.

——*Enlightenment Crossings: Pre- and Post-Modern Discourses – Anthropological* (Manchester: Manchester University Press, 1991).

Rowland, Herbert and Karl J. Fink (eds), *The Eighteenth Century German Book Review* (Heidelberg: Winter, 1995).

Rudolph, Gerhard, 'Hallers Lehre von der Irritabilität und Sensibilität', in Karl E. Rothschuh (ed.), *Von Boerhaave bis Berger: Die Entwicklung der Kontinentalen Physiologie im 18. und 19. Jahrhundert mit Besonderer Berücksichtigung der Neurophysiologie* (Stuttgart: Fischer, 1964), 14–34.

Ruestow, Edward G., *Physics at Seventeenth and Eighteenth-Century Leiden: Philosophy and the New Science in the University* (The Hague: Nijhoff, 1973; Archives Internationales d'Histoire des Idées, Series Minor, 11).

Rupke, Nicolaas A. (ed.), Vivisection in Historical Perspective (London and New York: Croom Helm, 1987; The Wellcome Institute Series in the History of Medicine).

Salomon-Bayet, Claire, *L'Institution de la Science et l'Expérience du Vivant. Méthode et Expérience à l'Académie Royale des Sciences 1666–1793* (Paris: Flammarion, 1978).

Sander, Sabine, *Handwerkschirurgen: Sozialgeschichte einer Verdrängten Berufsgruppe* (Göttingen: Vandenhoeck & Ruprecht, 1989).

Saussure, Raymond de, 'Haller and Mettrie', *Journal of the History of Medicine*, 4 (1949), 431–49.

Schmallenbach, Hans-Joachim, *Johann Georg Zimmermann und die Irritabilitätslehre* (Münster Univ. M.D. thesis, 1967).

Schneider, Ute, *Friedrich Nicolais Allgemeine Deutsche Bibliothek als Integrationsmedium der Gelehrtenrepublik* (Wiesbaden: Harrassowitz, 1995).

Schofield, Robert, *Mechanism and Materialism: British Natural Philosophy in an Age of Reason* (Princeton: Princeton University Press, 1970).

Schramm, Hans P. (ed.), *Johann Georg Zimmermann: Königlich Grossbritannischer Leibarzt (1728–1795)* (Wiesbaden: Harrassowitz, 1998; Wolfenbütteler Forschungen, 82).

Schulte, B.P.M., 'The Neurophysiology of Herman Boerhaave', in Karl Eduard Rothschuh (ed.), *Von Boerhaave bis Berger: Die Entwicklung der Kontinentalen Physiologie im 18. und 19. Jahrhundert* (Stuttgart: Fischer, 1964), 5–13.

Sgard, Jean (ed.), *Dictionnaire des Journaux, 1600–1789* (2 vols, Paris: Universitas, 1991).

Shapin, Steven and Schaffer, Simon, *Leviathan and the Air-Pump: Hobbes, Boyle, and the Experimental Life* (Princeton: Princeton University Press, 1985).

Singer, Charles and Underwood, E. Ashworth, *A Short History of Medicine*, (2nd edn, Oxford: Clarendon Press, 1962).

Sonntag, Otto, 'The Motivations of the Scientist: The Self-Image of Albrecht von Haller', *Isis*, 65 (1974), 336–51.

——'Albrecht von Haller on Academies and the Advancement of Science: The Case of Göttingen', *Annals of Science*, 32 (1975), 379–91.

——'The Mental and Temperamental Qualities of Haller's Scientist', *Physis: Rivista Internazionale di Storia della Scienza*, 19 (1977), 173–84.

Spillane, John David, *The Doctrine of the Nerves: Chapters in the History of the Neurology* (Oxford: Oxford University Press, 1981).

Sprengel, Kurt, *Versuch einer Pragmatischen Geschichte der Arzneikunde* (5 vols., Halle: Gebauer, 1792–1803).

Steinke, Hubert, ' "Die Ehre des Rechthabens": Experiment und Theorie im Streit um die Lehre von der Irritabilität', *Sudhoffs Archiv*, 82 (1998), 141–69.

——'Divergierende Resultate eines Forschungslabors des 18. Jahrhunderts: Zimmermann und Haller über Irritabilität', in Robert Seidel (ed.), *Wissen und Wissensvermittlung im 18. Jahrhundert: Beiträge zur Sozialgeschichte der Naturwissenschaften zur Zeit der Aufklärung* (Heidelberg: Palatina, 2001; Cardanus, Jahrbuch für Wissenschaftsgeschichte, 1, 2000), 93–115.

——'Tissot Traducteur de Haller: de l'Expérience à la Théorie', in Vincent Barras and Micheline Louis-Courvoisier (eds), *La Médecine des Lumières: Tout autour de Tissot* (Chêne-Bourg: Georg, 2001; Bibliothèque d'Histoire des Sciences, 3), 103–11.

——'Haller's Concept of Irritability and Sensibility and its Reception in France', *La Lettre de la Maison Française d'Oxford*, 14 (2001), 37–69 (volume on 'Mécanisme et Vitalisme').

——'Der Patron im Netz: Die Rolle des Briefwechsels in wissenschaftlichen Kontroversen', in Martin Stuber, Stefan Hächler and Luc Lienhard (eds), *Hallers Netz: Ein Gelehrtenbriefwechsel im Zeitalter der Aufklärung* (Basel: Schwabe, 2005; Studia Halleriana, 9) (in press).

——and Claudia Profos (eds), *Bibliographia Halleriana. Verzeichnis der Schriften von und über Albrecht von Haller* (Basel: Schwabe, 2004; Studia Halleriana, 8).

Stevenson, Lloyd G., 'Anatomical Reasoning in Physiological Thought', in Chandler McC. Brooks and Paul F. Cranefield (eds), *The Historical Development of Physiological Thought* (New York: Hafner, 1959), 27–38.

Stichweh, Rudolf, *Zur Entstehung des Modernen Systems Wissenschaftlicher Disziplinen: Physik in Deutschland, 1740–1890* (Frankfurt am Main: Suhrkamp, 1984).

Stuber, Martin, 'Journal and Letter: The Interaction between two Communication Media in the Correspondence of Albrecht von Haller', *Studies on Voltaire and the Eighteenth Century* (2004), 114–41.

Tansey, E.M., 'The Physiological Tradition', in W.F. Bynum and Roy Porter (eds), *Companion Encyclopedia of the History of Medicine* (2 vols., London and New York: Routledge, 1993), 120–52.

Tega, Walter, 'Mens Agitat Molem. L'Accademia delle Scienze di Bologna (1711–1804)', in Renzo Cremante and Walter Tega (eds), *Scienza e Letteratura nella Cultura Italiana del Settecento* (Bologna: Il Mulino, 1984), 65–108.

Temkin, Owsei, 'The Classical Roots of Glisson's Doctrine of Irritation', *Bulletin of the History of Medicine*, 38 (1964), 297–328.

——*The Double Face of Janus and Other Essays in the History of Medicine* (Baltimore and London: Johns Hopkins University Press, 1977).

Thode, Brita, *Die Göttinger Anatomie 1733–1828* (Göttingen Univ. M.D. thesis, 1979).

Thomson, Ann, 'La Mettrie, Lecteur et Traducteur de Boerhaave', *Dix-Huitième Siècle*, 23 (1991), 23–29.

Toellner, Richard, '"Anima et Irritabilitas": Hallers Abwehr von Animismus und Materialismus', *Sudhoffs Archiv*, 51 (1967), 129–44.

——'Die Verbindung von Lehre und Forschung an der Jungen Georgia Augusta zu Göttingen', *Hippokrates*, 39 (1968), 859–63.

——*Albrecht von Haller. Ueber die Einheit im Denken des Letzten Universalgelehrten* (Wiesbaden: Steiner, 1971; Sudhoffs Archiv, Beiheft 10).

——'Haller und Leibniz: Zwei Universalgelehrte der Aufklärung', *Studia Leibnitiana, Supplementa*, 1 (1973–5), 249–60.

——'Mechanismus – Vitalismus: Ein Paradigmawechsel? Testfall Haller', in Alwin Diemer (ed.), *Die Struktur Wissenschaftlicher Revolutionen und die Geschichte der Wissenschaften* (Meisenheim: Hain, 1977), 61–72.

——'Die Bedeutung des Physico-theologischen Gottesbeweises für die Nachcartesianische Physiologie im 18. Jahrhundert', *Berichte zur Wissenschafisgeschichte*, 5 (1982), 75–82.

——'Albrecht von Haller (1708–1777)', in Dietrich von Engelhardt and Fritz Hartmann (eds), *Klassiker der Medizin* (Munich: Beck, 1991), 245–61.

——'Principles and Forces of Life in Haller', in Guido Cimino and François Duchesneau (eds), *Vitalisms from Haller to the Cell Theory* (Florence: Olschki, 1997), 31–9.

Tröhler, Ulrich and Sabine Mildner-Mazzei, *Vom Medizinstudenten zum Doktor. Die Göttinger Medizinischen Promotionen im 18. Jahrhundert* (Göttingen: Vandenhoeck & Ruprecht, 1993; Göttinger Universitätsschriften, Serie C: Kataloge, 3).

Vila, Anne C., *Enlightenment and Pathology: Sensibility in the Literature and Medicine of Eighteenth-Century France* (Baltimore: Johns Hopkins University Press, 1998).

Waa, G. van der, '*De Irritabilitate*': *Een Onderzoek naar de Betekenis van het Irritabiliteitsbegrip in de Geschiedenis van de Achttiende-Eeuwse Nederlandse Fysiologie* (Rotterdam: Erasmus, 1992; Nieuwe Nederlandse Bijdragen tot de Geschiedenis der Geneeskunde en der Natuurwetenschappen, 41).

Wellman, Kathleen, *La Mettrie: Medicine, Philosophy, and Enlightenment* (Durham: Duke University Press, 1992).

Williams, Elizabeth A., *The Physical and the Moral: Anthropology, Physiology and Philosophical Medicine in France, 1750–1850* (Cambridge: Cambridge University Press, 1994).

——*A Cultural History of Medical Vitalism in Enlightenment Montpellier* (Aldershot: Ashgate, 2003).

Wilson, Philip, 'An Enlightenment Science?: Surgery and the Royal Society', in Roy Porter (ed.), *Medicine in the Enlightenment* (Amsterdam: Rodopi, 1995; Clio Medica, 29), 360–86.

Wittern, Renate (ed.), *Die Professoren und Dozenten der Friedrich–Alexander–Universität Erlangen: 1743–1960* (Erlangen: Universitäts-Bibliothek, 1993).

Wolman, Maria, *Entzündung: Studie zur Geschichte eines Biologischen Begriffs* (Heidelberg: Hüthig, 1962).

Wright, John P., 'Boerhaave on Minds, Human Beings and Mental Diseases', *Studies in Eighteenth-Century Culture*, 20 (1990), 289–302.

——'Metaphysics and Physiology: Mind, Body, and the Animal Economy in Eighteenth-Century Scotland', in M.A. Stewart (ed.), *Studies in the Philosophy of Scottish Enlightenment* (Oxford: Clarendon Press, 1990), 251–302.

——'Substance versus Function Dualism in Eighteenth-Century Medicine', in *idem* and Paul Potter (eds), *Psyche and Soma: Physicians and Metaphysicians on the Mind-Body Problem from Antiquity to Enlightenment* (Oxford: Clarendon Press, 2000), 237-54.

Wrigley, Richard, *The Origins of French Art Criticism: From the Ancien Régime to the Restoration* (Oxford: Clarendon Press, 1993).

Zwaag, P. van der, *Wouter van Doeveren, Medicus, 1730–1783* (Assen: Van Gorcum, 1970).

Appendix: The Spread of Experiment

This list indicates only experiments carried out on living animals, and not the equally numerous tests on surgical patients and anatomical demonstrations. Often we are not informed about witnesses who may have joined the performances.

Key

(p): physician; (s): surgeon; (ms): medical student; (o): other
I: experiments on irritability; S: experiments on sensibility
f: favouring Haller's position; c: contradicting Haller's position; a: ambiguous results

Place	initiator/experimenter/ author	Date	Reported witnesses		
Netherlands					
Groningen	Doeveren, Wouter van (p)	1754, 1758	J. Gummer (ms), M. van Geuns (p), J. Gout (p), L. Stenuis (p) and others	S	a
	Doeveren, Wouter van (p) Verschuir, Walther Forsten (ms)	1764–1765	Eilerts (ms), Kutsch (ms), Munnik (ms), Hoffmann (ms), Verster (ms), Stolte (ms), Dryfhout (ms)	I,S	c
Leiden	Doeveren, Wouter van (p)	1751–1752	A. van Royen (p), J. Tak (p), H.D. Gaub (p), F. Winter (p), F. Klanke (ms), J.F. Martinet (o), and many other physicians and students	I,S	c
	Musschenbroek, Pieter van (o)	1752	together with an unidentified professor	I	c
	Bos, Imam Jakob van den (p) Bikker, Lambertus (p)	1757	E.P. Visvliet (p)	I,S	c
Utrecht	Doeveren, Wouter van (p)	1753	J.D. Hahn (p), G. van Vianen (ms), J. Oosterdijk (p), and students	S	f

German Countries and Switzerland					
Basel	Ramspeck, Jakob Christoph (p)	1751		S	c
	Mieg, Achilles (p)	1758	J.R. Stähelin. (p)	I,S	f
	Müller, Johann Rudolf (ms)	1760	F.L. Watt (ms)	I	f
Berlin	Zinn, Johann Gottfried (p)				
	Meckel, Johann Friedrich (p)	1751		S	f
	Pallas, August Friedrich (p)	1760s	A.J. Güldenstädt (ms)	I	f
Bern	Haller, Albrecht von (p)	1756		I,S	f
Bremen	Runge, Johann Georg (p)	1756		I	f
Göttingen	Haller, Albrecht von (p) Zinn, Johann Gottfried (ms) Zimmermann, Johann Georg (ms) Castell, Peter (ms) Walstorff, Johann Dietrich (ms) Brunn Johann Heinrich von (ms)	1748– 1753	17 different pupils, S. Hollmann (o)	I,S	f
	Zinn, Johann Gottfried (p)	1755– 1756		I,S	f
Hamburg	Cropp, Friedrich Ludwig Christian (p) (identity uncertain)	1751		S	f
Tübingen	Andreae, Jakob Eberhard (ms)	1758		I	c
Great Britain					
Edinburgh	Whytt, Robert (p)	1740s		I	c
	Ramsay, Robert (p)	1760s	R. Whytt (p)	S	f
	Monro, Alexander (p)	1760s?		I	a
	Smith, Thomas (ms)	1765– 1767	J. White (p), W. Withering (p), J. Bostock (ms)	I	c
London	Brocklesby, Richard (p)	1755		I	f

France

Auxerre	Housset, Etienne-Jean-Pierre (p)	1756	Friniat (p), Martin (p)	S	f
Dijon	Hoin, Jean-Jacques (s)	1760 1762		S	f
Montpellier	Lamure, François B. de (p) Sauvages, François B. de (p)	1751– 1752	L.M. Girard de Villars (ms)	I,S	c
	Tandon, Antoine (p) Lamure, François B. de (p)	1755– 1756	F.B. de Sauvages (p), E.J.P. Housset (ms), J.P. de Jausserand (ms), and many others	I,S	c
	Housset, Etienne-Jean-Pierre (ms)	1756	J.J. Ménuret de Chambaud (ms), Brac (ms), Collin (ms), L. Vitet (ms), Guillemeau (ms)	S	f
	Jausserand, Jean Pierre de (ms)	1757		S	c
Paris	Grandclas, Claude-François (p)	1752	F.P.L. Poulletier de la Salle (p), J. Sue (s), A.F. Pallas (p), and others	I,S	c
	Lorry, Anne-Charles (p) Vandermonde, Charles-Aug. (p)	1753		I, S	c
	Doeveren, Wouter van (ms)	1753	C.-F. Grandclas (p), J. Tak (p), two French students	S	c
	Girard de Villars, Louis-Marie (ms)	1755– 1756	L.A. Lavirotte (p), A.C. Lorry (p), J.B.L. Chomel (p), Gervaise (p), A. Ferrein (p), J. Descemet (ms), J.C. Ramspeck (p), and others	I,S	c
	Bordenave, Toussaint (s)	1757		S	f
	Sigaud-Lafond, Joseph-Aignan (p)	1765– 1767	at least one anatomist	S	c
	Portal, Antoine (p)	1771	physiological course, various students	I,S	f
	Arthaud, Charles (ms)	1771		S	f
	Vicq d'Azyr, Félix (p)	1775– 1776		S	f
Rouen	Le Cat, Claude-Nicolas (s, p)	1751– 1755, 1761		S, I	c

Italy					
Bologna	Laghi, Tommaso (p)	1755/10	G. Galeazzi (p), G.M. Bacchetoni (p,s), L. Canuti (p), F.M. Galli Bibiena (p), G.G. Ballanti (p), and others	I,S	c
	Fontana, Felice (o)	1756–1758	G. Veratti (p), P.P. Molinelli (p,s), P. Fantoni (o), F. Algarotti (o), F.M. Zanotti (o)	I,S	f
	Caldani, Leopoldo Marcantonio (p)	1756–1760	Conte Malvezzi (o), G. Veratti (p), P.P. Molinelli (p,s), P. Fantoni (o), V. Menghini (p), F. Algarotti (o), G.A. Pujati (p), L. Bassi (o)	I,S	f
Brescia	Vandelli, Domenico (p)	1756	G. Salvadori (?), C. a Magistra (?), R. Gallina (?)	S	c
Ferrara	Vari, Ignazio (p)	1755	several professors of medicine and philosophy	S	f
Florence	Bianchi, Giuseppe (s)	1755	A. Nannoni (s), E. Audrichi (o), Fossi (o), four other physicians and two surgeons	S	f
	Pozzi, Caesareo Giuseppe (o)	1755	Conte P. Pierucci (o), G. Vespa (s), O. Cametti (o), C. Guadagni (o), A. Gatti (o), F. Fossi (o), X. Manetti (o), B. Pupigliani (p), G. Vannucci (p), F. Tozzetti (p), F. Pagnini (p), P. Molini (p), M. Bianconi (p), A. Cocchi (p), G. Targioni Tozzetti (p), C.A. Collini (o), B. Collini (p), and others	S	f
	Vannucci, Gesualdo (p)	1755		S	f
	Grima, Michel Angiolo (s)	1756	Barbette (p), Fabrini (p), Buonaparte (o), Conte P. Pierrucci (o), G. Vannucci (p), Ruffo (?), Spaneo (?), Ugolini (?), C.A. Guadagni (o), and others	S	c
	Fontana, Felice (o)	1766		I	f
Lucca	Marcuzzi, Gregorio (p,s)	1755	G.L. Graziani (p)	S	f
Modena	Morandi, Morando (p)	1755		S	f

Naples	Sanseverini, Domenico (p)	1755		S	c
	Sarcone, Michele (p)	1760s	several persons	S	f
Padua	Vandelli, Domenico (p)	1756	N. Mezzavia (s), G. Bertossi (ms), F. Trento (o); A. Pujati (p), Sografi (s), Rana (s), Moya (p), Ghirus (p), P. Arduino (o), and several students	S	c
	Caldani, Leopoldo Marcantonio (p)	1764–		I	f
Pisa	Fontana, Felice (o)	1758–1766		I	f
	Sichi, Luca (s)	1763–1764		I,S	f
Rome	Barbiellini, Camillo (p)	1755		S	c
	Tosetti, Urbano (o)	1755	G. Bassani (p), G.B. Balduini (s), N. Giraldi (s), G.V. Petrini (o)	S	f
	Palliani, Ludovico (s)	1755–1757		S	f
	Moretti, Giovanni Battista (o) Massimi, Lorenzo (p)	1756		S	f
	Piazza, Giovanni Battista (p,s)	1756?	several surgeons	S	f
Turin	Bianchi, Giambattista (p)	1755–1759	Plazza (o), Gallo (p), Marini (p), several other doctors and students	S	a
	Bertrandi, Giovanni Ambrogio (p)	1757		S	f
Venice	Pagani, Orazio Maria (p) Bonioli, Cammillo (s)	1757		S	f
	Caldani, Leopoldo Marcantonio (p)	1760–1763		I	f
Other countries					
Grodno	Gilibert, Emanuel (o)	1770s		I	f
Copenhagen	Oeder, Georg Christian (p)	1751		I	f
	Heuermann, Georg (p,s)	1752		S	f
Prague	Radniczky, Ignatius (ms)	1756	J.T. Klinkosch (p), W. Mac Neven (p), Du-Toy (p), and many students	S	c
	Klinkosch, Joseph Thaddäus (p)	1756		S	c
	Trzebiczky, Franz Xaver Caspar (ms)	1772		I	c

Index

J

K

L

M

S

T

Z